Current Topics in Microbiology and Immunology 166

Mechanisms in B-Cell Neoplasia 1990

Workshop
at the National Cancer Institute
National Institutes of Health
Bethesda, MD, USA, March 28-30, 1990

Organized and Edited by
M. Potter and F. Melchers

With 143 Figures

Springer-Verlag
Berlin Heidelberg NewYork London
Paris Tokyo Hong Kong Barcelona

MICHAEL POTTER, M.D.
Laboratory of Genetics, National Institutes of Health
National Cancer Institute
Bethesda, MD 20892,
USA

Prof. Dr. FRITZ MELCHERS
Basel Institute for Immunology
Grenzacherstraße 487
CH-4005 Basel

ISBN-13: 978-3-642-75891-1 e-ISBN-13: 978-3-642-75889-8
DOI: 10.1007/978-3-642-75889-8

Library of Congress Catalog Card Number 15-12910
Softcover reprint of the hardcover 1st edition 1990

2123/3020-543210 – Printed on acid-free paper.

Preface

The eighth workshop in this series on Mechanisms in B-Cell Neoplasia 1990 was held in Wilson Hall at the National Institutes of Health, Bethesda, Maryland on March 28-30. Five major topics formed the basis for the discussions : 1) progress in experimental models of B-cell tumorigenesis, 2) the role of IL-6 in plasma cell tumor formation with particular emphasis on human myeloma, 3) immortalization and regulation of mitosis in B-cells, 4) the myc gene in B-cell neoplasia, and 5) the role of EBV and other oncogenes in transformation of human B-lymphocytes. A meeting on the Epidemiology of Myeloma was held at the N.I.H. on the preceding day, and many of those interested in the clinical aspects of myeloma were also participants at the workshop.

Experimental Models of B-Cell Tumor Development

We have seen in the last eight years the steady growth of model experimental systems, many of which have been designed to be counterparts of the major forms of human B cell tumors, e.g., follicular lymphomas, Burkitt's lymphomas, acute B-cell leukemia and multiple myeloma. A variety of novel ways of inducing these tumors has been described. Advantage has been taken of the "experiments in nature" to identify critical genes that play a role in tumor pathogenesis. These genes have been identified by being near to viral insertion and chromosomal translocation sites, or by having been incorporated or transduced into a defective transforming retrovirus. Not all of these genes have transforming activity in vitro (e.g., bcl-2, IL-6, IL-6R). Thus, in B-cell neoplasia we must deal with certain consistent mutations that appear to be oncogenic, i.e., play a role in the pathogenesis of the tumor, but which do not necessarily involve a known proto-oncogene. The genes implicated in translocations and retroviral insertions have been isolated and reintroduced into engineered transforming retroviral elements that can be used to infect mice and cells. Many of the genes that play a role in B-cell tumorigenesis, e.g., myc, abl, bcl-2 and IL-6, have been reconstructed with new promoting and enhancing elements and introduced into the germ cells in mice to make transgenic strains of mice. One of the important advances in this field has been the discovery of a trangenic model for follicular lymphomas. Korsmeyer et al. (this book) introduced a bcl-2 minigene that lacks a large portion of intron-2 and several other modifications coupled to the regulatory elements of an IgG gene. Strasser et al. (this book) have also made a bcl-2 transgenic mouse wherein the bcl-2 gene was placed under the control of the Ig-heavy chain enhancer, but it contrasts to Korsmeyer's results, as these transgenic mice have not yet developed lymphomas. There are other transgenic strains that develop pre-B cell lymphomas (Eu-myc), plasmacytomas (Eu-abl) and Castelman-like disease (Eu-IL-6). Transgenic mice carrying one of the above mentioned genes can be mated to other transgenic strains to produce hybrids that carry two potential oncogenes. Hybrids carrying two oncogenes in some cases, e.g., Eu-myc and v-abl, pim-1 and myc leads to the more rapid development of tumors. A remarkable feature of tumor development in many of the transgenic mouse systems has been the development of clonal tumors. Only a few of the B-cells in transgenic mice are vulnerable to transformation, suggesting that additional events are required.

MODELS OF HUMAN B-CELL TUMORS

B-Cell Tumor	Experimental
Acute B-Cell Leukemia	bcr-abl transgenic mice; A-MuLV infected mice; Pre-B tumors in Eu-myc transgenic mice
Burkitt's Lymphoma	Transplantation of PBL to SCID mice Transplantation of LCL lines to nu/nu mice
Follicular lymphoma	bcl-2 transgenic mice
Monoclonal Gammopathy of Undetermined Origin	Spontaneous BMG in C57BL/Ka mice
Plasmacytoma, Multiple Myeloma (MGUS)	Spontaneous immunocytomas in rats Induced plasmacytomas (mice) Eu-abl transgenic mice Various transforming retroviruses in pristane treated mice: RIM(Eu-myc/v-Ha-ras); J3V1(v-myc/v-raf-1); ABL-MYC (v-abl/c-myc) Spontaneous marrow seeking plasma cell tumors in mice
Castelman's Disease	IL-6 transgenic mice Infection of stem cells with IL-6 containing retroviruses

Retrovectors carrying two oncogenes such as J3V1 (v-myc/v-raf1); RIM (Eu-myc/v-Ha-ras) have been previously shown to rapidly induce the formation of plasmacytomas in pristane-conditioned mice. An extremely potent new plasmactomagenic virus ABL-MYC has been developed in Rex Risser's laboratory (Largaespada et al., this book). ABL-MYC can induce plasmacytomas in virtually 100% of pristane-treated BALB/cAn mice with extremely short latent periods (Weissinger et al., this book). Interestingly this virus induces plasmacytomas in non-pristane-primed mice of various strains and thus overcomes the strain specific barrier. In some of these potent tumor inducing systems, such as with the ABL/MYC virus, polyclonal or oligoclonal tumors are sometimes induced.

Advances have been made in testing the tumorigenicity of human cells by transplantation to immunodeficient mice. The transplantation of the immortalized lymphoblastoid cell lines, unadapted peripheral blood lymphocytes (PBL) and Burkitt lymphoma cells, to nude mice has been studied in the past. Current studies have now used the SCID mouse as a recipient with surprising results. The transplantation of PBL from EBV-positive normal donors have resulted in the generation of human tumors in SCID mice [see Mosier et al.; Rowe et al. (this book)]. These mice appear to be a new model for testing tumorigenicity of cell lines and for studying tumor pathogenesis of normal cells.

One of the problems that has plagued cancer research for years is how to compare a tumor with its normal counterpart. While EBV-immortalized B-cell cell lines (LBL) have been available in man, there is as yet no similar counterpart in mice. One of the best sources of B-cells in mice have been resting or LPS-activated B-cells from the spleen. Some potentially valuable B-cells can now be cultured in vitro or isolated in abundance from in vivo sources that should be very useful in making important comparisons. A new development in this area has been the successful culture of early B cells in the mouse (F. Melchers). Murine pre-B cells in the process of rearranging their H-chain loci (DH-JH;VHDH-JH) before they have rearranged L chain loci can be grown as cell lines and clones on stromal cells under the stimulatory influence of IL-7 for several months. These are some of the longest lived proliferating normal cells in the B-cell lineage. They can re-populate the B-cell compartment of SCID mice and, therefore, can be considered to be committed stem cells. By changing the stimulatory environment these cells can be induced in vitro to undergo rearrangements of the light chain genes. These cells are potential targets for tranformation. Another new interesting source of normal cells are the plasma cells that appear in abundance in IL-6 transgenic mice or from C57BL-W/Wv congenitally anemic mice that have been reconstituted with cells infected with an IL-6-containing retrovirus (Brandt, this book). Immortalized cell lines can also be isolated from Bcl-2 transgenic mice.

POLYCLONAL B-CELL POPULATIONS

1. B cells from Eu-myc transgenic mice
2. Cultures of Pro-B and Pre-B cells (mice)
3. Immortalized B-cells from Bcl-2 transgenic mice
4. LPS blasts
5. EBV-Immortalized LCL lines (man)
6. Hyperplastic plasma cells (IL-6 transgenic mice)

Pre-neoplastic states

In many of the natural forms of B-cell neoplasia, pre-neoplastic cell types are demonstrable: cells in transformed follicles in the avian bursa; possibly the immortalized B-lymphocytes in Burkitt's lymphoma; the focal proliferating plasma cells in the oil granulomata of BALB/c mice; the smoldering stages of follicular lymphomas; cells in a few certain benign monoclonal gammopathies (MGUS). The relationship of oncogenic mutations to early stages in natural tumor development is a subject of considerable interest. Are such cells the equivalent of the initiation stages of epidermal carcinogenesis?

IL-6 in Plasma Cell Tumor Formation

Specific microenvironments appear to play an important role in B-cell tumorigenesis, and contact and close interactions with stromal cells contribute important growth factors for the developing neoplastic cell. IL-6 is known to be required as a factor for the in vitro growth of many pristane-induced plasmacytomas in mice. It has also been shown to be an important growth factor for some human myeloma cell lines in vitro. The critical question in both the human and mouse systems is what role IL-6 plays in the pathogenesis of plasma cell tumor development. Overproduction of IL-6 in IL-6 transgenic mice or by the introduction of retroviruses containing the IL-6 gene into stem cells (Brandt, this book) induces a Castelman-like disease in mice characterized by a vast over-production and hyperplasia of polyclonal plasma cells; thus far these cells lack tumorigenic properties. Kawano et al. in 1988 (Nature 332:83) reported that human multiple myeloma cells produce IL-6 in an autocrine fashion. Also very recently Blankenstein et al. (J. Exp. Med. 171:965,1990) have found that the MPC-11 plasmacytoma cell line has a transposed IAP (intracisternal A particle) gene inserted into the IL-6 structural gene and is producing IL-6 constitutively. Suematsu et al. (this book) have found that the plasmacytoma P3U1 [tissue culture line of MOPC21] has an IAP insert in the IL-6R gene. MOPC21 has been previously shown to have an IAP insert in the mos oncogene. These cells make increased numbers of IL-6 receptors. This is not in the strict sense an autocrine mechanism, although it can potentially facilitate the intake of IL-6 from the environment and hence is an autocrine-like mechanism. There may be others. Other workers favor the paracrine origin of IL-6 and have found evidence that IL-6 is produced by bone marrow stromal cells. In fact, rather dramatic levels of IL-6 appear in the serum of patients in advanced stages of myeloma and plasma cell leukemia (Bataille et al., J Clin Invest 84:2008,1989). The autocrine-paracrine argument awaits further sorting-out. Perhaps both mechanisms can supply a source of IL-6 although the paracrine source is probably the more common. We see in these discussions the beginnings of an understanding of the signalling pathways that are activated by IL-6 (Nakajima and Wall, this book). The dependence of plasmacytoma and myeloma cells for IL-6 in vitro has opened up new possibilities for controlling the growth of these tumors in vivo by interfering with the supply of IL-6.

Immortalization and Mitotic Cycling in B-Cells

B-cells are members of the hematopoietic family and have probably the most complex ontogeny of any cell in the system rivalled only by T-lymphocytes. Like other hematopoietic cells B-cells are continuously regenerated from pluripotenial hematopoietic stem cells and from B-lineage committed cells. Unlike neutrophils and erythrocytes, which complete their differentiation and development during a single burst of mitotic activity B-cells, undergo a protracted form of development in which the cells can exit and re-enter from mitotic cycling. Bursts of mitotic activity are triggered in extramedullary sites by contact with antigens, T-lymphocytes and cytokines. Immunoglobulin heavy chain switching and somatic mutational processes take place in extramedullary B-cells. Both of these functions are associated with rearrangements or substitutions in the germline Ig-gene. The principal value of exiting from mitotic cycling after exposure to antigen is that some B-cells can re-enter the recirculating pool as relatively long lived memory cells. Reactivation into mitotic activity can be triggered by new encounters with antigens or T-cell-derived cytokines. B-cell tumors in man and many other species are the most prevalent form of lymphoid neoplasm. The protracted development of B-cells provides opportunities to exit and enter mitotic cycling and to persist or hide in the body as partially transforming resting cells. These characteristics make cells of the B-cell lineage more prone to tumor development.

There was much discussion throughout the meeting on the role of stromal cells in B-cell tumor development. It is generally thought that stromal cells produce essential cytokines that can act over short distances, but in addition there is growing evidence that certain B-cell tumor cells (e.g., the mouse plasmacytomas) require intimate adhesion to a substratum for growth. Miyake and Kincade (this book) have developed a monoclonal antibody to Pgp-1/CD44 that blocks lymphopoiesis in bone marrow cultures and adhesion of plasma cells to stromal cells. DeGrassi et al. (this book) have developed a plasmacytoma cell line that requires intimate adhesion with a reticular fibroblast to grow. This stromal requirement cannot be replaced by IL-6. Further adaptation of human myeloma cells to tissue culture requires the use of bone marrow-derived stromal cells. What is the biological significance of this intimate association of B cell and fixed stromal cell? Is it simply a mechanism for bringing a plasma cell close to an important cytokine source, or are other essential signals or factors for replication transferred between the cells? (See discussion by B. Klein et al., this book.)

Genetic predisposition to tumor development as seen in pristane- or plastic-induced plasmacytoma development in BALB/cAn mice raises the question in what cell is genetic susceptibility expressed, the B-cell itself or in the host reactive tissue in which the plasmacytoma actually develops? The work of Silva, Wiener et al. now supports the possibility that this vulnerability is expressed in B-cells and not necessarily the tissue microenvironment. Silva and Wiener have constructed BALB/c-DBA/2 chimaeras and injected them with pristane.

Plasmacytomas were induced in chimaeric DBA/2 mice (a plasmacytoma-resistant strain) that have approximately 10% BALB/c cells. These plasmacytomas were shown to be of the donor (BALB/c) type, strongly suggesting that the genetic resistance of the DBA/2 mouse was not determined in the tissue microenvironment (or more specifically the induced oil granuloma microenvironment) but rather resided in the B-cell itself. Osmond et al. (this book) have found that an intra-peritoneal injection of pristane induces a proliferation of B-cell precursors in the bone marrow; they did find differences in the pre-B cells in plasmacytoma induction susceptible BALB/cAnPt and DBA/2 mice.

A fundamental issue in the pathogenesis of immunoglobulinsecreting tumors such as the plasmacytomas in mice or multiple myeloma in man is: in which cell does the first somatic mutational change occur? The consistent chromosomal translocations that involve the PVT-1 and IgL chain loci, which occur in the mouse plasmacytomas, suggest that this illegitimate recombination occurred during Ig light chain gene rearrangements that take place in the pre-B stages of development. Evidence obtained by Joshua Epstein (Epstein et al., Blood 71:861, 1988) in multiple myeloma indicates that some myelomas express markers associated with other hematopoietic lineages. This raises the possibility that the earliest changes may take place earlier than the pre-B stage at the level of a multipotential hematopoietic stem cell. In contrast one might argue that the target cell in plasmacytoma development can be a late B-lymphocyte, i.e., one that has encountered antigen and undergone heavy chain switching, somatic hypermutation in addition to Ig-gene rearrangement. The issue is unresolved.

The c-myc Gene in B-Cell Neoplasia

Dysregulation of c-myc transcription has been implicated in many forms of B-cell tumor development: Bursal lymphoma in chickens; Endemic and Sporadic Burkitt's lymphoma, AIDS associated B-cell lymphoma in man; induced plasmacytoma in mice and spontaneous immunocytoma of rats.

Surprisingly, the biological function of c-myc is not yet understood. In previous workshops in this series much of the work focused on the regulation of c-myc transcription. C-myc has a number of regulatory sites that extend at least 1 kb, 5' of the first exon. We are continuing to learn more about this complex problem of transcriptional regulation. Ultimately understanding the alterations of c-myc and its product will depend upon the elucidation of the functions of the normal gene and its products. A promising direction towards this end comes from the analysis the structure of the myc protein itself. Identification in myc of a leucine zipper (Landschulz et al., Science 243:1681,1989), an amphipathic DNA binding helix-loop-helix peptide motif found in regulatory proteins (Ig enhancer binding E12, MyoD and Drosophila daughterless proteins [Tapscott et al., Science 242:405,1988; Murre et al., Cell 56:777,1989) nuclear localization site, high content of proline, etc., has suggested that the c-myc protein has homologies with other nuclear factors that regulate gene expression, possibly at the transcriptional level. While the myc protein has not been shown to bind to a specific DNA sequence or to be a transcription factor, it may possibly be a component of a heterodimeric system. This remains to be investigated.

EBV and Other Oncogenes in the Neoplastic Transformation of Human B-lymphocytes

An exciting new development in the field of human B-cell lymphomagenesis has been the observation of Mosier et al. that the injection of PBL from EBV-positive normal volunteers into SCID mice results in the formation of B-cell tumors. In a few of the cases novel rearrangements of the c-myc gene have been discovered in these abnormal cell types. It has been known for some time that Burkitt lymphoma cells and EBV immortalized LBL transfected with myc or ras oncogenes grow as tumors in nu/nu mice. Wolf et al. (this book) have found evidence for recessive genes in Burkitt lymphoma cells by fusing BL/LCL lines. The hybrid cells contained both the degregulated myc genes and EBV but failed to grow progressively in nu/nu mice. Differences in the degree of tumorigenicity of Burkitt lymphoma cells in nu/nu hosts have been described (Gurtsevitch et al., Inter. J. Cancer 41:87,1987). The EBV immortalization step is as yet not understood but provides an experimental advantage for studying the stages and steps in neoplastic tranformation in a human system. There is renewed interest in studying the insertional properties of EBV (Hurley et al., Zimber-Strobl et al., this book); as yet there is no evidence that these inserts play a role in the oncogenic process.

Comments: Multiple Genetic Changes in Neoplastic Development; Oncogene Cooperation

It is generally thought that the process of neoplastic development depends upon a succession of mutational events. One hesitates to say steps, as it has been proposed by others that these events may not follow an ordered sequence but rather simply represent an accumulation of mutations in a clonal lineage. Nonetheless, the concept of steps where one depends upon the preceeding step is important in B-cell neoplasia because of the complex ontogeny of the B-cell. It is possible that one step or oncogenic mutational event may occur at one level of development and not create a growth disorder in that cell but rather set the stage for a second or third subsequent change that takes place later in development. Nonetheless, it is argued by some that even though an oncogenic mutation may not result in a phenotypically obvious uncontrolled growth, it may provide a survival advantage to the cell in which it occurs (e.g., as with the activation of bcl-2 in follicular lymphoma (see paper by Korsmeyer et al.).

One of the widely accepted concepts of neoplastic development is that the process depends upon multiple genetic changes. While there has been considerable success in identifying one oncogenic mutation associated with the pathogenesis of a given type of B-cell tumor, there is relatively little hard data that supports the multiple mutational concept. There are extensive data with rather artificial tumor induction systems (e.g., infection with transforming retroviruses that contain 2 transforming oncogenes; experiments with transgenic mice) that demonstrate that two oncogenic mutations can cooperate in inducing neoplastic growth. But these experiments must be viewed with caution in attempting to reconstruct a natural scenario of tumor development, principally because cooperating oncogenes have not yet been found with consistency in B-cell tumorigenesis. Transgenic mice provide valuable systems to look for cooperating mutations. One example which might be used as evidence of favor of cooperating oncogenic mutations is derived from studies in plasmacytoma formation in Eu-abl transgenic mice. These tumors have been shown to have rearranged c-myc genes presumably due to chromosomal translocations (see Harris et al., this book). Similarly B-cells infected with Abelson virus implanted into pristanecondi-tioned mice develop into plasmacytomas that also have chromosomal translocations which dysregulate myc transciption (see Wiener et al., this book). A more intense examination should now be made to identify cooperating oncogenic mutations in naturally occuring B-cell tumors as well in B-cell tumors that are induced by nontrans-forming viruses (e.g., mouse plasmactytomagenesis) or in systems where only a single oncogene has been introduced. For example, how exhaustively has c-abl (or a reasonable equivalent) been examined in plasmacytomagenesis of Burkitt lymphomagenesis?

There are some possible explanations for the lack of success in finding second, third and other consistent mutations in the pathogenetic history of such B-cell tumors. Trivially, the searches may have been thus far too limited. There may be biological reasons as well; second and third mutations may not involve any of the known major oncogenes. Also they may not be 'consistent', that is, a variety of different alterations may be able to cooperate effectively with a major mutation. For example, many of the major oncogenes are known to be tyrosine or ser/thr kinases which have multiple substrates. This suggests that mutations affecting the phosphorylation or de-phosphorylation of a single (though important) substrate could mimic, in part, the effect of the major oncogene. These smaller incremental types of mutations may have been elusive. Another difficulty in defining secondary and tertiary stages or steps may be encumbered by the inability to propagate or culture intermediate stage cells. Searches are now underway to find such stages as, for example, in the EBV-immortalized cell systems.

Epigenetic changes may possibly play a role in B-cell transformation. This is a poorly defined area of research, but the possibility cannot be dismissed. In one of the classic models of carcinogenesis the induction of epidermal tumors in mice, the initiation (induced by carcinogenic, genotoxic agents such the polycyclic aromatic hydrocarbons) and promotion phases (induced by phorbol esters and other agents) have been extensively studied. It is generally thought that promoting agents are not directly genotoxic or mutagenic. Are there equivalents of promotion in B-cell neoplasia? One candidate system may be the induction of plasmacytomas by paraffin oils or plastics. Another kind of epigenetic change is the demethylation of DNA. This has been associated with many kinds of tumors and has been discussed as a possible form of epigenetic change that can have heritable properties (Holliday and Jeggo, Cancer Surveys 4;557,1985).

One approach to understanding pathogenetic mechanism has been provided by the study of oncogene cooperation. The discovery of transforming retroviral elements that contain two oncogenes provided the "experiment in nature" for this. At least four transforming retroviruses with two oncogenes have been described; among these is the avian MH-2 virus that carries v-myc and v-mil (the avian counterpart of raf). Oncogene cooperation has been extensively studied in fibroblast transformation where two genes are essential for the acquisition of tumorigenicity. These fall into two complementarity groups of genes which have been separated on the basis of their localization in the cell, i.e., nuclear versus cytoplasmic (R.A.Weinberg, Science 230:770,1985). This distinction however, may be fortuitous and not reflect the true biological basis of cooperativity. A crucial question here is whether complementation has its origins in separate regulative signal transduction pathways. For example, the mitotic cycle from Go to mitosis may depend upon the activation of sets of genes and these events may be separated in space and time and further require the activation of different signal transduction pathways. Some of the earliest hypotheses on cooperation speculated that oncogenic changes blocked differentiation and stimulated cell cycling. Another form of cooperating oncogenic mutations may relate to the organization of cells; in this context mutations leading to the stimulation of mitosis may be complemented by a second set of mutations that modify the receptor landscape of the cell. The two types of mutations result in a disoriented cell that continues to cycle.

We are beginning to perceive dimly the outlines of pathogenetic mechanism in B-cell neoplasia, but such visions are often illusory and at this stage must be taken with great caution. However, it is very clear that progress will depend upon another form of cooperation, exchanges and collaborations among interested scientists from different laboratories with different perspectives. This workshop serves this purpose. We thank the Division of Cancer Biology, Diagnosis and Centers of the National Cancer Institute for funding this meeting and for the interest and support of Dr. Alan Rabson. This book was assembled and edited by Ms. Victoria Rogers to whom we are most grateful.

Michael Potter

Fritz Melchers

Table of Contents

Experimental Plasmacytomas and B-Cell Tumors

Growth Regulation to B-Cells: Immortalization

C-*myc* Genes in B-Cell Neoplasia

Role of EBV in B-Cell Neoplasia

Contributors

Their addresses can be found at the beginning of their chapters

IL-6 in Multiple Myeloma

IL-6 in Multiple Myeloma

IL-6 as a Growth Factor for Human Multiple Myeloma Cells – A Short Overview

K. Nilsson, H. Jernberg and M. Pettersson

Laboratory of Tumor Biology, Department of Pathology, University of Uppsala, S-751 85 Uppsala, Sweden

INTRODUCTION

The growth and differentiation of normal B-cells are under stringent control by complex cell-cell interactions, including cell-cell contacts and growth-and differentiation factor signalling. The activation of resting B-cells, and their subsequent growth and/or differentiation, have thus been found to be regulated by multiple, alternative cytokines acting in concert to ensure a fine-tuned control of the process. Malignant B-cells, best exemplified by chronic lymphocytic leukemia cells, have also been found to require such complex signalling for activation-, growth- and differentiation induction. (Nilsson 1989). It was therefore surprising when Kawano et al (1988) reported that the growth of human multiple myeloma (MM) biopsy cells *in vitro* could be stimulated by IL-6 and serum only, and that the growth could be inhibited by anti-IL-6 antibodies.

Human IL-6 is a pleiotropic, multifunctional cytokine produced by many cell types and was originally described to be a differentiation- but not a growth factor for B-cells (Kishimoto 1989). It can therefore be speculated that the growth stimulatory effect IL-6 of MM cells reflects some malignancy-associated defect(s) in the gene programs controlling the growth and differentiation of such cells.

The article by Kawano et al (1988) has been followed by a number of independent studies aimed at finding out the biological effects of IL-6 on MM cells, specifically at testing the possibility of an autocrine IL-6 loop operating in MM cells. In this introduction to the session on the biology of MM these studies, and as yet unpublished studies on a panel of MM cell lines performed in our own laboratory, will be summarized. In evaluating the role of IL-6 as a growth factor for human MM cells it is important to put together the excisting data from studies both on fresh biopsy cells and established cell lines, since both types of materials for studies on the biology of MM are to some extent artefactual. In the use of MM biopsy cell populations the admixture of non-MM bone-marrow cells, even after extensive purification, is an unavoidable problem. In long term cultured MM cell lines secondary genetic changes, affecting genes involved in the control of growth and differentiation, may occur, making the cells unrepresentative of MM cells *in vivo*. In the following we will make statements about IL-6 as a growth factor for MM cells and illustrate some of them with results from our own studies on our MM cell line panel.

Current Topics in Microbiology and Immunology, Vol. 166

THE RESPONSE TO EXOGENEOUS IL-6 IN MM BIOPSY CELLS AND MM CELL LINES I HETEROGENEOUS

Kawano et al (1988), in their original study on IL-6 as a growth facto for MM cells, found that Il-6 stimulated the growth of MM biopsy cells from 12/26 patients. In studies performed subsequently in other laboratories about the same fraction (40-60 %) of IL-6 dependent MM biopsy cells was found among a total of 105 tested cases (Asaoku et al 1988; Anderson et al 1989; Kawano et al 1989; Klein et al 1989; Tanabe et al 1989; Zhang et al 1989).

Table 1 shows our own results on the IL-6 dependency in a panel of established human MM cell lines. Three of the 7 lines responded to IL- The IL-6 responsive lines were strictly dependent on feeder cells or feeder cell conditioned medium for maintenance. The IL-6 independent cell lines had all been in continuous culture for 8 years or more and, as discussed below, the IL-6 independency observed in the majority of the cell lines may represent the result of a secondary phenotypic chan *in vitro*. Recently, Klein et al (to be published) have shown that not only the culture technique employing feeder cells as a source of IL-6 (Nilsson et al 1970; Shimizu et al 1989) but also IL-6 and serum containing tissue culture medium in the absence of feeder cells can supply MM cells with the nutrients and growth stimulatory signals necessary for long term growth *in vitro*.

THE EXPRESSION OF IL-6 AND IL-6 RECEPTOR IN MM CELLS IS HETEROGENEOUS

Il-6 mRNA may be detected by northern blot analyses in MM biopsies (Kawano et al 1988 and 1989; Freeman et al 1989). In our study (Table IL-6 was detectable only in the U-266 cell line using northern blot analyses and poly-A selected mRNA. This MM line, has also been found t express IL-6 mRNA by Kawano et al (1988) but was reported IL-6 mRNA negative by Klein et al (1989).

Kawano et al (1988) reported expression of IL-6 receptors (IL-6R) both in IL-6 responding and non-responding MM biopsy cells. Table 1 demonstrates that the IL-6R is expressed in 5/7 of the MM cell lines. It thus seems that the IL-6R may be expressed on IL-6 independent MM cells also during long term culture *in vitro*.

IL-6 IS THE MAJOR, BUT NOT THE ONLY, PARACRINE GROWTH FACTOR FOR MM CELLS DEPENDENT ON EXOGENEOUS IL-6

The studies by Kawano et al (1988) and Klein et al (1989) showed that anti-IL-6 antibodies inhibited the growth of MM biopsy cells. This finding was differently interpretated in the two papers - indicating autocrine IL-6 stimulation of MM cell growth to the former group but a paracrine regulation of MM growth to the latter. Our studies support that IL-6 is primarily a paracrine growth factor for MM cells. With the exception for the long term passaged U-266 1984 we did not find IL-6 production in any of the cell lines and in U-266 1984 the production is low and variable. All feeder cell dependent cell lines respond to IL-6 and this response can be blocked by anti-IL-6 antibodies.

Table 1. **Growth characteristics, expression of IL-6 and IL-6 receptor mRNA and response to rIL-6 of a panel of human myeloma cell lines.**

Designation	Reference	Time of continuous cultivation in vitro	Dependence of feeder cells/ conditioned medium	Population doubling time(hrs)	IL-6 mRNA expression	IL6R mRNA expression	Mitogenic response to exogenous rIL-6
U-266 1970	Nilsson et al 1970	2 years	yes	120	yes (weak)	yes (weak)	yes
U-266 1984	Hellman etal 1988	21 years	no	40	yes	yes	no
U-1958	Jernberg et al 1987	5 years	yes	100	no	yes	yes
U-1996	Jernberg et al 1987	5 years	yes	48	no	yes	yes
Karpas 707	Karpas etal 1982	8 years	no	40	no	yes	no
L363	Diehl et al 1978	12 years	no	48	no	no	no
FRAVEL	Miller et al 1982	10 years	no	48	no	no	no

Supernatant from the feeder cells contains IL-6, and the growth stimulatory effect by feeder cell supernatant can be completely inhibited by antibodies against IL-6. That IL-6 is the "feeder factor" is supported by the studies of Shimizu et al (1989) who could block th growth stimulatory effect of the macrophage monolayer, used to support the growth of two new IL-6 dependent established MM cell lines, by anti-IL-6 antibodies. Although IL-6 seems to be an essential growth factor for IL-6 dependent MM cell lines we have, however, not been ab to manintain such cells in serum free medium + IL-6 for more that 11 days. The addition of at least 0.5% fetal calf serum is a prereqiusite for long term growth. We have shown that serum free medium + IL-6 supplemented with the physiologic IL-6 carrier protein α2- macroglobul (Matsuda et al 1989) can not support long term survial and growth of M cells. This suggests that serum provides some factor(s) critical for maintenance of MM cells in long term culture.

Taken together, the above results suggest to us 1) that IL-6 acts primarily as a paracrine growth factor for MM, but only for a fraction of the cases and 2) that some other cytokines may be co-stimulatory wit IL-6 since serum is required to maintain MM cells in long term culture. Some studies have addressed the question whether other cytokines may stimulate growth in MM cells alone or in combination with IL-6. MM biopsy cells did not respond to exogeneous IL-1ß, IL-2, TNFß and IFNγ (Klein et al 1989) or IL-1α, IL-1ß Il-2, IL-4, G/M-CSF, G-CSF and M-CS (Anderson et al 1989). Since the production of IL-6 was shown both in cultures of IL-6 responding and non-responding biopsy cells (Kawano et al 1988) these negative findings suggest thus that none of the examine factors seems to represent MM growth factors co-stimulatory with IL-6. However, Andersson et al (1989) reported that MM biopsy cells from a fe patients responded to IL-3 and IL-5, alone or in combination with IL-6. In addition, Kawano et al (1989) found that IL-1ß stimulated growth in MM biopsy cells in 8/20 cases. Brenning (1985) and Bataille (this volume) has shown that INFα at low concentrations is growth stimulator for a fraction of MM biopsy cells. Finally, Bataille (this volume) has demonstrated that GM-CSF in combination with IL-6 stimulates MM cell growth.It was not shown in any of the studies whether or not the observed effect on the MM cell growth resulted from the direct interaction of the cytokines with the MM cells or indirect, via stimulation of contaminating non-MM to production of MM stimulatory cytokines, e.g. IL-6.

The studies on our panel of cell lines (free of contaminating non-MM cells) have clearly demonstrated that among the cytokines available as recombinant molecules only IL-6 significantly stimulates growth of MM cells. However, the possibility that long term cultured cells lines hav changed their growth factor requirement must always be kept in mind.

It is also possible that cytokines produced by the MM cells may regulat MM cell growth by autocrine mechanisms or by stimulating non-MM bone marrow cells to produce growth factors like IL-6, acting on MM cells by a paracrine mechanism. Table 2 summarizes studies on cytokine productio in cultured MM biopsy cells and cell lines. IL-1ß and TNFß seem to be the most frequent cytokines produced by MM cells. However, none of them seem to act as autocrine growth factors.

Table 2. **Cytokine expression (mRNA and/or protein) *in vitro* by cells from myeloma biopsies and in established myeloma cell lines**

Reference	Cytokine IL-1α	IL-1β	TNFα	TNFβ	IGF-1	TGFβ
Biopsies						
Cozzolino et al 1989	(+)	+	-	-	n.t.	n.t.
Lichtenstein et al 1989	-	+	+	-	n.t.	n.t.
Yamamoto et al 1989	-	+	-	-	n.t.	n.t.
Cell lines						
Garrett et al 1987	-	-	+*	+	n.t.	n.t.
Bataille et al 1989	-	-	-	+	n.t.	n.t.
Jernberg et al 1990	n.t.	+	-	+	+	+

n.t. = not tested; *mRNA+; protein-

IL-6 IS NOT A DIFFERENTIATION FACTOR FOR MM CELLS

In studies on IL-6 induced growth and Ig secretion Andersson et al (1989) and Tanabe et al (1989) in MM biopsy cells, and our own studies with the panel of MM cell lines, have demonstrated that Ig secretion is not induced by IL-6.

For normal B-cells IL-6 appears to be a differentiation factor (Hirano et al 1986; Maraguchi et al 1988) acting by transcriptional activation of Ig heavy and light chain genes of B-cells properly activated and co-stimulated by IL-2 (Raynal et al 1989; Splawski et al 1990). IL-6 is induced in normal B-cells and in B-type chronic lymphocytic B-cells induced to Ig secretion by IL-4 (Smeland et al 1989; Carlsson et al to be published) and may participate i the development of Ig secretory B-cells by an autocrine loop. Whether or not normal plasma cells can be induced to proliferation by IL-6 is not clear. It is widely assumed that plasma cells represent terminally differentiated, growth arrested cells. However, Klein et al (1989) report that plasma cells derived from inflammatory sites can be induced to proliferation by IL-6 and in the IL-6 transgenic mice IgG1 plasmacytosis develops (Suematsu et al 1989). It is thus possible that plasma cells in fact may be able to proliferate or that the cells studied by Klein et al (1989) and those growing in the transgenic mice do not represent fully mature plasma cells. Until the response of normal plasma cells to IL-6 has been more conclusively studied it is not possible to consider the growth induction in MM cells exposed to IL-6 as abnormal. What already, however, seems to be an abnormal response to IL-6 in MM cells is the lack of stimulation of the Ig secretion. This probably demonstrates a malignancy associated abnormality in the MM cell phenoype.

IFN-γ INHIBITS THE GROWTH STIMULATION OF IL-6 DEPENDENT MM CELLS BY EXOGENEOUS IL-6 BY AN UNKNOWN MECHANISM

Of all the cytokines tested on MM cells only IFN_{α} and IFNγ have been found to be growth inhibititory. IFNα can inhibit growth of biopsy cel from a fraction of patients with MM and of cells of some of the MM cel lines (Brenning et al 1985). This growth inhibition appears to be unrelated to sensitivity of the MM cells to IL-6. IFNγ, in contrast, inhibits the growth only of those MM cell lines and biopsy cells which respond to IL-6. The mechanism for this IFNγ effect on IL-6 responsive MM cells is unknown, but it seems not to be explainable by receptor cross-talk leading to IL-6R down-regulation, as is clear from our studies by northern blot analysis of IL-6R expression in IFNγ treated IL-6 dependent cell lines.

AUTOCRINE IL-6 STIMULATION OF MM CELLS IS RARE BUT MAY DEVELOP AS A CONSEQUENCE OF TUMOR PROGRESSION AS SUGGESTED BY STUDIES ON NEWLY ESTABLISHED AND LONG TERM CULTURED U-266 MM CELLS

As pointed out above the studies on MM biopsy cells do not conclusivel show that IL-6 might be an autocrine growth factor for MM cells. Also for mouse plasmacytomas there seems to be little evidence for autocrin IL-6 stimulation. However, Tohyama et al (1990) have demonstrated, that induction of constitutive secretion of IL-6 in a murine IL-6 dependent B-hybridoma by transfection of IL-6 cDNA can result in IL-6 independent growth apparently due to an autocrine growth stimulation. Il-6 may thus seem to have the potential to function as an autocrine growth factor for at least some B-cells.

We have examined the possibility that U-266 1984 may be stimulated by a autocrine IL-6 loop. We have found that U-266 1984 cells express both IL-6R and IL-6 mRNA and produce IL-6, although with considerable variability unrelated to the growth rate. IL-6 from the U-266 1984 cell can stimulate growth of human and mouse MM cells. The growth of the lin can not be blocked by anti-IL-6 antibodies of a batch known to inhibit growth of IL-6 dependent MM cell lines.IFNγ does not affect the growth of the U-266 1984 and anti-sense IL-6 mRNA experiments have been inconclusive. Taken together, the data do not demonstrate an autocrine loop in this cell line. However, recently we have found that an anti-IL 6R antibody has been found to inhibit the growth by 30 %. It is therefore possible that a mechanism of autocrine IL-6 stimulation is indeed operating in the U-266 1984 cell line. This is also supported by studies with this myeloma cell line at NIH (Rick Nordan personal communication).

The studies on IL-6 production in the U-266 cell line prompted us to test whether the expression of IL-6, IL-6R and response to IL-6 could undergo changes with time in established MM cell lines. This was plausible as only the recently established cell lines were dependent on IL-6. Table 3 demonstrates that several phenotypic changes have occurre during the 14 years that the U-266 cell line has been in continuous culture. The alterations include improved growth properties both *in vitro* (development of feeder cell and IL-6 independence) and *in vivo* (acquisition of tumorigenic potential in nude mice), a decrease in *bcl-* expression, loss of capacity to produce IGF-1 and an increase in the expression of IL-6R and IGF1R.

Table 3. Properties of a newly established (U-266 1970) and long term cultured (U-266 1984) human myeloma cell line

Characteristics	U-266 1970	U-266 1984
Growth properties		
Doubling time (hours)	108-140	40-45
Feeder-cell dependence	+	-
Growth in agarose	-	+
Tumorigenicity in nude mice	-	+
Secretion of IgE (fg/cell/24 hours)	4000	750
Secretion of IgA (fg/cell/24 hours)	0	20-200
IgA mRNA (differential splicing)	Neg.	Pos.
Growth factor response	IL-6 and serum dependency	IL-6 independency; serum dependency
Oncogene expression	c-*abl*, c-*myb*, L-*myc*, *bcl*-2(++)	c-*abl*, c-*myb*, L-*myc* *bcl*-2(+)
Cytokine mRNA expression	IL-6(+), IGF-1+, TGFß+, IL-1ß+,TNFα-, TNFß+,	IL-6(++), IGF-1- TGFß+, IL-1ß+, TNFα-, TNFß+
Cytokine receptor expression	IL-6R+, IGF-1R(+)	IL-6R++, IGF-1R++

It is apparent from detailed studies on this particular MM cell line, and from our experience with changes of growth properties in continuous cell lines in general, that there is usually an evolution towards independence of exogeneous growth factors. A general finding is that the need for serum diminishes as cell lines grow old *in vitro*, perhaps reflecting such an evolution. The possibility that MM cell lines escape the need for IL-6 is obvious. This can hypothetically be brought about by autocrine growth factor production, which in this case only rarely should be IL-6 but some other, as yet not defined factors (Klein et al 1987).

From the clinical studies performed so far it is not possible to deduce whether a similar evolution, from IL-6 dependency to independency might take place. The data from Montepellier (Zhang et al 1989) suggest that aggressive, fast growing MM, including plasma cell leukemia, remain IL-6 dependent while Asaoku et al (1989) report a decreased IL-6 response in advanced cases of MM.

Table 4. **The role of IL-6 as a growth factor for human myeloma cells - a summary.**

1. IL-6 is a major growth factor for a large fraction of human myelomas.
2. IL-6 is the growth factor produced by feeder cells (macrophages, fibroblasts, glial cells) in *in vitro* cultures.
3. IL-6 seems to act essentially as a paracrine growth factor.
4. Autocrine growth stimulation by IL-6 has not been proven in fresh biopsy cells but may occur in the human U-266 myeloma cell line.
5. IL-6 independency may develop as a result of tumor progression.
6. IL-6 is a growth factor but not a differentiation factor
7. IL-6 may synergize with GM-CSF, IL-3, IFNα and unknown factors in FCS, and possibly with factors produced by myeloma cells
8. IFNγ inhibits growth of IL-6 dependent human myeloma cell lines by an unknown mechanism.

ACKNOWLEDGEMENTS

This work was supported by a grant from the Swedish Cancer Society. We thank Mrs Eva Harryson for assistance in the preparation of this manuscript.

REFERENCES

Andersson K C, Jones R M, Morimoto C, Leavitt P, Barut B A (1989) Response patterns of purified myeloma cells to hematopoietic growth factors. Blood 73:1915-1924.

Asaoku H, Kawano M, Iwato K, Tanabe O, Tanaka H, Hirano T, Kishimoto T, Kuramoto A (1988) Decrease in BSF-2/IL-6 response in advanced cases of multiple myeloma. Blood 72:429-432.

Bataille R, Klein B, Jourdan M, Rossi J F, Durie B G M (1989) Spontaneous secretion of tumor necrosis factor-beta by human myeloma cell lines. Cancer 63:877-880.

Brenning G (1985) The *in vitro* effect of leucocyte-α interferon on human myeloma cells in a semisolid agar culture system. Scand J Haematol. 35:178-185.

Cozzolino F, Torcia M, Aldinucci D, Rubartelli A, Miliani A, Shaw A R, Lansdorp P M, Di Guglielmo R (1989) Production of interleukin-1 by bone marrow myeloma cells. Blood 74:380-387.

Diehl V, Schaadt M, Kirchner H, Hellriegel K-P, Gudat F, Fonatsch C, Lasekwitz E, Guggenheim R (1978) Long-term cultivation of plasma cell leukemia cells and autologous lymphoblasts (LCL) *in vitro*: a comparative study. Blut 36:331-338.

Freeman G J, Freedman A S, Rabinowe S N, Segil J M, Horowitz J, Rosen K, Whitman J F, Nadler L M (1989) Interleukin 6 gene expression in normal and neoplastic B cells. J Clin Invest 83:1512-1518.
Garrett I R, Durie B G M, Nedwin G E, Gillespie A, Bringman T, Sabatini M, Bertolini D R, Mundy G R (1987) Production of lymphotoxin, a bone-resorbing cytokine, by cultured human myeloma cells. N Engl J Med 317:526-532.
Hellman L, Josephson S, Jernberg H, Nilsson K, Pettersson U (1988) Immunoglobulin synthesis in the human myeloma cell line U-266; expression of two immunoglobulin heavy chain isotypes (ε and α) after long term cultivation *in vitro*. Eur J Immunol 18:905-910.
Hirano T, Yasukawa K, Harada H, Taga T, Watanabe Y, Matsuda T, Kashiwamura S-i, Nakajima K, Koyama K, Iwamatsu A, Tsunasawa S, Sakiyama F, Matsui H, Takahara Y, Taniguchi T, Kishimoto T (1986) Complementary DNA for a novel human interleukin (BSF-2) that induces B lymphocytes to produce immunoglobulin. Nature 324:73-76.
Jernberg H, Nilsson K, Zech L, Lutz D, Nowotny H, Scheirer W (1987) Establishment and phenotypic characterization of three new human myeloma cell lines (U-1957) and U-1995). Blood 69:1605-1612.
Karpas A, Fischer P, Swirsky D (1982) Human myeloma cell line carrying a Philadelphia chromosome. Science 216:997-999.
Kawano M, Hirano T, Matsuda T, Taga T, Horii Y, Iwato K, Asaoku H, Tang B, Tanabe O, Tanaka H, Kuramoto A, Kishimoto T (1988) Autocrine generation and requirement of BSF-2/IL-6 for human multiple myelomas. Nature 332:83-85.
Kawano M, TanakaH, Ishikawa H, Nobuyoshi M, Iwato K, Asaoku H, Tanabe O, Kuramoto A (1989) Interleukin-1 accelerates autocrine growth of myeloma cells through interleukin-6 in human myeloma. Blood 73:2145-2148.
Kishimoto T (1989) The biology of interleukin-6. Blood 74:1-10.
Klein B, Jourdan M, Vazques A, Dugas B, Bataille R (1987) Production of growth factors by human myeloma cells. Cancer Res 47:4856-4860.
Klein B, Zhang X-G, Jourdan M,. Content J, Hooussiau F, Aarden L, Piechaczyk M, Bataille R (1989) Paracrine rather than autocrine regulation of myeloma-cell growth and differentiation by interleukin-6. Blood 73:517-526.
Lichtensetein A, Berenson J, Norman D, Chang M-P, Carlile A (1989) Production of cytokines by bone marrow cells obtained from patients with multiple myeloma. Blood 74:1266-1273, 1989.
Matsuda T,Hirano T, Nagasawa S, Kishimoto T (1989) Identification of alpha2-macroglobulin as a carrier protein for IL-6. J Immunol 142:148-152.
Miller C H, Carbonell A, Peng R, Paglieroni T, MacKenzie M R (1982) A human plasma cell line. Induction and characterization. Cancer 49:2091-2096.
Muraguchi A, Hirano T, Tang B, Matsuda T, Horii Y, Nakajima K, Kishimoto T (1988) The essential role of B cell stimulatory factor 2 (BSF-2/IL-6) for the terminal differentiation of B cells. J Exp Med 167:332-344.
Nilsson K, Bennich H, Johansson S G O, Pontén J (1970) Established immunoglobulin producing myeloma (IgE) and lymphoblastoid (IgG) cell lines from an IgE myeloma patient. Clin Exp Immunol 7:477-489.
Nilsson K (1989) IN: F Melchers, M Potter (eds.) Mechanisms of B cell neoplasia. Editiones Roche, Basel, Switzerland, pp 229-235.
Raynal M-C, Liu Z, Hirano T, Mayer L, Kishimoto T, Chen-Kiang S (1989) Interleukin 6 induces secretion of IgG1 by coordinated transcriptional activation and differential mRNA accumulation. Proc Natl Acad Sci USA 86:8024-8028.
Shimizu S, Yoshioka R, Hirose Y, Sugai S, Tachibana J, Konda S (1989) Establishment of two interleukin 6 (B cell stimulatory factor

2/interferon ß2)-dependent human bone marrow-derived myeloma cell lines. J Exp Med 169:339-344.
Smeland EB, Kiil Blomhoff H, Funderud S, Shalaby M R, Espevik T (1989) Interleukin 4 induces selective production of interleukin 6 from normal human B lymphocytes. J Exp Med 170:1463-1468, 1989.
Splawski J B, McAnally L M, Lipsky P E (1990) IL-2 dependence of the promotion of human B cell differentiation by IL-6 (BSF-2). J Immunol 144:562-569.
Suematsu S, Matsuda T, Aozasa K, Akira S, Nakano N, Ohno S, Miyazaki J-I, Yamamura K-I, Hirano T, Kishimoto T (1989) Proc Natl Acad Sci USA 86:7547-7551.
Tanabe O, Kawano M, Tanaka H, Iwato K, Asaoku H, Ishikawa H, Nobuyoshi M, Hirano T, Kishimoto T, Kuramoto A (1989) BSF-2/IL-6 does not augment Ig secretion but stimulates proliferation in myeloma cells. Am J Hematol 31:258-262.
Tohyama N, Karasuyama H, Tada T (1990) Growth autonomy and tumorigenicity of interleukin 6-dependent B cells transfected with interleukin 6 cDNA. J Exp Med 171:389-400.
Yamamoto I, Kawano M, Sone T, Iwato K, Tanaka H, Ishikawa H, Kitamura N Lee K, Shigeno C, Konishi J, Asaoku H, Tanabe O, Nobuyoshi M, Ohmoto Y, Hirai Y, Higuchi M, Ohsawa T, Kuramoto A (1989) Production of interleukin 1ß, a potent bone resorbing cytokine, by cultured human myeloma cells. Cancer Res 49:4242-4246.
Zhang X G, Klein B, Bataille R (1989) Interleukin-6 is a potent myeloma cell growth factor in patients with aggressive multiple myeloma. Bloo 74:11-13.

Interleukin 6 (IL-6) and Its Receptor (IL-6R) in Myeloma/Plasmacytoma

S. Suematsu, M. Hibi, T. Sugita, M. Saito, M. Murakami, T. Matsusaka, T. Matsuda, T. Hirano, T. Taga, and T. Kishimoto.

Institute for Molecular and Cellular Biology, Osaka University, 1-3, Yamada-oka, Suita, Osaka 565, Japan

INTRODUCTION

Interleukin 6 (IL-6) was originally identified as a B cell differentiation factor (BCDF/BSF2) which induces final maturation of B cells into antibody producing cells (Hirano *et al* 1986). However, subsequent studies with recombinant molecules revealed that IL-6 has a wide variety of biological functions on various tissues and cells (Kishimoto 1989). As shown in Fig. 1, IL-6 acts not only on B cells but also on hematopoietic progenitors and hepatocytes and is involved in hematopoiesis and acute phase reactions. It also acts on nerve cells, epidermal keratinocytes and kidney mesangium cells. One of the most interesting activities is the induction of the growth of myeloma/plasmacytoma cells. IL-6 is a potent growth factor for myelomas/plasmacytomas and only 0.002 ng/ml of rIL-6 could induce 50% of the maximum proliferation in a human myeloma cell line (Muraguchi *et al* 1988). This concentration of IL-6 is 100-fold less than that required for the immunoglobulin induction in B cells.

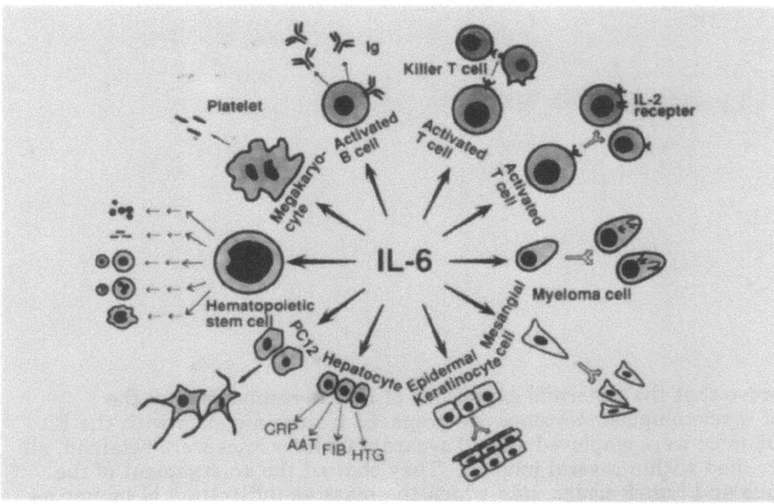

Fig. 1. Multifunction of IL-6

Current Topics in Microbiology and Immunology, Vol. 166

IL-6 AND MYELOMA/PLASMACYTOMA

Involvement of IL-6 in human multiple myelomas

The result suggests to us that deregulation of the IL-6 expression may be involved in the generation of myelomas/plasmacytomas. This possibility was examined with 24 cases of human multiple myelomas and the result obtained with freshly isolated myeloma cells was as follows; i) myeloma cells produce IL-6, ii) myeloma cells express IL-6 receptors, iii) IL-6 augments the *in vitro* growth of myeloma cells and iv) anti-IL-6 antibody inhibits the *in vivo* growth of myeloma cells. These results suggest that the autocrine mechanism of IL-6 may operate in the generation of human multiple myelomas (Kawano *et al* 1988).

γ_1 plasmacytosis in Eμ-IL-6 transgenic mice:

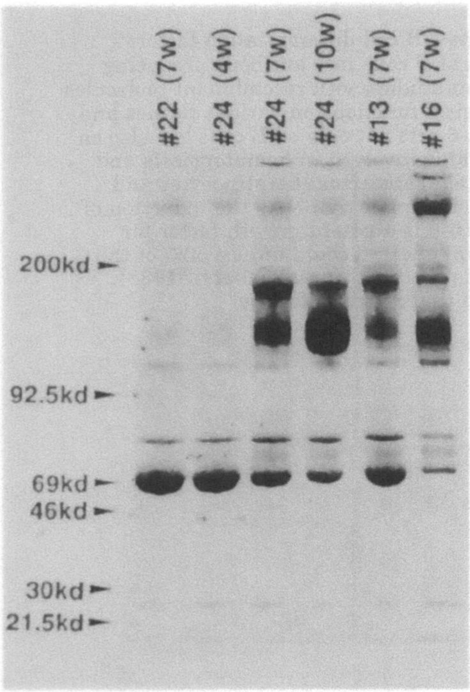

Fig. 2. SDS-PAGE analysis of the sera of IL-6 transgenic mice. #22 is a non-transgenic littermate.

In order to prove that the abnormal expression of IL-6 is responsible for the generation of myeloma/plasmacytoma, we prepared transgenic mice with the Eμ-IL-6 gene. B6 mice were employed and 10 separate founder mice were obtained, all of which were died within several months. They showed the enlargement of the spleen, thymus and lymph nodes, into which the massive infiltration of immature plasma cells were observed (Suematsu *et al* 1989). As shown in Fig. 2, SDS-PAGE analysis of the sera from the transgenic mice showed an increase in γ-globulin and a decrease in albumin. As described in Table 1, ELISA assay of the isotypes of immunoglobulins in their sera demonstrated the preferential increase in IgG_1 isotype but not other isotypes. However, Southern blot analysis of the DNA from the lymph nodes with the J_H probe revealed no monoclonal or oligoclonal bands of the Ig gene rearrangements. Moreover, plasma cells from the spleen or lymphnodes could not be transfered into the syngeneic recipients, indicating that tumors generated in the Eμ-IL-6 transgenic mice were not malignant monoclonal plasmacytomas.

Table 1. The preferential increase in IgG_1 in the sera of IL-6 transgenic mice.

No.(W)	IgM (mg/ml)	IgA	IgG1	IgG2a	IgG2b	IgG3	HuIL-6 (ng/ml)
13 (19)	0.4	0.5	170	0.1	2.0	0.4	0.8
16 (10)	0.6	1.0	75	0.4	0.4	0.3	20
24 (10)	2.0	1.3	240	0.4	3.0	0.2	1.9
4 (12)	0.3	1.0	140	0.1	1.7	0.03	1.4
11 (10)	3.4	1.7	140	0.1	2.4	0.2	2.4
18 (9)	5.4	5.4	61	0.1	2.8	0.02	12
26 (10)	0.4	0.5	35	0.1	2.5	0.05	3.4
33 (7)	0.6	1.3	60	0.1	1.5	0.1	8.8
39 (6)	6.4	2.6	175	0.1	1.4	0.01	4.0
46 (9)	4.6	2.8	60	0.1	1.5	0.2	1.0
50 (9)	12	0.8	150	0.2	4.5	0.3	N.D.
N (10)	0.8	1.2	0.6	0.2	1.7	0.2	<0.1
12 (10)	0.2	0.7	0.6	0.1	0.5	0.2	<0.1

Introduction of the BALB/c background into the B6 transgenic mice:

More than 20 years ago, Potter and his colleagues succeeded in the induction of plasmacytomas by intraperitoneal injection of pristane into BALB/c mice (Potter and Boyce 1962). Pristane induced granulomas which produced a large amount of a plasmacytoma growth factor and plasmacytomas were generated exclusively from the granuloma tissues. Recently, a murine plasmacytoma growth factor has been isolated and the result showed that a plasmacytoma growth factor was identical with IL-6. Therefore, the results strongly suggest that a paracrine mechanism of IL-6 may operate in pristane-induced plasmacytoma generation. All plasmacytomas induced by pristane showed the chromosomal translocation including the c-myc gene and they could be inducible only in BALB/c mice (Ohno *et al* 1979).

Therefore, attempt was made to introduce the genetic background of BALB/c mice into the IL-6 transgenic mice of B6 origin. In order to develop plasmacytosis more gradually in transgenic mice for the generation of their progeny, transgenic mice were prepared with the IL-6 gene fused with the L^d promoter. As shown in Fig. 3, the sera of the transgenic mice at 20 weeks did not show an increase in γ-globulin. On the other hand, BCF_1 mice between the IL-6 transgenic (B6) and BALB/c showed a large increase in γ-globulin at the same week. ELISA assay demonstrated that a preferential increase in IgG_1 isotype in the F_1 mice. Continuous backcrossing may provide an information about the genetic background for the generation of plasmacytomas or chromosomal translocation including the c-myc gene.

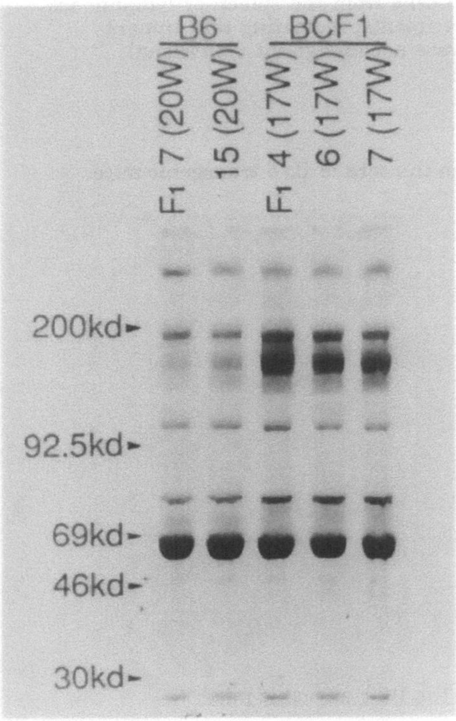

Fig. 3. Increase in γ-globulin in the sera of BCF_1 mice between L^dIL-6 transgenic and normal BALB/c mice.

Anti-IL-6 antibody prevents γ_1-plasmacytosis in IL-6 transgenic mice:

In order to prove that the generation of γ_1-plasmacytosis is due to the abnormal production of IL-6 in the transgenic mice, anti-human IL-6 monoclonal antibody was administered into the transgenic mice. An antibody (100 µg/mouse) was administered intraperitoneally three times a week starting at 4 weeks after birth. As shown in Table 2, an increase of the serum IgG_1 was prevented and the survival of the mice was prolonged by the administration of the antibody, although the effect was not complete, confirming that the abnormal production of IL-6 generates γ_1 plasmacytosis in transgenic mice.

Table 2. Prevention of IgG_1 plasmacytosis in IL-6 transgenic mice by the administration of anti-IL-6 antibody.

			IgG_1 (mg/ml)
Tg	Ab-	Eμ11-6 (8W)	23.3
		7 (8W)	71.7
	Ab+	Eμ11-9 (8W)	8.8
		10 (8W)	1.2
non-Tg		Eμ11-5 (8W)	0.7
		8 (8W)	0.5

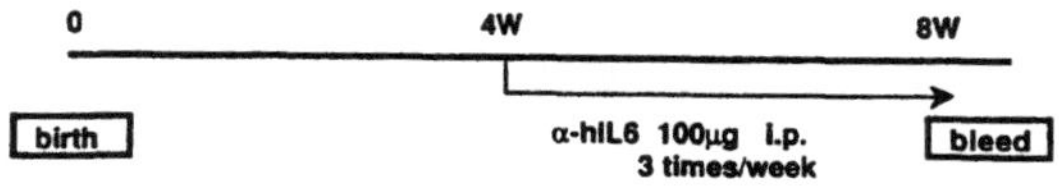

IL-6 RECEPTOR AND SIGNAL TRANSDUCTION

<u>IL-6 receptor belongs to a cytokine receptor family as well as to an immunoglobulin superfamily:</u>

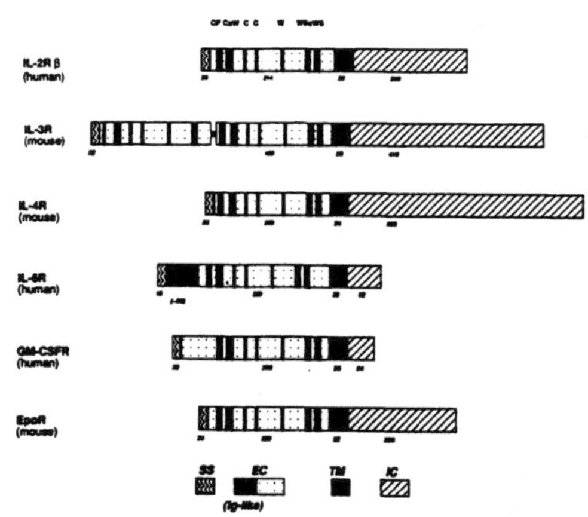

Fig. 4. Comparison of the structures of the receptors belonging to the cytokine receptor family.

IL-6 receptor (IL-6R) has been molecularly cloned by the expression cloning method with CDM8 vector (Yamasaki *et al* 1988). The receptor consists of 468 amino acid residues and has one immunoglobulin-like domain in the amino-terminal part of the extracellular portion. This is in contrast to the IL-1 receptor in which the entire extracellular portion consists of three-immunoglobulin-like domains (Sims *et al* 1988). Moreover, the Ig-like domain in IL-6 receptor was shown not to be involved in the ligand-binding. The rest of its extracellular portion was later found to belong to a cytokine-receptor family. From the cloning of the cDNAs for IL-2R β chain, IL-3, IL-4, IL-7, Epo and GM-CSF receptors, it was revealed that all these receptors possess certain common structural features in their extracellular portions. Four cysteine residues in the amino-terminal part and a Trp-Ser-X-Trp-Ser motif near the transmembrane portion and some other amino acids were commonly observed in analogous positions in these receptors (Fig. 4). These conserved amino acids may be important for constructing the basic structure of cytokine receptor molecules (Bazan 1989).

Interaction of IL-6R and its signal transducer (gp130):

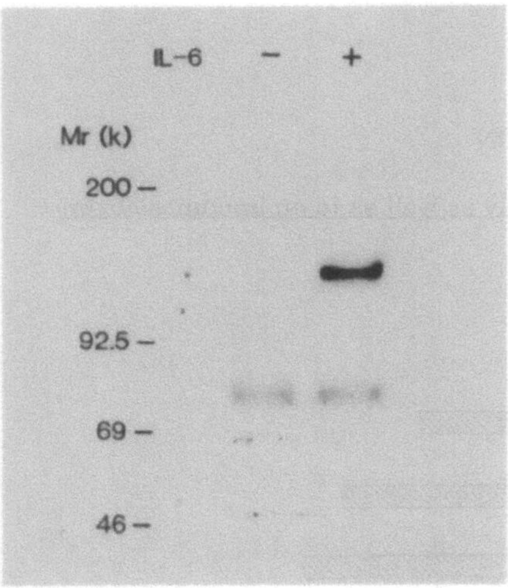

Fig. 5. Association of IL-6R and gp130 by IL-6 stimulation.

The precipitation of the cell lysates from a myeloma cell line, U266, with anti-IL-6R antibody showed that IL-6R consists of a highly glycosylated single polypeptide chain with a molecular weight of 80 kd (Hirata *et al* 1989). However, the intracytoplasmic portion of IL-6R has only 82 amino acid residues without any unique sequence for the signal transduction, suggesting the presence of another associated molecule responsible for the signal transduction. As shown in Fig. 5, when U266 myeloma cells were incubated with IL-6 for 15 - 30 min at 37°C and then their cell lysates were prepared in digitonin buffer, anti-IL-6R antibody could precipitate not only 80 kd IL-6R but also a second polypeptide chain with a MW of 130 kd (Taga *et al* 1989). The result suggests that the binding of IL-6 with IL-6R triggers the association of a polypeptide chain with a MW of 130 kd (gp130) with IL-6R. When mutated cDNA of IL-6R without its intractyoplasmic portion was transfected into a murine B lymphoma cell line, M12 which does not express IL-6R, anti-IL-6R could precipitate a murine gp130 together with a human IL-6R. The result indicates that the interaction between IL-6R and gp130 can occur between human and mouse and the intracytoplasmic portion of IL-6R is not required for its association. The result strongly suggests that the intra-cytoplasmic portion of IL-6R is not involved in its signal transduction and this is indeed the case. A murine myeloid leukemia cell line, M1, is responsive to IL-6 and IL-6 inhibits its growth and induces the differentiation into macrophages. When a cDNA for human IL-6R was transfected into M1 cells and the number of IL-6R was increased, M1 cells became more sensitive to IL-6 and much lower concentrations of IL-6 could induce the differentiation of M1 cells. Exactly the same dose response curve of IL-6 was observed when M1 cells transfected with the mutated cDNA without its intracytoplasmic portion were employed.

Soluble IL-6R showed agonistic effect:

Recombinant soluble IL-6R (rSIL-6R) without transmembrane and intracytoplasmic portions were expressed in CHO cells. rSIL-6R could bind IL-6 as efficiently as membrane-bound IL-6R. When rSIL-6R was incubated with a surface-iodinated B lymphoma line, M12 cells, in the presence of IL-6, anti-IL-6R antibody could precipitate a murine gp130. The result indicates that the complex between rSIL-6R and IL-6 can bind with a gp130 in its extracellular portion and confirmed the interaction between IL-6R and gp130 in its extracellular portion in the presence of IL-6.

The binding of the complex of rSIL-6R and IL-6 with gp130 could provide a signal into the cells through gp130. The incubation of M1 cells with rSIL-6R in the presence of IL-6 could inhibit their growth but rSIL-6R without IL-6 did not show any growth inhibitory effect on M1 cells.

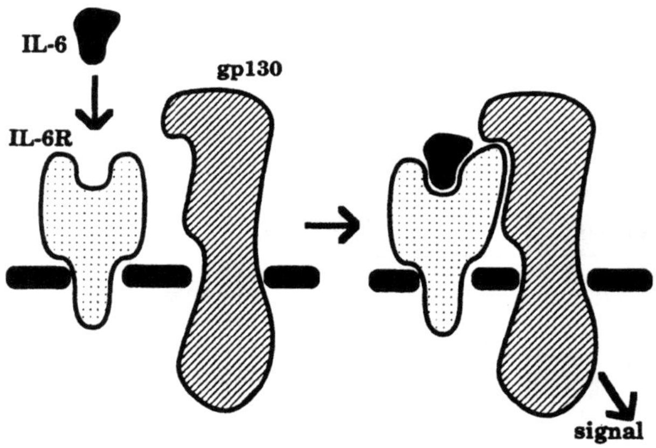

Fig. 6. Scheme for the IL-6R system.

The result was schematically summarized in Fig. 6. The IL-6 receptor system consists of two polypeptide chains, a legand binding molecule (80 kIL-6R) and a signal transducer (gp130). The binding of IL-6 with 80 kIL-6R triggers the association of gp130 and the signal is transduced through gp130. This receptor system may be applied to other cytokine receptor systems. Human and mouse IFNγ receptors were molecularly cloned and the study with the cDNA suggested that there might be a species-specific signal transducing molecules associating with the IFNγ receptor, since the expression of the human IFNγ receptor in mouse cells or the mouse IFNγ receptor in human cells using the cloned cDNAs was not sufficient for generating signals of human or mouse IFNγ, respectively (Agnet *et al* 1988; Gray *et al* 1989). GM-CSF receptor possesses a short intracytoplasmic portion compared with the other members of a cytokine receptor family. It has only 54 amino acid residues in the cytoplasm (Gearing *et al* 1989). The possible existence of the associated molecule might be expected from the result reported.

Rearrangement of the IL-6R gene in a plasmacytoma cell line:

The association of chromosomal aberrations with many murine and human tumors suggests that DNA rearrangements may constitute a general mechanism for tumor induction. The slowly transforming retroviruses were shown in some cases to activate transcription of a particular cellular *onc* gene. The intracisternal A particle (IAP) gene is a member of endogenous proretroviral-like elements present in ~1,000 copies per haploid genome of *Mus musculus*. IAP genes function as movable elements in the mouse genome. The possible role of the IAP genes in oncogenesis was suggested in several studies (Canaani *et al* 1983; Schrader and Crapper 1983).

We isolated two species of the cDNAs encoding murine IL-6R; one is abnormal and the other authentic, from a plasmacytoma cell line and normal spleen cells, respectively. As shown in Fig. 7, the extracellular portion of both murine IL-6Rs isolated from plasmacytoma cells and normal spleen cells were completely identical and their overall homology with human IL-6R was 66 and 48% at the DNA and the protein levels, respectively.

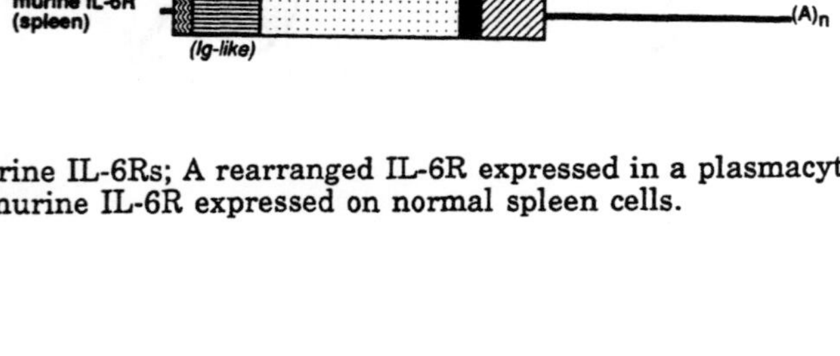

Fig. 7. Murine IL-6Rs; A rearranged IL-6R expressed in a plasmacytoma and an authentic murine IL-6R expressed on normal spleen cells.

The domain of an Ig superfamily found in the human IL-6R was also identified in the molecule encoded by the cloned cDNA. All cysteine residues present in the human IL-6R were found to be conserved in this murine molecule, except for Cys^{211} (human). Furthermore, Trp^{134} (human) residue and Trp^{303}-Ser-Xxx-Trp-Ser^{307} (human) sequence were also found at positions 130 and 300-304, respectively. However, the intracytoplasmic portions of these two IL-6Rs were completely different and only intracytoplasmic portion of the receptor isolated from normal spleen cells showed a significant homology with that of human IL-6R. It was found that the nucleotide sequence of the cDNA corresponding to the intracytoplasmic domain of the receptor isolated from plasmacytoma cells showed a striking homology (94%) with a part of an LTR of an IAP gene (Sugita et al 1990).

Northern blot analysis utilizing the cDNA fragment specific to the extracytoplasmic domain detected two species of mRNA (1.8 kb and 5.5 kb) in a plasmacytoma cell line (P3U1). One major mRNA species had a compatible size (1.8 kb) with the insert cDNA cloned from a plasmacytoma line. The other mRNA species, the expression of which was much lower than the former one, had a larger size (5.5 kb), compatible with that of human IL-6R and murine IL-6R from normal spleen cells. Furthermore, the former mRNA species was shown to be hybridized with an IAP-LTR specific cDNA fragment but the latter mRNA was not. Moreover, only the second species of the mRNA (5.5 kb) was expressed in the spleen, thymus, and liver of BALB/c mouse.

To examine whether the IL-6R with its intracytoplasmic portion encoded by a part of the IAP-LTR gene can transduce the signals, normal or abnormal IL-6R cDNA was transfected into an IL-6-dependent human T cell line, KT-3. Human IL-6 induced cell growth of KT3 cells, but murine IL-6 did not. However, KT-3 cells transfected with either normal or abnormal cDNA responded to murine IL-6 to proliferate. The data indicated that the IL-6R, in which an intracytoplasmic domain was replaced with the molecule encoded by a part of the IAP-LTR gene, could function and transduce the signals. This was in complete agreement with our previous observation with human IL-6R that an intracytoplasmic domain was not required for the signal transduction. Since this replacement could be responsible for the overexpression of IL-6R on a plasmacytoma cell line, the integration of the IAP into the receptor gene may function as a positive selection element for the development of certain plasmacytomas.

ACKNOWLEDGEMENT

This work was supported by Grants-in-Aid for Specially Promoted Research from the Japanese Ministry of Education, Science and Culture. The preparation of the transgenic mice was done by the collaboration with Dr. K. Yamamura, Kumamoto University. We thank Ms. M. Harayama for her editorial assistance.

REFERENCES

Aguet M, Dembic Z, Merlin G (1988) Molecular cloning and expression of the human interferon-γ receptor. Cell 55: 273-280.

Bazan JF (1989) A novel family of growth factor receptors: a common binding domain in the growth hormone, prolactin, erythropoietin and IL-6 receptors, and the p75 IL-2 receptor β chain. Biochem Biophys Res Commun 164: 788-795.

Canaani E, Dreaxen O, Klar A, Rechavi G, Ram D, Cohen JB, Givol D. (1983) Activation of the c-mos oncogene in a mouse plasmacytoma by insertion of an endogenous intracisternal A-particle genome. Proc Natl Acad Sci USA 80: 7118-722.

Gearing DP, King JA, Gough NM, Nicola NA (1989) Expression cloning of a receptor for human granulocyte=macrophage colony-stimulating factor. EMBO J 8: 3667-3676.

Gray PW, Leong S, Fennie EH, Farrar MA, Pingel JT, Fernandez-Luna J, Schreiber RD (1989) Cloning and expression of the cDNA for the murine interferon γ receptor. Proc Natl Acad Sci USA 86: 8497-8501.

Hirano T, Yasukawa K, Harada H, Taga T, Watanabe Y, Matsuda T, Kashiwamura S, Nakajima K, Koyama K, Iwamatu A, Tsunasawa S, Sakiyama F, Matsui H, Takahara Y, Taniguchi T, Kishimoto T (186) Complementary cDNA for a novel human interleukin (BSF-2) that induces B lymphocytes to produce immunoglobulin. Nature 324: 73-76.

Hirata Y, Taga T, Hibi M, Nakano N, Hirano T, Kishimoto T (1989) Characterization of IL-6 receptor expression by monoclonal and polyclonal antibodies. J Immunol 143: 2900-2906.

Kawano M, Hirano T, Matsuda T, Taga T, Horii Y, Iwato K, Asaoku H, Tang B, Tanabe O, Tanaka H, Kuramoto A, Kishimoto T (1988) Autocrine generation and essential requirements of BSF-2/IL-6 for human multiple myeloma. Nature 332: 83-85.

Kishimoto T (1989) The biology of interleukin 6. Blood 74: 1-10.

Muraguchi A, Hirano T, Tang B, Matsuda T, Horii Y, Nakajima K, Kishimoto T (1988) The essential role of B cell stimulatory factor 2 (BSF-2/IL-6) for the terminal differentiation of B cells. J Exp Med 167: 332-344.

Ohno S, Babonits M, Wiener F, Spira J, Klein G, Potter M (1979) Non-random chromosome changes involving the Ig-gene carrying chromosomes 12 and 6 in pristane-induced mouse plasmacytomas. Cell 18: 1001-1007.

Potter M, Boyce C (1962) Induction of plasma-cell neoplasms in strain BALB/c mice with mineral oil and mineral oil adjuvants. Nature 193: 1086-1087.

Schrader JW, Crapper RM (1983) Autogenous production of ahemopoietic growth factor, persisting-cell-stimulating factor, as a mechanism for transformation of bone marrow-derived cells. Proc Natl Acad Sci USA 80: 6892-6896.

Sims JE, March CJ, Cosman D, Widmer MB, MacDonald HR, McMahan CJ, Grubin CE, Wignall JM, Jackson JL, Call SM, Friend D, Alpert AR, Gillis S, Urdal DL, Dower SK (1988) cDNA expression cloning fo the IL-1 receptor, a member of the immunoglobulin superfamily. Science 241: 585-589.

Suematsu S, Matsuda T, Aozasa K, Akira S, Nakano N, Ohno S, Miyazaki J, Yamamura K, Hirano T, Kishimoto T (1989) IgG1 plasmacytosis in interleukin 6 transgenic mice. Proc Natl Acad Sci USA 86: 7547-7551.

Sugita T, Totsuka T, Saito M, Yamasaki K, Taga T, Hirano T, Kishimoto T (to be published) Functional murine interleukin 6 receptor with the intracisternal a particle gene product at its cytoplasmic domain. J Exp Med

Taga T, Hibi M, Hirata Y, Yamasaki K, Yasukawa K, Matsuda T, Hirata T, Kishimoto T (1989) Interleukin-6 triggers the association of its receptor with a possible signal transducer, gp130. Cell 58: 573-581.

Yamasaki K, Taga T, Hirata Y, Yawata H, Kawanishi Y, Seed B, Taniguchi T, Hirano T, Kishimoto T (1988) Cloning and expression of the human interleukin-6 (BSF-2/IFNβ) receptor. Science 241: 825-828.

Interleukin-6 is a Major Myeloma Cell Growth Factor In Vitro and In Vivo Especially in Patients with Terminal Disease

B. Klein, X.G. Zhang, M. Jourdan, M. Portier, and R. Bataille

INSERM U291, 99 rue Puech Villa, 34100- Montpellier, France

INTRODUCTION

Multiple myeloma (MM) is a B cell malignancy characterized by the slow proliferation, primarily in the bone marrow, of a clone of malignant plasma cells that destroy the bone tissue. This disease is lethal in any case and little therapeutic progress has been achieved these last 20 years. Although myeloma cells poorly proliferate in vivo, the in vivo labelling index of the myeloma cells (LI = percentage of myeloma cells in the S-phase) constitutes one of the best prognostic factors (Durie 1980). This emphasizes clearly the importance of knowing the cytokines that control tumoral growth for further improving the therapeutic control of this disease.

The aim of this report is to review the recent studies of our group demonstrating that, as regard to the growth and differentiation factors, human MM resembles closely murine plasmacytomas: i.e. interleukin-6 (IL-6) is a major growth factor for the tumor cells in vitro and in vivo, and is produced in large amounts by the tumoral environment.

IL-6 IS A MAJOR MYELOMA-CELL GROWTH FACTOR IN VITRO AND IN VIVO

IL-6 is a major myeloma-cell growth factor in vitro

We and others (Klein 1989a, Kawano 1988) have previously shown that IL-6 is a potent myeloma-cell growth factor in vitro. This demonstration is now greatly facilitated by the recent development of strongly neutralizing anti-IL-6 monoclonal antibodies (mab).

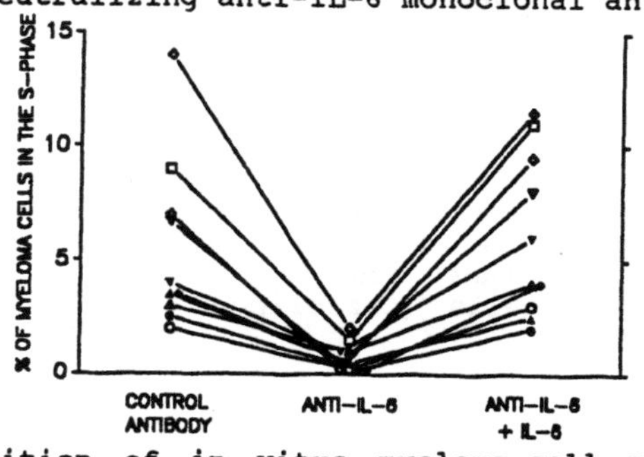

Fig. 1. Inhibition of in vitro myeloma-cell proliferation by an anti-IL-6 monoclonal antibody.

As indicated in Fig. 1, when bone-marrow cells from MM patients are cultured for 5 days in vitro, the spontaneous myeloma-cell proliferation occurring in the cultures is strongly inhibited by a potent anti-IL-6 mab (BE-8, John Wijdenes). The inhibitory effect of the BE-8 mab is reversed by adding an excess of recombinant (r) IL-6.

IL-6 is produced by the tumoral environment in vitro and in vivo

In a previous report (Klein 1989a), we have indicated that there was little evidence for supporting the attractive hypothesis of an autocrine production of IL-6 by myeloma cells as initially claimed by Kawano et al. (1988): these authors have in fact only shown that populations enriched in myeloma cells (still containing non-myeloma cells) produced a minor IL-6 activity in comparison to the huge amounts produced by the tumoral environment.

We have now confirmed our initial paracrine hypothesis by three lines of evidence: 1) there is a significant IL-6 mRNA expression in the bone marrow cells from most of the myeloma patients in vivo (13 out of 19 patients) and **this IL-6 mRNA is expressed mainly by monocytic and myeloid cells and not by purified myeloma cells** (Portier 1990) 2) we can now reproducibly get myeloma cell lines of which proliferation is completely dependent on exogenous IL-6 and which are not producing IL-6 (Klein 1989b and see in the following) 3) we have studied a MM patient with a pleural effusion. Pleural cells were found to comprise 95% myeloma cells and to express high levels of IL-6 mRNA. Purified myeloma cells (>98% myeloma cells) obtained after removal of monocytes, T cells and adherent cells did not express IL-6 mRNA. When pleural cells were put into culture, non-adherent myeloma cells proliferated closely in association with a feeder layer of autologous stromal adherent cells producing large amounts of IL-6. After one month of culturing, the adherent fibroblast-like cells died and a myeloma-cell line of which proliferation was completely dependent on exogenous IL-6 was obtained (Zhang, in preparation). **This last observation suggests that myeloma cells proliferate in close contact and probably bound to IL-6-producing accessory cells in vivo and allows to understand the difficulty to obtain pure myeloma-cell populations devoided of IL-6-producing accessory cells.**

In a recent study by Tohyama (1990), it was shown that injections of an exogenous IL-6-dependent hybridoma B cell line to mice resulted in a development of mixed tumors composed of hybridoma cells and host stromal cells and in the subsequent death of mice. Cultures of these mixed tumors yielded to the reappearance of the initial exogenous-IL-6-dependent hybridoma clone. Human MM looks very similar to this model and it is possible that **human myeloma cells produce mediators inducing the bone marrow environment to produce IL-6.** The identification of these mediators may be critical for blocking the interactions between tumor cells and its environment. IL-1β could be one of these mediators as it has been shown to be produced by myeloma cells (Cozzolino 1989, Kawano 1989).

IL-6 is involved in myeloma-cell proliferation in vivo

As indicated above, we have found an in-vivo expression of IL-6 mRNA in the bone marrows of myeloma patients unlike bone-marrow cells of healthy donors. We have also reported **increased serum IL-6 levels in patients with MM, the highest levels being obtained in patients with terminal disease and plasma-cell leukemia** (Bataille 1989). Moreover, a significant correlation between the in-vivo myeloma-cell proliferation and the serum IL-6 levels was noticed (Bataille 1989). This constitutes a first argument in favor of an involvement of IL-6 in myeloma-cell proliferation in vivo. An other argument is that **the best myeloma-cell responses to IL-6 in vitro are obtained with myeloma cells from patients with active disease and proliferating myeloma cells in vivo** (Zhang 1989). Myeloma cells which are not proliferating in vivo are generally not responding to IL-6 in vitro (Zhang 1989).

Our results absolutely challenge a recent report emphazising a decreased responsiveness to IL-6 in advanced stages of MM (Asaoku 1988). We have now studied 30 myeloma patients with advanced disease (comprising 14 plasma-cell leukemias) and for each patient the in-vitro myeloma-cell proliferation was completely dependent on exogenous IL-6 (Zhang, in preparation). Taken together, these observations emphazise the use of anti-IL-6 therapeutics in patients with advanced disease and resistant to standard chemotherapy.

A monoclonal antibody (mab) to IL-6 blocks myeloma-cell proliferation in vivo.

We have treated a patient with a primary plasma-cell leukemia resistant to chemotherapy and presenting with anemia, hypercalcemia, fever, in vivo proliferating myeloma cells (LI = 4.5 %), and high serum CRP levels (70 mg/L). Before treating the patient, we have checked that the myeloma-cell proliferation was completely blocked by the anti-IL-6 mab in vitro. The patient was treated for 2 months with a 10 mg daily dose of BE-8 mab prepared by J. Wijdenes (CRTS, Besançon, France) for human use with the following observations:

- no side effects related to mab treatment were noticed during the 2-month treatment.
- body temperature, which was stabilized at 38°C before treatment, returned to normal levels all along the treatment (about 37°C).
- the CRP levels were rapidly reduced within the 10 first days of treatment, remained undetectable (< 1mg/L) all along the treatment, and rose again 8 days after the cessation of the treatment (Fig. 2).
- the hypercalcemia was significantly reduced during the 2-month treatment.
- no immnunization against the anti-IL-6 mab was detectable all along the treatment.
- serum IL-6 levels were undetectable before and all along the treatment but rose rapidly 5 days after cessation of treatment.
- No in-vivo myeloma-cell proliferation was detectable during one month and an half of treatment (Fig. 2). A significant myeloma-cell proliferation was again detected at the end of treatment, indicating that the patient was progressively escaping the treatment (Fig. 2). In agreement, the serum myeloma protein was reduced by 25% during one month and an half of treatment and rose again at the end of treatment. Of interest, the in-vitro proliferation of the myeloma

cells harvested after 2 months of treatment (when proliferating myeloma cells were again detected in vivo) was still completely dependent of IL-6 and blocked by the anti-IL-6 BE-8 mab.

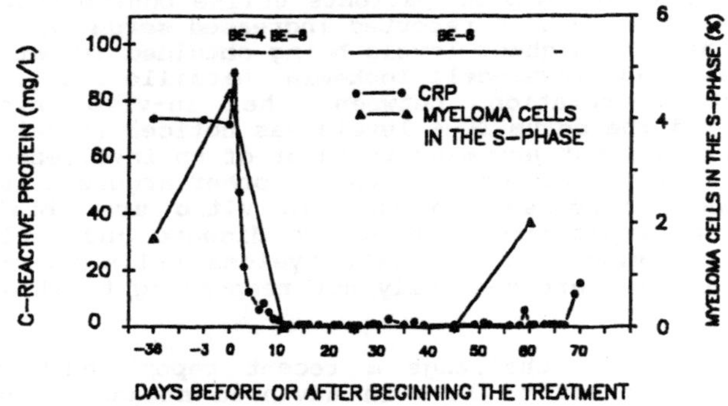

Fig. 2: In-vivo myeloma-cell LIs and serum CRP levels during treatment with the anti-IL-6 mab.

This first anti-IL-6 clinical trial allows to conclude that IL-6 is effectively the major myeloma-cell growth factor in vivo and that IL-6 completely controls the CRP production by human hepatocytes in vivo in agreement with in-vitro data (Heinrich 1990). It emphasizes the further use of anti-IL-6 therapeutics which apparently can be safety administrated without detectable side effects. It is important to understand why the patient has escaped the anti-IL-6 treatment. Indeed, at the end of treatment, no immnunization was detected, no biologically active IL-6 was available for the hepatocytes (undetectable CRP levels), and the in-vitro myeloma-cell proliferation was still completely dependent on IL-6. One possibility is that the anti-IL-6 mab was no longer efficient to block the IL-6 activity in the tumoral environment unlike the liver environment due either to a selection of an IL-6-hypersensitive myeloma-cell clone, and/or to the release of soluble IL-6 receptors in the bone-marrow environment able to compete with the anti-IL-6 mab and to activate myeloma cells (Novick 1989), and/or an increased IL-6 production by the tumoral environment. We are now investigating these different possibilities.

OTHER CYTOKINES CONTROLLING MYELOMA-CELL PROLIFERATION

Granulocyte-macrophage colony stimulating factor (GM-CSF) and interleukin-3 (IL-3) are potent stimulators of the myeloma-cell response to IL-6

As the obtention of GM-CSF-dependent plasmacytoma cell lines have been reported in mice (Vink 1988) , we have studied the effect of this hematopoietic growth factor in human MM. We have demonstrated that **GM-CSF is a potent myeloma-cell growth factor, but that its stimulatory action is completely inhibited by specific anti-IL-6 mabs** (Klein 1989c and Zhang, submitted for publication). GM-CSF is spontaneously produced in the cultures of myeloma bone marrow and contributes to the spontaneous myeloma-cell proliferation occurring in these cultures. We have no evidence that GM-CSF acts by modulating the endogenous IL-6 production since additions of either rGM-CSF or mab to GM-CSF do not affect this production (Klein 1989c and Zhang, submitted). In fact, GM-CSF acts by upregulating the response of fresh myeloma cells to IL-6 as evidenced by the recent appearance of IL-6-dependent myeloma cell lines (Klein 1989b and Zhang, submitted for publication). By using Scatchard analyses, we have presently found no upregulation of the first chain of the IL-6 receptor by GM-CSF.

IL-3 is also able to stimulate myeloma-cell proliferation by upregulating their response to IL-6 without having any detectable effect on the IL-6 production (Zhang, in preparation). These results are in agreement with a recent report by Bergui et al. (1989) showing that IL-3 and IL-6 together can induce a major differentiation of circulating myeloma-cell precursors.

These results are of major importance considering the possible therapeutical use of these cytokines for improving hematopoietic recovery in myeloma patients. We are now rapidly investigating whether other hematopoietic cytokines share the same myeloma-cell growth factor activities.

Reproducible generation of myeloma cell lines of which proliferation is completely dependent on exogenous IL-6.

By using a combination of Il-6 and GM-CSF, we have now obtained five different myeloma cell lines (termed XG-1 to XG-5) from fresh myeloma cells isolated from 5 out of 6 patients with extramedullary proliferation studied (Klein 1989b and Zhang, submitted for publication). No cell line has been presently obtained from bone-marrow cells of patients with only a bone-marrow tumoral invasion. As outlined in the following Fig. 3, no cell line could be obtained with IL-6 alone.

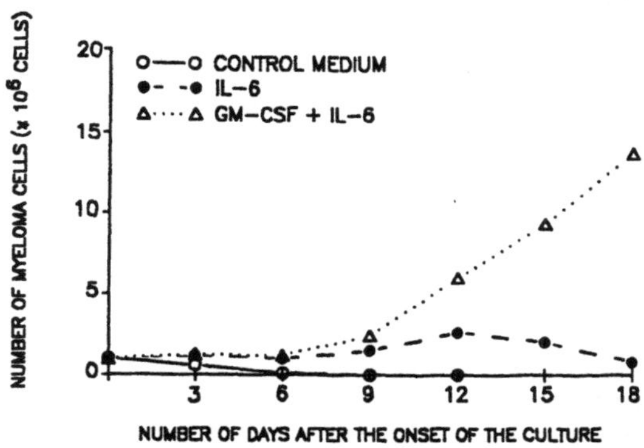

Fig. 3. Circulating fresh myeloma cells from a patient with a plasma-cell leukemia were cultured with either culture medium alone, or IL-6 (1000 U/ml), or IL-6 (1000 U/ml) and GM-CSF (10 ng/ml) together. Results are the estimated total myeloma-cell numbers generated in the different culture groups.

These myeloma cell lines are composed only of myeloma cells expressing the same immunoglobulin heavy and light genes rearrangements than the fresh tumor cells. After one or two months

of culturing, these myeloma cells could be grown with IL-6 alone unlike GM-CSF alone.

The reproducible and rapid ability to obtain exogenous-IL-6-dependent myeloma cell lines demonstrates that at least in the patients studied the myeloma stem cell, capable of self renewal, has a plasma-cell phenotype. It confirms our previous studies (Klein 1989a, Portier 1990) showing that IL-6 is a major myeloma-cell growth factor and that the great majority of myeloma cells are not producing IL-6 in vivo.

Selection for autocrine IL-6 producing myeloma cell subclones by interferon-alpha (IFN-α) or TNFα in vitro.

After several months of culture, our first IL-6-dependent myeloma cell line (termed XG-1) has increased by 10 fold its proliferative potential in response to IL-6 and is no longer sensitive to GM-CSF. The XG-1-cell proliferation is always completely dependent on exogenous IL-6, and XG-1 cells die within about 10 days without IL-6 (Klein 1989b and Zhang, submitted). Recently, we have found that **IFN-α or TNFα are also able to significantly stimulate XG-1 growth without addition of exogenous IL-6; further, XG-1 subclones growing with IFN-α or TNFα alone for several months have been easily obtained. Of major interest, the proliferation of these IFN-α- or TNFα-dependent XG-1 subclones was completely abrogated by anti-IL-6 mabs and these XG-1 subclones were found to produce their own autocrine IL-6 as an autocrine growth factor** (Jourdan 1990a and in preparation). In addition, the proliferation of these XG-1 sub-clones, obtained with IFN-α or TNFα, was no longer absolutely dependent on exogenous cytokines (although still increased by about 10 fold by addition of IFN-α , TNFα or IL-6) and autonomously growing XG-1 subclones could be obtained. **These XG-1 subclones were found to also produce IL-6 as an autocrine growth factor, without the need of exogenous IFN-α or TNFα stimulation** (Jourdan 1990 and in preparation). We are now investigating the mechanisms underlying this autocrine IL-6 production. An attractive hypothesis is that these autonomously-growing XG-1 subclones are producing autocrine IFN-α or TNFα which secondarily induced an autocrine IL-6 production.

To reconcile these recent data with the previous data demonstrating a paracrine production of IL-6, we can hypothesize that the great majority of myeloma cells do not produce IL-6 in vivo but that there might exist a minority of myeloma cells producing an autocrine IL-6. As IL-6 is produced in large amounts by the tumoral environment in vivo, there is no real selective advantage for the autocrine-IL-6-producing subclones. When these cells are cultured without adding exogenous IL-6 in vitro, the IL-6 produced by the autocrine IL-6-producing subclones must be rapidly absorbed by the IL-6-non-producing myeloma cells which are in a great majority and all myeloma cells (producing or not IL-6) rapidly die. In the presence of cytokines like IFN-α or TNFα, an autocrine production of IL-6 must be induced in many myeloma cells allowing the progressive emergence of the subclones initially producing an autocrine IL-6.

IFN-α has recently been shown to increase the plateau phase duration in patients responsive to chemotherapy (Mandelli 1989). These in-vivo results are not necessarily in contradiction with the present results as we have shown that myeloma cells from patients with inactive disease and no proliferating myeloma cells in vivo

were not responding to IL-6 in vitro (Zhang 1989). However, we can fear that in patients in early relapse a continuous injection of IFN-α might rapidly allow the emergence of an autocrine-IL-6-producing myeloma subclone and thus aggravate the disease. Such a fulminating relapse following IFN-α treatment and dependent on the IFN-α treatment has recently been observed by Bladé et al.(personnal communication). It seems therefore reasonable that to exclude such a scenario in vivo.

Interferon-gamma (IFN-gamma) is a potent inhibitor of myeloma-cell proliferation.

We have found that IFN-gamma was able to completely inhibit the IL-6-dependent myeloma cell proliferation in all the 25 patients studied (Zhang, in preparation). IFN-gamma did not affect the endogenous IL-6 production but was directly acting on myeloma cells. In addition, IFN-gamma also completely inhibited the proliferation of our 5 IL-6-dependent myeloma cell lines (Zhang, in preparation). Preliminary data suggest that IFN-gamma is able to down-regulate the expression of the first chain of the IL-6 receptor complex. These results emphasize a potential therapeutical use of IFN-gamma in MM as this cytokine also inhibits the cytokine-mediated bone resorption which is one of the major clinical problem in MM.

CONCLUSION

As schematized in Fig. 4, we have reviewed our recent results showing that IL-6 is a major and central myeloma-cell growth factor in vitro and in vivo. Thus human MM looks very similar to the murine plasmacytoma model, probably because IL-6 is a major growth factor for murine (Suetmatsu 1989) and human (Jourdan 1990b) normal plasma cells.

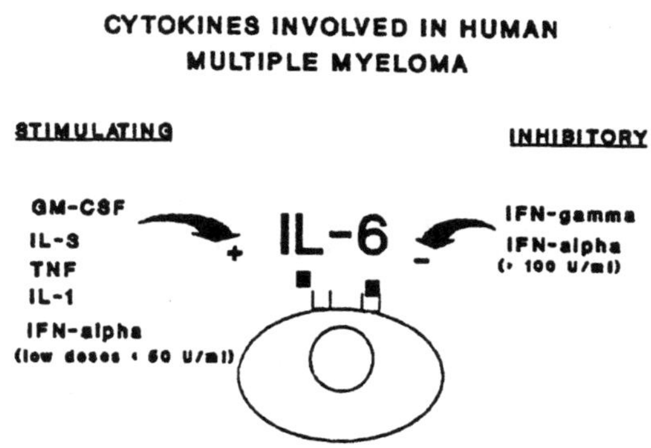

Fig. 4. Cytokines involved in human multiple myeloma

As in the murine plasmacytoma model, IL-6 is produced by the tumoral environment in vivo and not by the great majority of fresh myeloma cells. Other cytokines can stimulate myeloma-cell growth but until now, all these cytokines have been found to mediate their effects through IL-6.

GM-CSF and IL-3 increase the response to IL-6 of myeloma cells. IFN-α and TNFα induce an autocrine production of IL-6 by myeloma cells. Finally, IFN-gamma and high doses of IFN-α inhibit myeloma-cell proliferation probably by antagonizing the response to IL-6. The discovery of the major cytokines involved in MM has immediate clinical applications in this disease for which very little therapeutical progress has been achieved in the last 20 years. In particular, we have shown that anti-IL-6 therapeutics could efficiently block myeloma-cell proliferation in vivo. Moreover, the fact that several cytokines presently used in patients with MM in vivo are actually potent in-vitro tumoral growth factors emphasize the necessity to develop serious and extensive biological studies in patients receiving these cytokines before to generalize their clinical use.

ACKNOWLEDGEMENTS

This work was supported by grants from l'Association de Recherches contre le Cancer et la ligue Nationale de Lutte contre le Cancer (Paris, France

REFERENCES

Asaoku H, Kawano M, Iwato K, Tanbe O, Hirano T, Kishimoto T, Kuramoto A (1988) Decrease in BSF-2/IL-6 response in advanced cases of multiple myeloma. Blood 72:429-432.

Bataille R, Jourdan M, Zhang XG, Klein B (1989) Serum levels of interleukin-6, a potent myeloma cell growth factor, as a reflect of disease severity in plasma cell dyscrasias. J. Clin. Invest. 84:2008-2011.

Bergui L, Schena M, Gaidano G, Riva M, Caligaris-Cappio F (1989) Interleukin 3 and interleukin 6 synergistically promote the proliferation and differentiation of malignant plasma cell precursors in multiple myeloma. J Exp Med 170:613-618.

Cozzolino F, Torcia M, Aldinucci D, Rubatelli A, Miliani A, Shaw AR, Landsdorp P, Di Gugliemo R (1989) Production of interleukin 1 by bone marrow myeloma cells. Blood 74:380-387.

Durie BGM, Salmon SE, Moon TE (1980) Pretreatment tumor mass, cell kinetics, and prognosis in multiple myeloma. Blood 55:364-372.

Heinrich PC, Castell JV, Andus T (1990) Interleukin-6 and the acute phase response. Biochem. J. 265:621-636.

Kawano M, Hirano T, Matsuda T, Taga T, Iwato K, Asaoku H, Tang B, Tanabe O,, Tanka H, Kuramoto A, Kishimoto T (1988) Autocrine generation and requirement of BSF-2/IL-6 for human multiple myeloma. Nature 332:83-85.

Kawano M, Yamamoto I, Iwato K, Tanaka H, Asaoku H, Tanabe O, Ishikawa H, Nobuyoshi M, Ohmoto Y, Hirai Y, Kuramoto A (1989) Interleukin-1 beta rather than lymphotoxin as the major bone resorbing activity in human multiple myeloma. Blood 73:1646-1649.

Jourdan M, Zhang XG, Portier M, Bataille R, Klein B (1990a) Selection of autocrine IL-6-producing myeloma cell subclones by interferon-alpha in vitro. French Society of Immunology, Abstr., September 27-28, 1990.

Jourdan M, Bataille R, seguin J, Zhang XG, Chaptal PA, Klein B (1990b) Constitutive production of interleukin-6 and immunologic features in cardiac myxomas. Arth. and Rheum 33:398-403.

Klein B, Zhang XG, Jourdan M, Content J, Houssiau F, Aarden L, Piechaczyck, Bataille R (1989a) Paracrine rather than autocrine regulation of myeloma-cell growth and differentiation by interleukin-6. Blood 73:517-526.

Klein B, zhang XG, Jourdan M, Duperray C, portier M, Cantaloube JF, Haagen I, Bataille R (1989b) An IL-6 dependent human myeloma cell line: crucial role of hematopoietic growth factors in its rapid generation. Blood 74(suppl 1):749

Klein B, Zhang XG, Jourdan M, Bataille R (1989c) GM-CSF synergizes with IL-6 in supporting the proliferation of human myeloma cells. Blood 74(suppl 1): 898

Mandelli F, Avvisati G, Petrucci MT (1989) Interferon: a new strategy in the treatment of multiple myeloma. Eur J Haematol suppl 51, 43:129-134.

Novick D, Engelmann H, Wallach D, Rubinstein M (1989) Soluble cytokine receptors are present in normal human urine. J Exp med 170:1409-1414.

Suematsu S, Matsuda T, Aozasa K, Akira S, Nakano N, Ohno S, Miyazaki J, Yamamura K, Hirano T, Kishimoto T (1989) IgG1 plasmacytosis in interleukin-6 transgenic mice. 86:7547-7551.

Tohyama N, Karasuyama H, Tada T (1990) Growth autonomy and tumorigenicity of interleukin-6-dependent B cells transfected with interleukin 6 cDNA. J Exp med 171:389-400.

Zhang XG, Klein B, Bataille R (1989) Interleukin-6 is a potent myeloma-cell growth factor in patients with agressive multiple myeloma. Blood 74:11-13.

Macrophages as an Important Source of Paracrine IL6 in Myeloma Bone Marrow

Brian G.M. Durie, Elizabeth E. Vela, Yvette Frutiger*.

Department of Haematology, Charing Cross and Westminster Medical School, University of London, Charing Cross Hospital, Fulham Palace Road, London W6 8RF
The Arizona Cancer Center, College of Medicine, 1515 North Campbell Avenue, Tucson, Arizona 85724*.

INTRODUCTION

Interleukin 6 (IL6) is a pleiotropic cytokine which has been shown to be a very potent growth factor for myeloma cells, particularly cells derived from patients with aggressive clinical disease [1,2]. Although it has been documented that myeloma cells have large numbers of IL-6 receptors and that both the bone marrow and serum of patients with active myeloma contain increased levels of IL6, the source(s) and regulation of IL6 levels are not fully understood. In some studies it has been indicated that the IL6 growth stimulatory loop appears to be autocrine in nature [3]: whereas other data show that a paracrine loop is more likely with the predominant source of IL6 being accessory cells, such as macrophages, fibroblasts and probably other bone marrow cells [4]. The current report is a preliminary analysis of a large study in which both cell separation and anti IL6 blocking techniques (both IL6 and IL6 receptor) were used to evaluate the sources and functions of IL6 in myeloma bone marrow specimens. The data indicate that the paracrine mechanism is crucial.

MATERIALS AND METHODS

Bone marrow samples were obtained in preservative free heparin, from 93 patients with multiple myeloma or monclonal gammopathy of undetermined significance (MGUS). The cells were processed and cultured by a variety of techniques, (all previously published), including soft agar, methylcellulose and liquid culture methods [5]. Preliminary T lymphocyte depletion using neuraminidase treated sheep red blood cells was carried out to ensure rigorous T cell lymphocyte removal prior to culturing. Monocyte/macrophages were selectively removed using L-leucine methyl ester (LME)[6].

Current Topics in Microbiology and Immunology, Vol. 166

RESULTS

TABLE 1

MYELOMA COLONY GROWTH

	Myeloma growth	***Colonies***
Unfractionated in agar	3/21 (14%)	0/21 (0%)
Nonadherent with BALB/CCM* in agar	11/37 (30%)	0/37 (0%)
T-depleted in methylcellulose (MeC)	28/40 (70%)	12/40 (30%)
T-depleted plus MRC-5** (MeC)	22/25 (88%)	15/25 (60%)

*BALB/C mouse, conditioned media. **Human, foetal lung fibroblast line

The overall results of myeloma culture are summarised in Table 1. T lymphocyte depletion significantly improved the myeloma growth, (70% versus 14-30% without T depletion: $p < 0.05$) . The use of the MRC 5 fibroblast feeder layer (a source of IL-6) further enhanced growth. The addition of exogenous IL6 to T-depleted marrows (Genetics Institute 500 - 2500 units/ml) significantly increased CD38 positive myeloma growth, (Figure 1). This increase in growth was both in terms of cell number and proliferation (BU-1/Ki-67). Growth was significantly increased in 9 of 28 (32%) specimens. The median pre-culture labelling index for the IL6 sensitive group was 2% with all being > 1%. An additional 7 patients (25%), had an increase in the percentage of myeloma cells in culture, without an increase in total cell number. The remaining 12 patients (43%), included 3 with MGUS and had no increase in plasma cells, but did have stimulation of myeloid and monocytic components.

The addition of IL6 blocking antibody and macrophage depletion were evaluated and are illustrated in figure 2 . There was a dramatic reduction in myeloma growth associated with both IL6 blocking antibody addition and with macrophage depletion. Each of these manipulations were additive in the substantial reduction in subsequent myeloma growth. Conversely following macrophage depletion, myeloma growth was largely restored with the addition of exogenous IL6. Complete phenotypic studies of the IL6 sensitive and resistant plasma cells is currently being conducted, including quantitative evaluation of the of the IL6 receptor using monoclonal antibody (Kishimoto). Initial correlations indicate that the presence of a labelling index of > 1% plus phenotypic aberrancy, including positivity for CALLA and myelomonocytic antigens, predict for IL6 sensitivity.

FIGURE 1

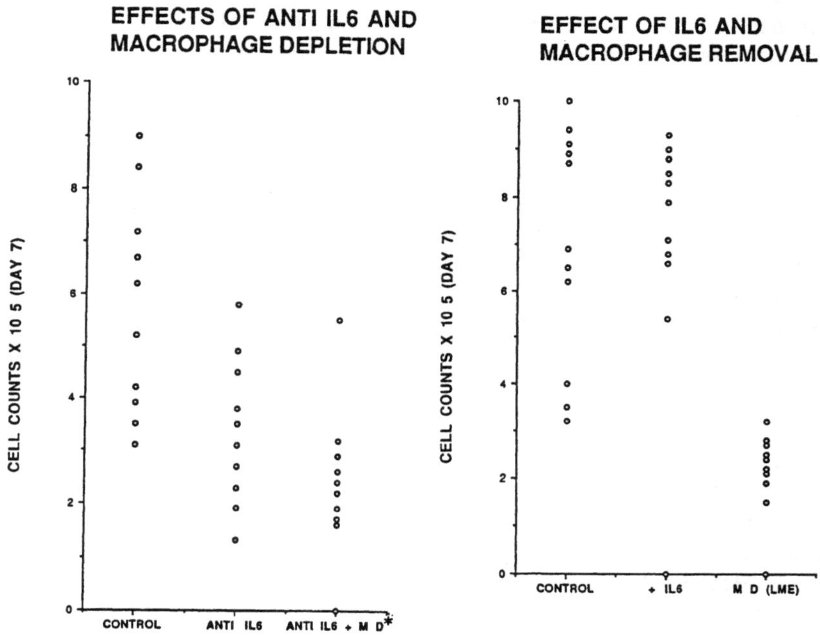

* MD=macrophage depletion ; LME = L-leucine methyl ester (see methods).

DISCUSSION

The data indicate that exogenous IL-6 significantly increases the proliferation of myeloma cells obtained from patients with pre culture bone marrow myeloma cell labelling indices of > 1%. Further, since addition of IL-6 blocking antibody reversed this effect one can infer that IL6 is a part of the in vivo proliferative stimulation for the myeloma cells from this subset of patients. This same subset of patients also had other markers of active disease, including elevated serum B2 microglobulin and phenotypic aberrancy. The importance of the IL6 mechanism was also illustrated by the dramatic inhibition with the addition of blocking anti IL-6 antibodies in all experiments. The role of endogenous macrophages was illustrated by the similar effects produced by depletion of activated macrophages and the effect of the anti IL-6 antibodies without macrophage depletion.

The results raise a number of questions including: **1. why do myeloma cells from certain patients become sensitive to IL-6 ?** and **2. why are macrophages activated to produce excess IL-6 ?** The role of macrophages is supported by morphologic studies indicating close contact between bone marrow macrophages and myeloma cells (7). There is obviously the possibility of a positive feedback loop with stimulation of the macrophages (endogenous [eg by myeloma cells via cytokines such as IL-1b and TNF B] or exogenous), production of IL6 and stimulation of myeloma growth with then further macrohage stimulation. A variety of infectious and immunologic stimuli (including within the idiotypic/anti-idiotypic network) could enhance this feedback loop system. It is important to elucidate the exact trigger mechanisms involved both to better understand the basic biology as well as to develop methods to potentially inhibit the enhancement of such trigger mechanisms of which IL-6 is such a crucial and central part.

REFERENCES

1. Kishimoto T (1989): The biology of interleukin-6. Blood 74:1.
2. Xhang X.G., Klein B., Bataille R (1989): Interleukin-6 is a potent myeloma-cell growth factor in patients with aggressive multiple myeloma.
3. Kwano M, Hirano T, Matsuda T, Taga T, Horii Y, Iwatok, Asaoku H, Tang B, Tanabe O, Tanaka H, Kuramoto A, Kishimoto T (1988): Autocrine generation and essential requirement of BSF-2/IL-6 for human multiple myelomas. Nature 332:83.
4. Klein B, Zhang XG, Jourdan M, Content J, Houssiau F, Aarden L, Piechaczyck M, Bataille R (1989): A paracrine rather than autocrine regulation of myeloma cell growth and differentiation by interleukin-6. Blood 73:517.
5. Durie B.G.M., Vela E, Baum V (1985): Establishment of two new myeloma cell lines from bilateral pleural effusions: Evidence for sequential in vivo clonal change. Blood 66:548.
6. Blom J (1977): The ultrastructure of macrophages found in contact with plasma cells in the bone marrow of patients with multiple myeloma. Acta Path Microbiol. Scan. Sect. A, 85:335.
7. Thiele D.L., Lipsky P.E. (1986): The immunosuppressive activity of $_L$-Leucyl-$_L$-Leucine Methyl Ester: Selective ablation of cytotoxic lymphocytes and monocytes. J. Immunol. 136:1038

Retroviral-Mediated Transfer of Interleukin-6 Into Hematopoietic Cells of Mice Results in a Syndrome Resembling Castleman's Disease

S.J. Brandt, D.M. Bodine, C.E. Dunbar, and A.W. Nienhuis

Clinical Hematology Branch, National Heart, Lung, and Blood Institute, National Institutes of Health, Bethesda, MD 20892

INTRODUCTION

Interleukin-6 (IL-6) is a multifunctional cytokine believed to have an important role in host defense, particularly in the acute phase response to infection or injury (Kishimoto 1989). IL-6 has prominent actions on cells of the immune system, including stimulation of immunoglobulin synthesis by activated B-cells and differentiation of cytotoxic T-cells. In addition, it induces hematopoietic stem cells to enter cell cycle, committed myeloid progenitors to differentiate into granulocytes and macrophages, and hepatocytes to synthesize a variety of plasma proteins. IL-6 is synthesized in response to a number of stimuli that include viruses and other cytokines, such as interleukin-1 and tumor necrosis factor, which may have overlapping biological activities.

Considerable, although largely circumstantial, evidence links dysregulated IL-6 expression with a number of diseases, including psoriasis (Grossman et al 1989), mesangioproliferative glomerulonephritis (Horii et al 1989), rheumatoid arthritis (Guerne et al 1989), and Castleman's disease (Yoshizaki et al 1989). IL-6 may act as an autocrine or paracrine growth factor in multiple myeloma (Kawano et al 1988; Klein et al 1989), myeloid leukemia (Hoang et al 1988), lymphoma (Yee et al 1989), and renal cell carcinoma (Miki et al 1989).

To investigate the consequences of dysregulated synthesis of IL-6 _in vivo_, we inserted the coding sequences of murine IL-6 into the N2 retroviral vector, cocultured bone marrow from 5-fluorouracil-treated C57BL/6 donors with a high titer psi-2 producer clone, and transplanted these bone marrow cells into congenitally anemic W/Wv recipients.

RESULTS

A high titer psi-2 clone producing the N2-IL6 retrovirus (Fig. 1) was used for these experiments. Five animals transplanted with cells infected with the parental N2 virus served as controls.

Fifteen of 21 W/W^v mice (71%) receiving marrow cocultured with the N2-IL6 producer line developed a syndrome characterized by anemia, leukocytosis, hypoalbuminemia, and polyclonal hypergammaglobulinemia. Cellulose acetate electrophoresis of plasma from these animals revealed a diffusely increased gamma globulin fraction and, in many, hypoalbuminemia (Fig. 2). Plasma immunoglobulin measurements made 10 weeks post-transplant showed IgG, IgA, and IgM levels to be increased approximately ten-fold, three-fold, and two-fold, respectively, over controls. Immunoelectrophoresis of plasma from one IL-6 animal demonstrated increased precipitin arcs for the IgG_1 and IgG_{2b} subclasses.

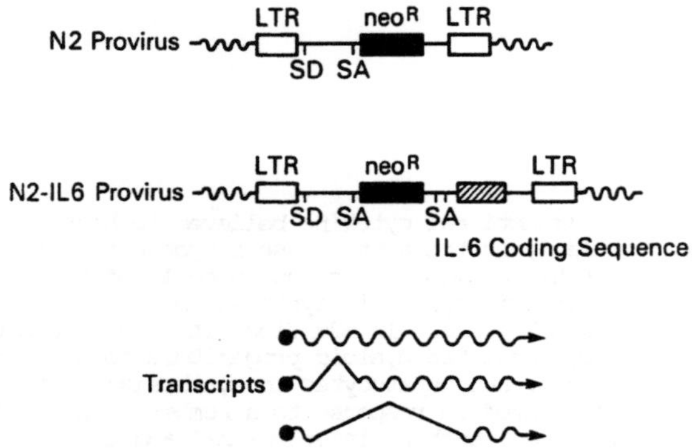

Fig. 1. Structures of proviral forms of N2 and N2-IL6 retroviruses and expected transcripts. SD and SA denote splice donor and acceptor sequences, respectively.

Animals receiving marrow transduced with the N2-IL6 virus developed leukocytosis of varying duration, with total white blood cell counts of 100,000-500,000/ul, beginning 3 weeks after transplant. The majority of cells were mature granulocytes, with no cells more immature than band forms noted. These animals also developed a hypochromic, microcytic anemia. The mean hematocrit of one cohort of IL-6 mice 10 weeks post-transplant was 29% compared to 49% for a group of concurrent controls that received marrow infected with the parental N2 virus. The mean platelet counts of IL-6 animals were approximately half those of a group of concurrent controls.

The elevated levels of circulating IL-6 activity that were measured in these animals appeared to have significant systemic effects. The majority of animals with the syndrome were inactive, and many appeared dyspneic. Subcutaneous fat was markedly decreased, and most animals developed palpable splenomegaly. Five of the first 14 animals studied died within 15 weeks following transplant.

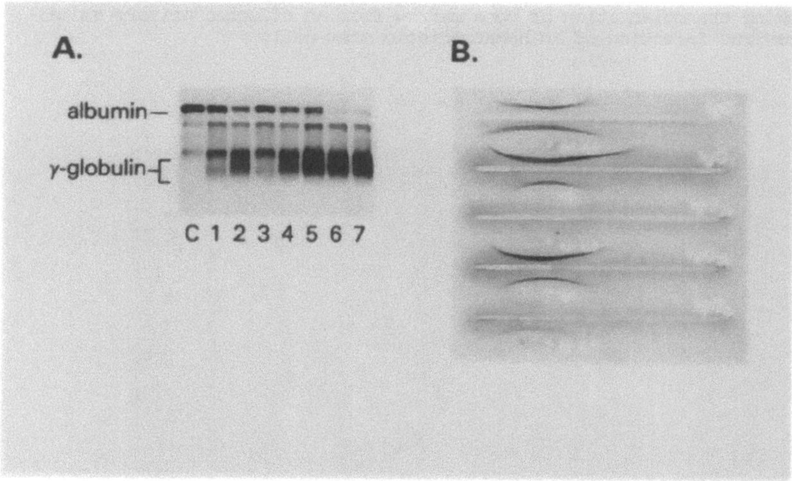

Fig. 2. Immunoglobulin analysis of IL-6 mice. A. Cellulose acetate electrophoresis of plasma from control (C) and from individual animals 8 weeks (1-4) and 14 weeks (5-7) following transplantation with bone marrow cells infected with the N2-IL6 virus. Marked hypoalbuminemia, in addition to polyclonal hypergammaglobulinemia, is apparent in animals 6 and 7. B. Immunoelectrophoresis of plasma obtained 8 weeks post-transplant from an IL-6 animal and a concurrent N2 control.

Postmortem examination revealed massive splenomegaly and peripheral lymph node enlargement without significant mediastinal or retroperitoneal adenopathy. Histological analysis showed almost complete replacement of nodes by mature plasma cells; sinuses were dilated and follicles markedly atretic (Fig. 3). Marked expansion of the splenic red pulp by plasma cells was noted, with smaller numbers also found in kidney, portal zones of liver, bone marrow, and thymus. A pleomorphic infiltrate that included plasma cells and neutrophils was seen in the alveolar septae of lung. In addition to the pathological changes produced in organs by plasma cell infiltration, glomerular enlargement from mesangial cell hyperplasia was noted in stained sections of kidney.

No distinct monoclonal or oligoclonal immunoglobulin gene rearrangements were detected by Southern blot analysis of DNA extracted from tissues rich in plasma cells using a heavy chain joining region probe, consistent with the results of plasma protein analysis demonstrating polyclonal hypergammaglobulinemia. The identity of proviral integration patterns in different hematopoietic and lymphoid organs in individual animals and in secondary recipients that developed the syndrome following transplantation of bone marrow from an affected primary animal confirmed retrovirus infection of an hematopoietic stem cell.

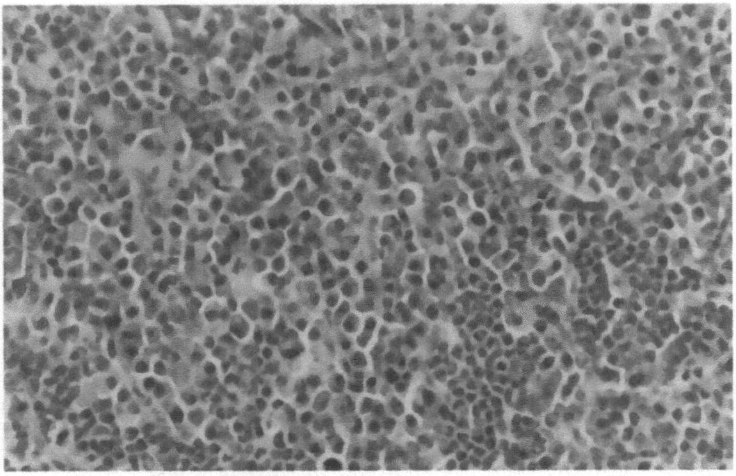

Fig. 3. Lymph node histology of IL-6 mice. Hematoxylin and eosin-stained section of lymph node demonstrates diffuse infiltrate of mature plasma cells.

DISCUSSION

W/W^v mice transplanted with bone marrow cells infected with a retrovirus bearing the coding sequences for murine IL-6 developed a syndrome characterized by anemia, transient leukocytosis, hypoalbuminemia, and polyclonal hypergammaglobulinemia, with splenomegaly and peripheral lymphadenopathy. These findings are strikingly similar to those of patients with multicentric Castleman's disease, previously considered to be a hyperplastic lymphoproliferative disorder of unknown etiology (Frizzera et al 1985). A recent study from Kishimoto and colleagues (Yoshizaki et al 1989) reported that the lymph nodes of two patients with the plasma cell variant of Castleman's disease could elaborate large amounts of IL-6 in culture, and they suggested that this represented the source of the elevated levels of circulating IL-6 measured. Our data provide a direct test of their hypothesis that dysregulated synthesis of IL-6 has a causative role in this disorder.

Similar findings to those resulting from retrovirus-mediated transfer of murine

IL-6 sequences into hematopoietic cells have also been noted in transgenic mice carrying the human IL-6 gene linked to the immunoglobulin heavy chain enhancer, although no hematologic data were reported (Suematsu et al 1989). Taken together, these studies indicate that dysregulated IL-6 production can account for the polyclonal hypergammaglobulinemia, changes in plasma proteins, and anemia seen in multicentric Castleman's disease and allied disorders. Although not detected in either type of mice, clonal immunoglobulin gene rearrangements have been noted in some patients with multicentric Castleman's disease (Hanson et al 1988), and a proportion of these patients ultimately develop monoclonal lymphoproliferative disorders, including multiple myeloma. It is likely that additional genetic changes must occur to allow evolution of a neoplastic clone.

REFERENCES

Frizzera G, Peterson BA, Bayrd ED, Goldman A (1985) A systemic lymphoproliferative disorder with morphologic features of Castleman's disease: Clinical findings and clinicopathologic correlations in 15 patients. J Clin Oncol 3:1202-1216

Grossman RM, Krueger J, Yourish D, Granelli-Piperno A, Murphy DP, May LT, Kupper TS, Sehgal PB, Gottlieb AB (1989) Interleukin 6 is expressed in high levels in psoriatic skin and stimulates proliferation of cultured human keratinocytes. Proc Natl Acad Sci USA 86:6367-6371

Guerne P-A, Zuraw BL, Vaughan JH, Carson DA, Lotz M (1989) Synovium as a source of interleukin 6 in vitro. Contribution to local and systemic manifestations of arthritis. J Clin Invest 83:585-592

Hanson CA, Frizzera G, Patton DF, Peterson BA, McClain KL, Gajl-Peczalska KJ, Kersey JH (1988) Clonal rearrangement for immunoglobulin and T-cell receptor genes in systemic Castleman's disease. Association with Epstein-Barr virus. Am J Pathol 131:84-91

Hoang T, Haman A, Goncalves O, Wong GG, Clark SC (1988) Interleukin-6 enhances growth factor-dependent proliferation of the blast cells of acute myeloblastic leukemia. Blood 72:823-826

Horii Y, Muraguchi A, Iwano M, Matsuda T, Hirayama T, Yamada H, Fujii Y, Dohi K, Ishikawa H, Ohmoto Y, Yoshizaki K, Hirano T, Kishimoto T (1989) Involvement of IL-6 in mesangial proliferative glomerulonephritis. J Immunol 143:3949-3955

Kawano M, Hirano T, Matsuda T, Taga T, Horii Y, Iwato K, Asaoku H, Tang B, Tanabe O, Tanaka H, Kuramoto A, Kishimoto T (1988) Autocrine generation and requirement of BSF-2/IL-6 for human multiple myelomas. Nature 332:83-85

Kishimoto T (1989) The biology of interleukin-6. Blood 74:1-10

Klein B, Zhang X-G, Jourdan M, Content J, Houssiau F, Aarden L, Piechaczyk M, Bataille R (1989) Paracrine rather than autocrine regulation of myeloma-cell growth and differentiation by interleukin-6. Blood 73:517-526.

Miki S, Iwano M, Miki Y, Yamamoto M, Tang B, Yokokawa K, Sonoda T, Hirano T, Kishimoto T (1989) Interleukin-6 (IL-6) functions as an in vitro autocrine growth factor in renal cell carcinomas. FEBS Lett 250:607-610

Suematsu S, Matsuda T, Aozasa K, Akira S, Nakano N, Ohno S, Miyazaki J-I, Yamamura K-I, Hirano T, Kishimoto T (1989) IgG1 plasmacytosis in interleukin 6 transgenic mice. Proc Natl Acad Sci USA 86:7547-7551

Yee C, Biondi A, Wang XH, Iscove NN, de Sousa J, Aarden LA, Wong GG, Clark SC, Messner HA, Minden MD (1989) A possible autocrine role for interleukin-6 in two lymphoma cell lines. Blood 74:798-804

Yoshizaki K, Matsuda T, Nishimoto N, Kuritani T, Taeho L, Aozasa K, Nakahata T, Kawai H, Tagoh H, Komori T, Kishimoto S, Hirano T, Kishimoto T (1989) Pathogenic significance of interleukin-6 (IL-6/BSF-2) in Castleman's disease. Blood 74:1360-1367

A Monoclonal Antibody Specific for the Murine IL-6-Receptor Inhibits the Growth of a Mouse Plasmacytoma In Vivo

P.G. Coulie, A. Vink, and J. Van Snick.

Ludwig Institute for Cancer Research, 75 avenue Hippocrate, B-1200 Brussels, Belgium.

INTRODUCTION

Antibodies specific for the human IL-6 receptor have been obtained by immunization with cells transfected with the cDNA for the 80kDa protein that was identified by Yamasaki as the human IL-6 receptor (Yamasaki, 1988; Hirata, 1989)). Unfortunately, similar reagents do not yet exist for the mouse IL-6 receptor.

A monoclonal antibody capable of blocking the binding of IL-6 to its receptor would be of interest to evaluate the importance of IL-6 in the in vivo growth of mouse plasmacytomas (Nordan, 1986; Van Snick, 1987). Here we desribe the derivation of such a reagent and demonstrate its inhibitory action on the growth of a transplantable plasmacytoma.

IDENTIFICATION OF THE 15A7 MONOCLONAL ANTIBODY.

Lewis rats were immunized with OKT4 cells and their spleen cells were fused with the SP2.neo cell line (Van Snick, 1985). Culture supernatants of hybridomas were screened for the inhibition of IL-6-binding to 2F4 cells (a murine B-cell hybridoma expressing a high number of IL-6 receptors). Of several hundred hybridomas tested, one, designated 15A7, was found to secrete a monoclonal antibody competing for the binding of ^{125}I-labeled IL-6. The antibody, which was typed as an IgG2b (courtesy of Pr H. Bazin, Experimental Immunology Unit, Université Catholique de Louvain), was purified by affinity chromatography with the anti-rat kappa chain monoclonal antibody MARK.1 (a gift of Pr H. Bazin).

In order to perform reciprocal inhibitions of binding with murine IL-6, the antibody was iodine-labeled by the chloramine T method, and IL-6 was labeled with the Bolton and Hunter reagent (Coulie, 1989).

Figure 1 shows the inhibitions of IL-6- and 15A7-binding to 2F4 cells by several concentrations of purified 15A7 or of murine recombinant IL-6. Half-maximal inhibition is obtained with 5 nM of 15A7. The binding of 15A7 is also competed for by IL-6.

Current Topics in Microbiology and Immunology, Vol. 166

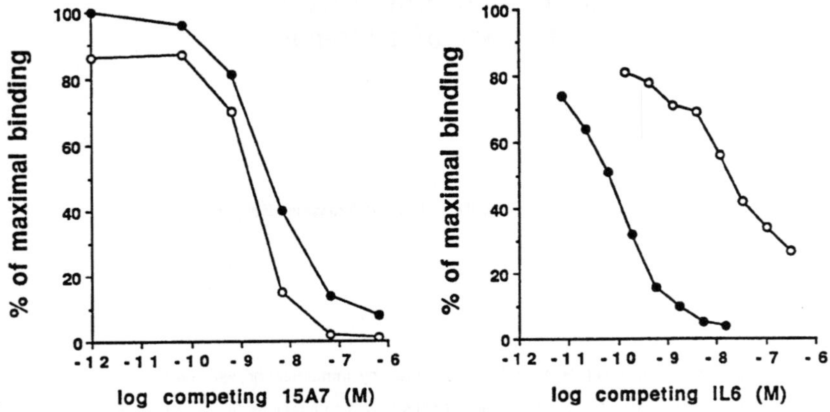

Fig. 1. Inhibitions of the bindings of ^{125}I-IL-6 or ^{125}I-15A7 by unlabeled 15A7 or IL-6. 2F4 cells (10^6) were incubated for 3h at 4°C in 140 µl of RPMI-1640 medium supplemented with 10% FCS and 20 mM sodium azide, containing 3.2×10^{-11} M labeled IL-6 (closed symbols) or 3.1×10^{-11} M labeled 15A7 (open symbols), and the indicated concentrations of competing 15A7 or murine recombinant IL-6. Bound and free ligands were separated with the phtalate method. Each point is the mean of triplicate determinations.

INHIBITION OF IL-6-DEPENDENT PLASMACYTOMA GROWTH BY 15A7.

Antibody 15A7 completely inhibited the in vitro growth of IL-6-dependent plasmacytoma line T1033C2, but did not significantly affect the growth of an IL-6-independent variant derived from the same plasmacytoma line (Fig. 2).

The inhibition observed in vitro prompted us to test the effect of the antibody on the in vivo growth of the same tumor. To this end, hybridoma 15A7 was grown in nude mice and ascitic fluid was collected under sterile conditions to avoid contamination with lipopolysaccharide. Increasing doses of the ascites were then injected s.c. in the back of normal BALB/c mice 24h before i.p. challenge of the animals with 10^6 IL-6-dependent T1033C2 cells. Control animals were injected with saline or with an isotype-matched anti-CD4 ascitis (GK1.5). As shown in Fig. 3, a dose-dependent inhibition of tumor growth was observed. With a relatively high dose of antibody (1 ml of ascites contained 0.5 mg/ml of antibody), 7/8 mice injected with 15A7 were protected while none of those injected with a similar dose of GK1.5 ascites survived.

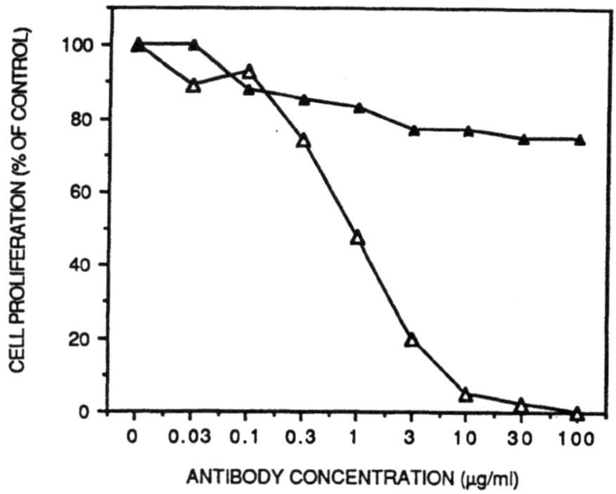

Fig. 2. Inhibition of plasmacytoma growth by antibody 15A7 in vitro. IL-6 dependent plasmacytoma T1033C2 (open symbols) and IL-6 independent variant (T33Bi) derived from the same cell line (closed symbols) were grown in microtiter plates in the presence of increasing concentrations of antibody 15A7. Cultures of IL-6 dependent cells were supplemented with 150 U/ml of mouse recombinant IL-6.

Experiments carried out with monoclonal anti-IL-6 antibody 6B4 (Vink, 1988) yielded similar inhibitions. These results demonstrate that plasmacytoma T1033C2 requires IL-6 for growth in vivo.
If these findings can be generalized to other plasmacytomas, reagents capable of blocking the binding of IL-6 to its receptor may become interesting candidates for the control of plasmacytoma growth.

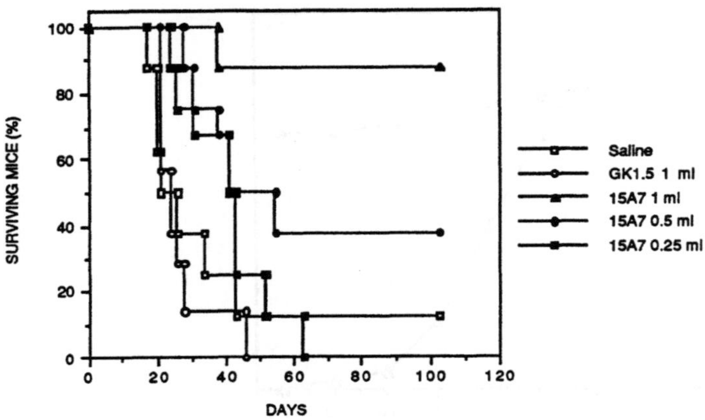

Fig. 3. Anti-tumoral effect of antibody 15A7 in BALB/c mice injected with plasmacytoma T1033C2.

REFERENCES

Coulie PG, Stevens M, Van Snick J (1989) High- and low-affinity receptors for murine interleukin 6. Distinct distributions on B and T cells. Eur. J. Immunol. 19:2107.

Hirata Y, Taga T, Hibi M, Nakano N, Hirano T, Kishimoto T (1989) Characterization of IL-6 receptor expression by monoclonal and polyclonal antibodies. J. Immunol. 143:2900.

Nordan R, Potter M (1986) A macrophage-derived factor required by plasmacytomas for survival and proliferation in vitro. Science 233:566.

Van Snick J, De Plaen E, Boon T (1985) A neomycin-resistant cell line for improved production of monoclonal antibodies to cell surface antigens. Eur. J. Immunol. 15:1148.

Van Snick J, Vink A, Cayphas S, Uyttenhove C (1987) Interleukin-HP1, a T cell-derived hybridoma growth factor that supports the in vitro growth of murine plasmacytomas. J. Exp. Med. 165:641.

Vink A, Coulie PG, Wauters P, Nordan RP, Van Snick J (1988) B cell growth and differentiation activity of interleukin-HP1 and related murine plasmacytoma growth factors. Synergy with interleukin 1. Eur. J. Immunol. 18:607.

Yamasaki K, Taga T, Hirata Y, Yawata H, Kawanishi Y, Seed B, Taniguchi T, Hirano T, Kishimoto T (1988) Cloning and expression of the human interleukin-6 (BSF-2/IFNß 2) receptor. Science 241:825.

Production of IL-2 in CD25 Positive Malignant Lymphomas

D. Emilie, M. Peuchmaur and P. Galanaud

INSERM U 131, 32 rue des Carnets, 92140 Clamart, France.

INTRODUCTION

An immune reaction develops in direct contact with lymphomatous cells in lymph nodes invaded by malignant lymphomas. This antitumoral immune response, which may slow down the progression of the disease, is characterized by the presence of activated T cells. We recently showed that production of interleukins by these reactive T cells is one of the mechanisms potentially involved in this antineoplastic response (Peuchmaur et al 1990a). However, several interleukins including IL-2 are potent lymphocyte growth factors, and malignant lymphomatous cells may express the corresponding receptor. This suggests that the immune reaction directed against lymphomatous cells may in some cases lead to the production of the growth factor(s) required for the proliferation of malignant cells.

PRODUCTION OF IL-2 BY REACTIVE T CELLS IN LARGE CELL LYMPHOMAS

We here show that an IL-2 dependent paracrin growth may indeed be involved in CD25 positive malignant lymphomas.

IL-2 production in CD25 positive lymphomas

Using *in situ* hybridization with an IL-2 specific probe, we analyzed the production of IL-2 in lymphomatous lymph nodes from 12 patients suffering from a large cell lymphoma (LCL). In all these cases, lymphomatous cells expressed the α chain of the IL-2 receptor (CD25, Tac antigen). Five of these lymphomas displayed a B cell phenotype and 5 a T cell phenotype. Two cases expressed no B or T cell markers. None of these patients was infected by HTLV-1. We observed IL-2 producing cells in all cases. The amount of IL-2 producing T cells was in the range of that observed in malignant follicular lymphomas (Peuchmaur et al 1990a) and

Current Topics in Microbiology and Immunology, Vol. 166

in infected lymph nodes (Emilie et al 1990a). It was also comparable to that observed in 9 cases of CD25 negative LCL studied in parallel.

IL-2 is produced by reactive T cells and not by malignant cells

To determine whether IL-2 was produced by lymphomatous cells or reactive cells, we performed double labelling experiments combining *in situ* hybridization with an IL-2 probe and immunohistochemistry with antibodies recognizing antigens expressed by either malignant cells (CD30 and EMA) or reactive cells (CD3 and CD2 in cases of non T lymphomas). We observed that IL-2 producing cells were CD3 and CD2 positive but expressed none of the antigens present on malignant cells. Therefore, malignant cells do not contribute to the *in situ* production of IL-2. This was true for T cell lymphomas as well as for non T lymphomas.

CD25 positive malignant cells express both chains of the IL-2 receptor

We next performed in 9 cases of CD25 positive LCL immunohistochemistry studies using a monoclonal antibody directed against the p75 chain (β chain) of the IL-2 receptor. We observed that in all cases the β chain could be detected in lymphomatous cells. As a mean, the level of expression of p75 by individual cells appeared to be higher in LCL expressing a T cell phenotype than in LCL expressing B cell antigens or in non T non B LCL. Interestingly, virtually all malignant cells expressed the β chain as well as the α chain of the IL-2 receptor in the 9 cases studied. This finding suggests that malignant cells express a functional IL-2 receptor in CD25 positive LCL. Double labelling experiments combining an anti-p55 antibody and an anti-p75 antibody confirmed that both chains were simultaneously expressed by most individual malignant cells. However, the staining of these cells by the anti-p75 antibody displayed a pattern clearly different from that obtained with the anti-p55 antibody: whereas this latter labelled only the surface of malignant cells, p75 was found to be essentially located inside the cytoplasm of these cells. Whether this peculiar pattern reflects an internalization of p75 following its transient expression at the surface of lymphomatous cells remains to be determined.

CONCLUSION

Our results show that IL-2 is produced by reactive T cells in lymph nodes invaded by CD25 positive LCL. Moreover, lymphomatous cells may express a functional IL-2 receptor, as both p55 and p75 are detected in these cells. IL-2 may thus act as a paracrin growth factor in CD25 positive LCL (Peuchmaur et al 1990b). This finding may have therapeutic consequences: monoclonal antibodies directed against the IL-2 receptor may stop lymphomatous progression in CD25 positive LCL by disrupting this putative paracrin loop (Emilie et al 1990b). In this case, such antibodies may improve the course of HTLV-1 negative CD25 positive LCL, as previously described in HTLV-1 induced adult T cell leukemia (Waldmann et al 1988).

REFERENCES

Emilie D, Peuchmaur M, Maillot MC, Crevon MC, Brousse N, Delfraissy JF, Dormont J, Galanaud P (1990a) Production of interleukins in HIV-1 replicating lymph nodes. J. Clin. Invest. In press

Emilie D, Peuchmaur M, Solal-Céligny P, Boué F, Delfraissy JF, Tertian G, Dormont J, Galanaud P (1990b) Anti IL-2 receptor antibody treatment for lymphoproliferative disease : transient positive response in a case of Ki-1^{+} anaplastic large cell lymphoma. Submitted for publication

Peuchmaur M, Emilie D, Crevon MC, Brousse N, Gaulard P, D'Agay MF, Galanaud P, Solal-Céligny P (1990a) IL-2 and Interferon gamma production in follicular malignant lymphomas. Am. J. Clin. Pathol. In press.

Peuchmaur M, Emilie D, Crevon MC, Solal-Céligny MC, Maillot MC, Lemaigre G, Galanaud P (1990b) IL-2 mRNA expression in Tac positive malignant lymphomas. Am. J. Pathol 136: 383-390

Waldmann TA, Goldman CK, Bongiovanni KF, Sharrow SO, Davey MP, Cease KB, Greenberg SJ, Longo DL (1988) Therapy of patients with human T cell lymphotrophic virus I-induced adult T-cell leukemia with anti-Tac, a monoclonal antibody to the receptor for interleukin-2. Blood 72: 1805-1816

IL-6 Induces Hybridoma Cell Growth Through a Novel Signalling Pathway

Koichi Nakajima and Randolph Wall

Department of Microbiology and Immunology, UCLA School of Medicine and the Molecular Biology Institute, University of California Los Angeles, California 90024

INTRODUCTION

With the successful cloning of Interleukin-6 (IL-6) (Haegeman et al. 1986; Hirano et al. 1986; Zilberstein et al. 1986), it became evident that a number of diverse physiological responses in different cell lineages were mediated by this single cytokine (for reviews, see (Kishimoto et al. 1988; Wong et al. 1988). IL-6 regulates the proliferation and the terminal differentiation of cells at different stages in the B cell lineage. IL-6 supports the growth of certain murine hybridoma and plasmacytoma cells (Nordan et al. 1986; Van Damme et al. 1987; Van Snick et al. 1986) as well as some human myeloma cells (Kawano et al. 1988). IL-6 also stimulates activated normal B cells and selected B cell lines to differentiate to immunoglobulin secretion without inducing growth (Hirano et al. 1985; Muraguchi et al. 1988; Raynal et al. 1989).

We have characterized the intracellular signalling events in IL-6 induced hybridoma and plasmacytoma cell growth with the aim of determining if these same events occur in IL-6 induced B cell differentiation. This report summarizes our findings which indicate that IL-6-induced hybridoma or plasmacytoma cell growth activates a previously unknown intracellular signalling pathway that does not utilize any major protein kinase or second messenger system, either alone or in combination. IL-6 activated signalling is initiated by the rapid tyrosine kinase phosphorylation of 160 kd protein followed by the activation of a second protein kinase sensitive to the potent kinase inhibitor, H7. Both of these protein kinase activities are required for the activation of two primary response genes, TIS11 (Lim et al. 1987; Varnum et al. 1989A) and the transcription factor, *junB* (Ryder et al. 1988). These findings define a new intracellular protein kinase cascade linking the IL-6 cell surface receptor complex to the expression of two nuclear-acting primary response genes implicated in gene control.

RESULTS

IL-6-Induced Hybridoma Cell Growth Does Not Use Any Known Signalling Pathway

The murine B cell hybridoma, MH60.BSF-2, was chosen for these studies on IL-6 induced signal transduction because its growth is strictly dependent on this cytokine. No other factors tested, (including; IL-1α, IL-1β, IL-2, IL-3, IL-4, IL-5, IFN-β, IFN-γ, or G-CSF) supported the survival and growth of MH60.BSF-2 cells (Matsuda et al. 1988). To test whether activators of known signal transduction pathways could bypass the IL-6-dependency of MH60.BSF-2 B cells, we examinined the

Current Topics in Microbiology and Immunology, Vol. 166

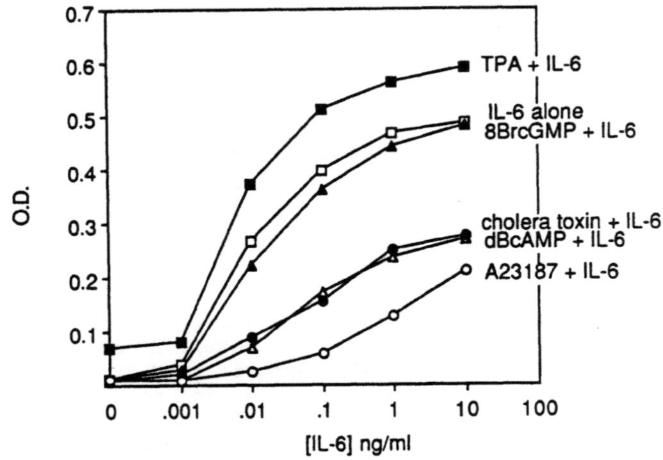

Fig. 1. Activators of protein kinases fail to relieve the IL-6-dependency of MH60.BSF-2 hybridoma cells for growth. MH60.BSF-2 cells deprived of IL-6 for 15 hr, were stimulated with the following activators with or without varying concentration of IL-6 for 30 hr. The activators and ranges of concentration tested are as follows: 12-O-tetradecanoyl phorbol-13-acetate (TPA), 10 ng/ml - 200 ng/ml. N^6, 2'-O-dibutyryl adenosine 3':5'-cyclic monophosphate (dBcAMP), 1 µM - 1 mM. Choleratoxin, 10 ng/ml - 1 µg/ml. Calcium ionophore, A23187, 100 µM - 300 µM. 8-bromoguanosine 3':5' cyclic monophosphate (8-Brc-GMP), 1 mM. The proliferative response was measured by the colorimetric assay using MTT {3-(4,5-dimethylthiazol-2-yl)-2,5-diphenyl tetrazolium bromide}. Since the stimulatory effect of TPA and the inhibitory effects of cBcAMP, cholera toxin, and A23187 are in dose-dependent manners, the representative results of TPA at 50 ng/ml, dBcAMP at 1 mM, cholera toxin at 1 µg/ml, 8-BrcGMP at 1 mM, and A23187 at 300 µM are shown. Each point represents the mean of triplicates with less than 15% variation.

effects of protein kinase activators in the presence and absence of IL-6, on the growth of IL-6-deprived MH60.BSF-2 cells. Cell proliferation was assayed by a colorimetric method using {3-(4,5-dimethylthiazol-2-yl)-2,5-diphenyl tetrazolium bromide}. The activators used were: TPA for protein kinase C (PKC), dibutyryl cAMP for cAMP dependent kinase (PKA), Cholera toxin, an activator of adenylate cyclase, a cGMP analogue 8-bromoguanosine, 3:5 cyclic monophosphate (8-BrcGMP) for cGMP-dependent kinase, and the calcium ionophore, A23187, for Ca/calmodulin dependent kinases. MH60.BSF-2 cells retained > 95% viability following deprivation of IL-6 for 15-17 hr. Complete loss of viability was observed after 48 hr without IL-6 (data not shown). As shown in Figure 1, TPA moderately enhanced the IL-6-dependent growth even when cells were stimulated maximally with 10 ng/ml of IL-6 and increased slightly the number of surviving cells in the absence of IL-6. None of the other activators of cyclic nucleotide- or Ca/calmodulin-dependent protein kinases maintained the survival or promoted the growth of MH60.BSF-2 cells in the absence of IL-6. The failure of these activators to abrogate the IL-6-dependency of MH60.BSF-2 cells suggests that the growth promoting activity of IL-6 for this B cell hybridoma cannot be attributed to the activation of PKC, PKA, Ca/calmodulin kinases or cGMP dependent kinase.

It is possible that IL-6 may simultaneously activate several signal transduction pathways as seen in the cases of platelet derived growth factor (PDGF) (see review by Williams 1989), epidermal growth factor (EGF), or colony stimulating factor-1 (CSF-1) (Yarden et al. 1988). To test this possibility, the effects of combinations of TPA, A23187 and dibutyryl cAMP were examined on growth of MH60.BSF2 cells in the absence or presence of IL-6. No tested combination maintained cell survival or enhanced the IL-6-induced cell growth to the extent observed with 50 ng/ml TPA plus 10 ng/ml IL-6 (data not shown). From these results, it is unlikely that IL-6 acts via multiple kinase pathways using second messenger systems.

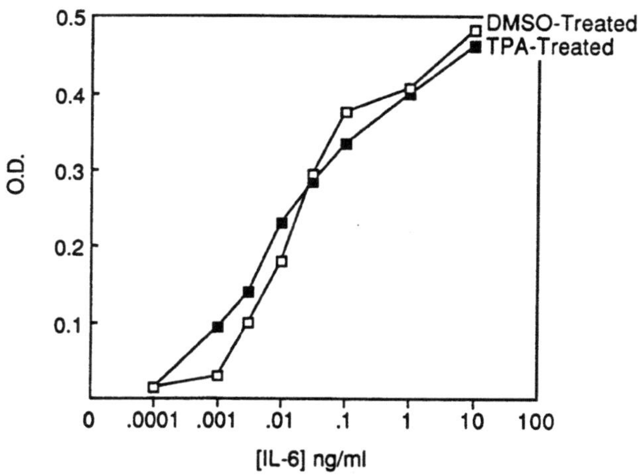

Figure 2. PKC-depletion does not affect IL-6-induced MH60.BSF-2. MH60.BSF-2 cells were treated with 1 µg/ml TPA or 0.1% DMSO in the presence of 1 ng/ml IL-6 for 25 hr, washed to remove IL-6 and treated with TPA or DMSO for a further 15 hr. The proliferative response of TPA-treated and DMSO-treated (untreated) cells to IL-6 were assayed using MTT as described in Figure 1 legend. TPA did not enhance the proliferative response of TPA-treated cells to IL-6 (not shown), and did not induce TIS11 and *junB* gene expression in TPA-treated cells as described in text.

Since TPA had a moderately stimulatory effect on IL-6-dependent cell growth, we next examined whether the IL6-induced cell growth required the activation of PKC. For this purpose, long-term treatment with TPA was performed to deplete PKC activity. Hybridoma cells were treated with 1 μg/ml TPA or 0.1% DMSO in the presence of 1 ng/ml IL-6 for 25 hr, washed three times to remove IL6 and incubated further with 1 μg/ml TPA or 0.1% DMSO, without IL-6, for 15 hr. The effectiveness of this PKC depletion method was confirmed by showing that the TPA-inducible genes, TIS11 (Lim et al. 1987) and *junB* (Ryder et al. 1988), were not activated by 50 or 200 ng/ml TPA in TPA-treated cells (results not shown). The depletion of PKC by pretreating cells with TPA had no effect on the IL6-dependent growth of MH60.BSF2 cells (Figure 2). This confirms that the activation of PKC is not required for IL-6-stimulated hybridoma cell growth. These combined results suggest that none of the major signaling pathways using well-characterized second messenger systems (PKC, PKA, cGMP dependent kinase, Ca/calmodulin dependent kinases) are involved in the IL-6-activated signal transduction pathway leading to the survival and growth of B cell hybridoma line, MH60.BSF-2.

<u>IL6 Rapidly Induces a Tyrosine Kinase Activity Followed by the Activation of the Primary Response Genes, TIS 11 and *junB*.</u>

To test the possible involvement of tyrosine kinases in IL-6 induced hybridoma cell proliferation, we analyzed MH-60.BSF-2 cells rapidly following IL-6 stimulation for tyrosine phosphorylated proteins using anti-phosphotyrosine monoclonal antibody (Huhn et al. 1987). IL-6-deprived MH60.BSF-2 cells were labeled with {^{32}P} orthophosphate in phosphate-free medium for 2 hr, and then stimulated with IL-6 (300 ng/ml). Phosphotyrosine containing proteins were immunoprecipitated from labeled cell lysates with monoclonal anti-phosphotyrosine antibody (1G2) and analyzed on SDS-PAGE under reducing conditions. IL-6 rapidly induced the transient appearance of a single new labelled cellular phosphoprotein of 160 kd in IL-6 stimulated MH60.BSF-2 cells. This tyrosine phosphorylated 160 kd protein appeared within 2 min., reached a maximal level at 5 min, and thereafter rapidly declined (Figure 3). Such a rapid and transient stimulation of tyrosine phosphorylation is a typical response seen with different growth factors.

We also tested the IL-6 induced responses of nine immediate early genes including; *c-fos* (Curran et al. 1982; Van Beveren et al. 1983), *c-jun* (Angel et al. 1988; Bohmann et al. 1987), *junB* (Ryder et al. 1988), Krox20 (Chavier et al. 1988) TIS11 (Varnum et al. 1989A), TIS1 (also called nur77, NGFIB) (Hazel et al. 1988; Milbrandt 1988), TIS7 (PC4) (Tirone et al. 1989; Varnum et al. 1989B), TIS8 (also called Egr-1, NGFIA, Zif286, Krox24) (Christy et al. 1988; Lemaire et al. 1988; Milbrandt 1987; Sukhatme et al. 1988), and TIS10 (Lim et al. 1987). Only TIS11 and *junB* mRNAs were induced following IL-6 stimulation of IL-6 starved MH60.BSF-2 hybridoma cells (Figure 3). Both TIS11 and *junB* mRNA levels increased 10-20 fold between 30-60 min following IL-6 stimulation and then declined gradually out to 120 min. The inductions of TIS11 and *junB* mRNA were not affected by cycloheximide indicating that protein synthesis is not required for their activation by IL-6 (results not shown). IL-6 induced TIS11 and *junB* gene expression with similar kinetics in the

IL-6-dependent plasmacytoma cell line, T1165 (data not shown). Interestingly, both the primary response genes activated by IL-6 are nuclear-acting genes implicated in gene control.

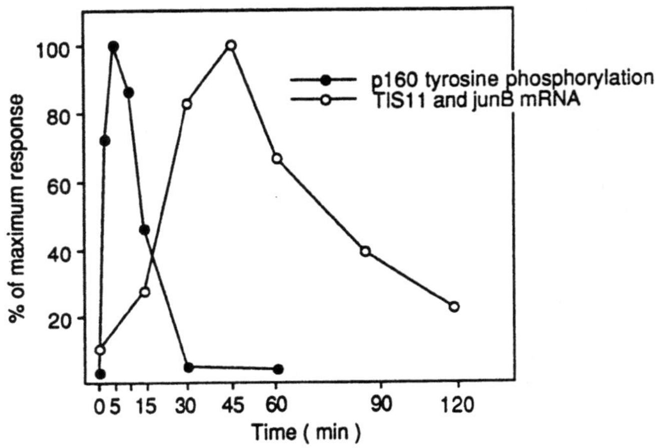

Figure 3. Time course of IL-6-induced p160 tyrosine phosphorylation and expression of primary response genes. IL-6-deprived MH60.BSF-2 cells were labeled with 1 mCi/ml{^{32}p} orthophosphate in phosphate-free medium containing 10% dialyzed FCS for 2 hr and stimulated with 300 ng/ml IL-6 for indicated period. Phosphotyrosine containing proteins were immunoprecipitated from cell lysates with monoclonal anti-phosphotyrosine antibody, 1G2, coupled to sepharose 4B and analyzed by 7.5% SDS-PAGE under reducing condition. For mRNA inductions, IL-6-deprived NH60.BSF-2 cells were stimulated with 10 ng/ml IL-6 for the indicated times. Total RNA samples were analyzed by electrophoresis in 1% agarose MOPS-formaldehyde gels. Blots were probed with ^{32}p-labeled DNA fragments of TIS11, *junB*, or CHO-B. CHO-B is a ubiquitous gene used as an internal control gene for normalization of samples (Harphold et al., 1979). Other immediately early genes which showed no response to IL-6 were c-fos, TIS1 (*nur*77/NGFIB), TIS7 (PC4), TIS 8 (Egr1/NGFIA/Aif286/Krox24), TIS10, *c-jun*, and Krox20.

Of the protein kinase activators tested for effects on MH-60.BSF-2 cell growth (see Figure 1) only TPA activated TIS11 and *junB* mRNA expression (results not shown). However, TIS11 and *junB* inductions by IL-6 were not altered in PKC-depleted cells (Table 1). These results establish that IL-6 activated expression of TIS11 and *junB* genes occurs via a pathway distinct from PKC, PKA, cGMP-dependent kinase, or Ca/calmodulin-dependent kinases. Accordingly, the induction of TIS11 and *junB* primary response gene expression shows the same requirements as IL-6 stimulated cell proliferation (see Figure 1).

Both Tyrosine Kinase and H7-sensitive Kinase Activities Are Required for IL-6-induced TIS11 and *junB* Gene Expression.

We tested the effects of various protein kinase inhibitors on IL-6-stimulated p160 tyrosine phosphorylation and on TIS11 and *junB* gene induction. MH-60.BSF-2 cells were treated with inhibitors for 30 min prior to IL-6 addition. Cell extracts were prepared and analyzed after IL-6 stimulation either for 5 min (for p160 tyrosine phosphorylation) or for 45 min (for TIS11 and *junB* mRNA).

Two highly specific tyrosine kinase inhibitors with different modes of action were used, tyrphostin (RG50863) (Yaish et al. 1988) and genistein (Akiyama et al. 1987). We also tested the potent, non-specific protein kinase inhibitor, H7, {1-(5-isoquinolinesulfonyl)-2-methylpiperazine} which is known to inhibit PKC, PKA and cGMP-dependent protein kinases (Hidaka et al. 1984). Both tyrphostin (RG50863) and genistein drastically inhibited IL-6 activated p160 tyrosine phosphorylation ($\geq$ 90% reduction) at concentrations known to have minimal effects on serine/threonine kinases (e.g., PKC, PKA, phosphorylase kinase) (summarized in Table 1). In contrast, H7 did not inhibit tyrosine phosphorylation of p160 indicating that this early event following IL-6 stimulation does not require prior activity of any H7-sensitive protein kinase(s).

Table 1. Two Kinase Activities Are Required for IL-6-Induced *junB* and TIS11 Gene Expression

Inhibitor	Tyrosine Kinase Phosphorylation of p160	*junB* and TIS11 Expression
Genistein[a]	inhibited	ND
Tyrphostin[a]	inhibited	inhibited
H7[b]	not inhibited	inhibited
Sphingosine[c]	ND	not inhibited
W7[d]	ND	not inhibited
PKC-Depletion[e]	ND	not inhibited

[a]specific tyrosine kinase inhibitor
[b]PKC, PKA, cGMP-dependent kinase inhibitor
[c]specific PKC inhibitor
[d]calmodulin antagonist
[e]PKC depleted by 40 hr TPA treatment
ND = not done

Tryphostin (RG50863) and H7 also were tested on IL-6-induced TIS11 and *junB* mRNA induction under the same conditions used to inhibit p160 tyrosine phosphorylation (except that a 45 min incubation was used for mRNA studies). The tyrphostin (RG50863) inhibition of IL-6-activated TIS11 and *junB* mRNA expression was comparable (≥ 90%) to the reduction in tyrosine phosphorylation of p160 produced by this inhibitor (summarized in Table 1). This finding indicates that tyrosine kinase phosphorylation of p160 is necessary for subsequent TIS11 and *junB* induction.

The potent protein kinase inhibitor, H7, completely blocked IL-6-induced TIS11 and *junB* gene expression (Table 1). This suggests that a second kinase activity intercedes between the early tyrosine kinase activity and TIS11 and *junB* induction. The nature of this kinase is not known. However, this H7-sensitive kinase does not correspond to any well characterized second messenger system. For example, sphingosine (a selective inhibitor of PKC, (Hannun et al. 1986) had no affect on IL-6-stimulated TIS11 and *junB* gene expression (Table 1). Likewise, PKC-depletion by long-term TPA treatment also did not effect on IL-6-induced TIS11 and *junB* expression. The calmodulin antagonist, W7, N-(6-aminohexyl)-5-chloro-1-naphthalenesulfonamide (Hidaka et al. 1981) also did not inhibit IL-6-induced TIS11 and *junB* mRNA expression. Thus, the activation of these two immediate early response genes by IL-6 apparently requires the activities of a tyrosine kinase as well as a second kinase sensitive to the general protein kinase inhibitor, H7, but distinct from any major signal transduction pathways utilizing known mediators such as diacylglycerol, Ca^{++}, cAMP or cGMP. Figure 4 proposes a model for the intracellular signalling events revealed in our studies on IL-6 signal transduction in hybridoma and plasmacytoma cell growth.

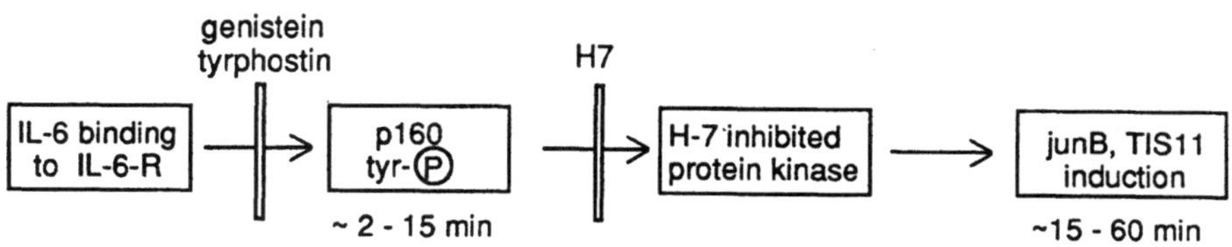

Figure 4. A model for the early signaling events in IL-6-induced hybridoma cell growth. Vertical bars denote blocks in the pathway by the specific tyrosine kinase inhibitors, genistein and tyrphostin, and by the protein kinase inhibitor, H7. Combined results of studies with kinase activators and inhibitors show that this signaling pathway does not involve PKC, PKA, cGMP-dependent or Ca^{++}/calmodulin-dependent kinases.

DISCUSSION

IL-6-induced signalling in MH-60 hybridoma cells evidently does not involve any known signal transduction pathway utilizing well-characterized second messengers (i.e., PKC, PKA, cGMP-dependent kinase, Ca/calmodulin kinases). Instead, IL-6-induced hybridoma cell growth is signalled through a novel protein kinase cascade initiated by tyrosine kinase phosphorylation of p160 rapidly following IL-6 receptor binding. This tyrosine phosphorylation plays a pivotal role in activating an H7-sensitive kinase(s) leading to the subsequent activation of the primary response genes, TIS11 and *junB*. The rapid kinetics of tyrosine phosphorylation of p160 and the subsequent activation of primary response gene expression following IL-6 induction are consistent with this model. This temporal sequence of steps is also supported by our studies using highly specific protein tyrosine kinase inhibitors. Tyrphostin (RG50863) efficiently blocked IL-6 induced tyrosine phosphorylation of p160 as well as activation of TIS11 and *junB* transcription. The existence of another protein kinase activity intervening between these steps is supported by the finding that the general protein kinase inhibitor, H7, blocked TIS11 and *junB* induction, but did not affect p160 tyrosine phosphorylation. These results are most consistent with a single signal transduction pathway leading to TIS11 and *junB* induction (Figure 4).

These findings raise a number of intriguing questions for further studies. The identities of the tyrosine kinase and H7 sensitive kinase are not known. The cloned 80 kd subunit of the IL-6 receptor does not contain a kinase domain and therefore does not appear to function directly in signal transduction (Yamasaki et al. 1988). IL-6 binding to the 80 kd receptor molecule results in the rapid formation (i.e., in 1-2 min) of a cell surface complex with a 130 kd non-ligand binding glycoprotein (gp130) (Taga et al. 1989). It is postulated that this latter molecule may function in signal transduction following IL-6 binding. The relationship of gp130 and the tyrosine-phosphorylated p160 detected within 2 min after IL-6 binding remains to be established. The activation of TIS11 and *junB* expression is not affected by cycloheximide, indicating that the induction of these genes is mediated through post-translational mechanisms. The H7-sensitive kinase activated by IL-6 may activate primary response gene expression through the phosphorylation of existing transcription factors. The *junB* gene is a member of the AP-1 family of transcription factors (Angel et al. 1988; Angel et al. 1987; Bohmann et al. 1987; Lee et al. 1987; Ryder et al. 1988). It is likely that some of the secondary response genes induced by IL-6 (i.e., genes requiring protein synthesis for expression) may have AP-1 or AP-1-related sequences in their promoters. Finally, it will be extremely interesting to compare the signalling steps defined here in IL-6-induced hybridoma gell growth to those in IL-6-stimulated B cell differentiation and in IL-6-induced responses in other cell lineages.

ACKNOWLEDGEMENTS

Recombinant IL-6 was generously provided by L. Souza and AMGEN. Tyrphostin (RG50863) was supplied by Rorer Central Research. We thank D. Nathans and H. Herschman for providing cDNA clones and A. R. Frackelton, Jr. for the 1G2 hybridoma. Koichi Nakajima was supported by a Fellowship from the University of California Cancer Research Coordinating Committee (CCRC). This research was supported by an NIH Program Project CA12800 and NIH Research Grant, GM40185.

REFERENCES

Akiyama, T., J. Ishida, S. Nakagawa, H. Ogawara, S. Watanabe, N. Itoh, M. Shibuya and Y. Fukami. (1987). Genistein, a specific inhibitor of tyrosine-specific protein kinases. J. Biol. Chem. 262: 5592-5595.

Angel, P., E. A. Allegretto, S. T. Okino, K. Hattori, W. J. Boyle, T. Hunter and M. Karin. (1988). Oncogene *jun* encodes a sequence-specific trans-activator similar to AP-1. Nature. 332: 166-171.

Angel, P., M. Imagawa, R. Chiu, B. Stein, R. J. Imbra, H. J. Rahmsdorf, C. Jonat, P. Herrlich and M. Karin. (1987). Phorbol ester-inducible genes contain a common *cis* element recognized by a TPA-modulated trans-acting factor. Cell. 49: 729-739.

Bohmann, D., T. Bos, A. Admon, T. Nishimura, P. K. Vogt and R. Tjian. (1987). Human proto-oncogene *c-jun* encodes a DNA binding protein with structural and functional properties of transcription factor AP-1. Science. 238: 1386-1392.

Chavier, P., M. Zerial, P. Lemaire, J. Almendral, R. Bravo and P. Charnay. (1988). A gene encoding a protein with zinc fingers is activated during G0/G1 transition in cultured cells. EMBO J. 7: 29-35.

Christy, B. A., L. F. Lau and D. Nathans. (1988). A gene activated in mouse 3T3 cells by serum growth factors encodes a protein with 'zinc finger' sequences. Proc. Natl. Acad. Sci. USA. 85: 7857-7861.

Curran, T., G. Peters, C. Van Bevereren, N. M. Teich and I. M. Verma. (1982). FBJ murine osteosarcoma virus: identification and molecular cloning of biologically active proviral DNA. J. Virol. 44: 674-682.

Haegeman, G., J. Content, G. Volckaert, R. Derynck, J. Tavernier and W. Fiers. (1986). Structural analysis of the sequence coding for an inducible 26-kd protein in human fibroblasts. Eur. J. Biochem. 159: 625-632.

Hannun, Y. A., C. R. Loomis, J. A. H. Merrill and R. M. Bell. (1986). Sphingosine inhibition of protein kinase C activity and of phorbol dibutyrate binding *in vitro* and in human plateletes. J. Biol. Chem. 261: 12604-12609.

Hazel, T. G., D. Nathans and L. F. Lau. (1988). A gene inducible by serum growth factors encodes a member of the steroid and thyroid hormone receptor superfamily. Proc. Natl. Acad. Sci. USA. 85: 8444-8448.

Hidaka, H., M. Inagaki, S. Kawamoto and Y. Sasaki. (1984). Isoquinolinesulfonamides novel and potent inhibitors of cyclic nucleotide dependent protein kinase and protein kinase C. Biochem. 23: 5036-5041.

Hidaka, H., Y. Sasaki, T. Tanaka, T. Endo, S. Ohno, Y. Fuji and T. Nagata. (1981). N-(6-aminohexyl)-5-chloro napthalenesulfonamide, a calmodulin antagonist, inhibits cell proliferation. Proc. Natl. Acad. Sci. USA. 78: 4354-4357.

Hirano, T., T. Taga, N. Nakano, K. Yasukawa, S. Kashiwamura, K. Shimizu, K. Nakajima, K. H. Pyun and T. Kishimoto. (1985). Purification to homogeneity and characterization of human B-cell differentiation factor (BCDF or BSFp-2). Proc. Natl. Acad. Sci. 82: 5490-5494.

Hirano, T., K. Yasukawa, H. Harada, T. Taga, Y. Watanabe, T. Matsuda, S. Kashiwamura, K. Nakajima, K. Koyama, A. Iwamatsu, S. Tsunasawa, F. Sakiyama, H. Matsui, Y. Takahara and T. Taniguchi. (1986).

Complementary DNA for a novel human interleukin (BSF-2) that induces B lymphocytes to produce immunoglobulin. Nature. 324: 73-76.
Huhn, R. D., M. R. Posner, S. I. Rayter, J. G. Foulkes and J. A. R. Frackelton. (1987). Cell lines and peripheral blood leukocytes derived from individuals with chronic myelogenous leukemia display virtually identical proteins phosphorylated on tyrosine residues. Proc. Natl. Acad. Sci. USA. 84: 4408-4412.
Kawano, M., T. Hirano, T. Matsuda, T. T., Y. Horii, K. Iwato, H. Asaoku, B. Tang, O. Tanabe, H. Tanaka, A. Kuramoto and T. Kishimoto. (1988). Autocrine generation and requirement of BSF-2/IL-6 for human multiple myelomas. Nature. 332: 83-85.
Kishimoto, T. and T. Hirano. (1988). Molecular regulation of B lymphocyte response. Ann. Rev. Immunol. 6: 485-512.
Lee, W., P. Mitchell and R. Tjian. (1987). Purified transcription factor AP-1 interacts with TPA-inducible enhancer elements. Cell. 49: 741-752.
Lemaire, P., O. Relevant, R. Bravo and P. Charnay. (1988). Two genes encoding potential transcription factors with identical DNA-binding domains are activated by growth factors in cultured cells. Proc. Natl. Acad. Sci. USA. 85: 4691-4695.
Lim, R. W., B. C. Varnum and H. R. Herschman. (1987). Cloning of tetradecanoyl phorbol ester-induced "primary response" sequences and their expression in density-arrested Swiss 3T3 cells and a TPA nonproliferative variant. Oncogene. 1: 263-270.
Matsuda, T., T. Hirano and T. Kishimoto. (1988). Establishment of interleukin 6 (IL-6) B cell stimulatory factor 2 (BSF-2)-dependent cell line and preparation of anti-IL-6/BSF-2 monoclonal antibodies. Eur. J. Immunol. 18: 951-956.
Milbrandt, J. (1987). A nerve growth factor-induced gene encodes a possible transcriptional regulatory factor. Science. 238: 797-799.
Milbrandt, J. (1988). Nerve growth factor induces a gene homologous to the glucocorticoid receptor gene. Neuron. 1: 183-188.
Muraguchi, A., T. Hirano, B. Tang, T. Matsuda, Y. Horii, K. Nakajima and T. Kishimoto. (1988). The essential role of B cell stimulatory factor 2 (BSF-2/IL-6) for the terminal differentiation of B cells. J. Exp. Med. 167: 332-344.
Nordan, R. P. and M. Potter. (1986). A macrophage-derived factor required by plasmacytomas for survival and proliferation *in vitro*. Science. 233: 566-569.
Raynal, M., Z. Liu, T. Hirano, L. Mayer, T. Kishimoto and S. Chen-Kiang. (1989). Interleukin 6 induces secretion of IgG1 by coordinnated transcriptional activation and differentional mRNA accumulation. Proc. Natl. Acad. Sci. USA. 86: 8024-8028.
Ryder, K., L. F. Lau and D. Nathans. (1988). A gene activated by growth factors is related to the oncogene *v-jun*. Proc. Natl. Acad. Sci. USA. 85: 1487-1491.
Sukhatme, V. P., X. Cao, L. C. Chang, C. H. Tsai-Morris, D. Stamenkovich, P. D. P. Ferreira, D. R. Cohen, S. A. Edwards, T. B. Shows, T. Curran, M. M. Le Beau and E. D. Adamson. (1988). A zinc finger-encoding gene coregulated with *c-fos* during growth and differentiation, and after cellular depolarization. Cell. 53: 37-43.
Taga, T., M. Higi, Y. Hirata, K. Yamasaki, K. Yasukawa, T. Matsuda, T. Hirano and T. Kishimoto. (1989). Interleukin-6 triggers the association of its receptor with a possible signal transducer, gp130. Cell. 58: 573-581.

Tirone, F. and E. M. Shooter. (1989). Early gene regulation by nerve growth factor in PC12 cells: induction of an interferon-related gene. Proc. Natl. Acad. Sci. USA. 86: 2088-2092.
Van Beveren, C., F. Van Staalen, T. Curran, R. Müller and I. M. Verma. (1983). Analysis of FBJ-MuSV provirus and *c-fos* (mouse) gene reveals that viral and cellular *fos* gene products have different carboxyl termini. Cell. 32: 1241-1255.
Van Damme, J., G. Opdenakker, R. J. Simpson, M. R. Rubira, S. Cayphas, A. Vink, A. Billiau and J. Van Snick. (1987). Identification of the human 26 kd protein, interferon β2 (IFN-β2), as a B cell hybridoma/plasmacyhtoma growth factor induced by interleukin 1 and tumor necrosis factor. J. Exp. Med. 165: 914-919.
Van Snick, J., S. Cayphas, A. Vink, C. Uyttenhove, P. G. Coulie, M. R. Rubira and R. J. Simpson. (1986). Purification and NH2-terminal amino acid sequence of a T-cell-derived lymphokine with growth factor activity for B-cell hybridomas. Proc. Natl. Acad. Sci. USA. 83: 9679-9683.
Varnum, B., R. Lum, V. Sukhatme and H. Herschman. (1989A). Nucleotide sequence of a cDNA encoding TIS11, a message induced in Swiss 3T3 cells by the tumor promoter tetradecanoyl phorbol acetate. Oncogene. 4: 119-120.
Varnum, B. C., R. W. Lim and H. R. Herschman. (1989B). Characterization of TIS7, a gene induced in Swiss 3T3 cells by the tumor promoter tetradecanoyl phorbol acetate. Oncogene. 4: 1263.
Williams, L. T. (1989). Signal transduction by the platelet-drived growth factor receptor. Science. 243: 1564-1570.
Wong, G. C. and S. C. Clark. (1988). Multiple actions of interleukin-6 within a cytokine network. Today. 5: 137-139.
Yaish, P., A. Gazit, C. Gilon and C. Levitzki. (1988). Blocking of EGF-dependent cell proliferation by EGF receptor kinase inhibitors. Science. 242: 933-935.
Yamasaki, K., T. Taga, Y. Hirata, H. Yawata, Y. Kawanishi, B. Seed, T. Taniguchi, T. Hirano and T. Kishimoto. (1988). Cloning and expression of the human interleukin-6 (BSF-2/IFN-β2) receptor. Science. 241: 825-828.
Yarden, Y. and A. Ullrich. (1988). Growth factor receptor tyrosine kinases. Ann. Rev. Biochem. 57: 443-478.
Zilberstein, A., R. Ruggieri, J. H. Korn and M. Revel. (1986). Structure and expression of cDNA and genes for human interferon-β2, a distinct species inducible by growth-stimulatory cytokines. EMBO J. 5: 2529-2537.

Selective Killing of IL6 Receptor Bearing Myeloma Cells Using Recombinant IL6-Pseudomonas Toxin

Clay B. Siegall, David J. FitzGerald and Ira Pastan

Laboratory of Molecular Biology, Division of Cancer Biology and Diagnosis and Centers, National Cancer Institute, National Institutes of Health
Bethesda, MD 20892

Chemotherapy is widely used in the treatment of human cancer but unfortunately is not successful against many cancers and has undesirable side effects. If the chemotherapeutic agent was only delivered to the tumor cell, it should be possible to eliminate cancer cells without killing normal cell populations. Recently, a new field has emerged which addresses the possibility that one can selectively eliminate malignant cell populations. This new discipline takes advantage of the fact that many cancer cells display higher numbers of growth factor receptor on their surface than normal cells. By taking advantage of this increase, one can target these receptors with cytotoxic reagents such as protein toxins. These reagents are cytotoxic to target cells because a receptor specific ligand is linked to a potent toxin (reviewed in Pastan 1986; Vitetta 1987; FitzGerald and Pastan 1989). In this review, we will describe specific reagents which can target and kill multiple myeloma cells.

Pseudomonas exotoxin A (PE) is a bacterial toxin which is composed of three domains as demonstrated by structural analysis (Allured 1986). Functional analysis has shown that domain I is responsible for binding to the PE receptor, domain II for translocating the toxin across a cell membrane and domain III for ADP ribosylating elongation factor 2 thereby halting protein synthesis and producing cell death (Hwang 1987; Jinno 1988; Siegall 1989, Jinno 1989). Domain III also encodes sequences at its carboxyl terminus which are required for translocation (Chaudhary 1990). By removing domain I, we produce a molecule termed PE40 (Mr 40,000 Da) which retains the functions of domains II and III (Kondo 1988). The linking to PE40 of a ligand which binds to specific cellular receptors produces a cell-specific cytotoxic reagent (Pastan and FitzGerald 1989).

Interleukin-6 (IL6) acts on a variety of cell types, most notably B cells activated to induce immunoglobulin production (Hirano 1988). It also is an essential growth factor for many hybridomas and plasmacytomas (Nordan and Potter 1986; Van Snick 1987), and induces hepatocytes to synthesize

Current Topics in Microbiology and Immunology, Vol. 166

acute-phase proteins (Gauldie 1987). It has been demonstrated that IL6 receptors are expressed in high numbers on some human multiple myeloma cells (Taga 1987; Kawano 1988; Asaoku 1988). Evidence of aberrant production of IL6 and its receptor in human myeloma have indicated the functioning of an autocrine loop in the progression of the disease (Kawano 1988; Schwab submitted). Reports that proliferation of myeloma cells from patients or from established cell lines can be inhibited by the addition of anti-IL6 antibodies suggests that if one can interfere with autocrine stimulation, one can potentially inhibit the growth of multiple myeloma. Therefore, specific inhibitors of IL6 and IL6 receptor bearing cells may be useful as therapeutic reagents in the treatment of this disease.

To kill myeloma cells with high numbers of IL6 receptors, we developed a toxic IL6 molecule composed of IL6 fused to PE40 (IL6-PE40) (Siegall 1988; Fig. 1). This chimeric toxin was constructed by fusing the gene encoding IL6 with a DNA fragment encoding PE40. The gene fusion was expressed under control of the bacteriophage T7 late promoter in *E. coli* by induction of a rapidly growing culture with isopropyl-β-D-thiogalactopyranoside (Studier and Moffatt 1986). The fusion product which was localized to the inclusion bodies, was solubilized by denaturation in 7M guanidine-HCl followed by rapid renaturation in 80 volumes of phosphate-buffered saline. IL6-PE40 was then purified from the inclusion bodies by successive anion exchange and gel filtration chromatography. The fusion protein was shown to be authentic IL6-PE40 based on its molecular size on SDS-PAGE, immunoblotting using both anti-IL6 and anti-PE antibodies, and NH_2-terminal amino acid sequencing.

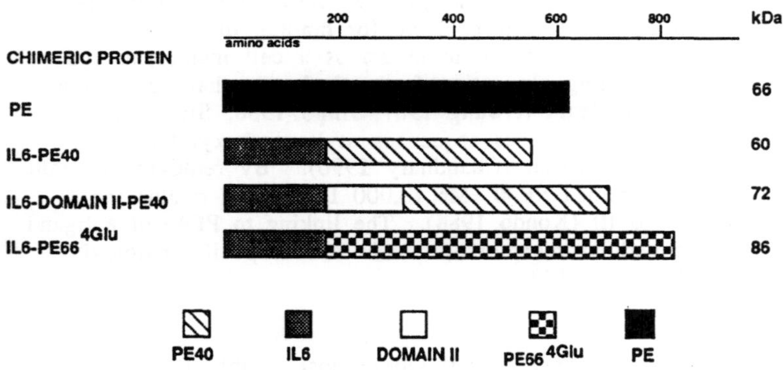

Fig. 1. Schematic diagram of IL6-PE40 and derivatives.

To determine the cytotoxic potential of IL6-PE40, we used the myeloma cell line U266 which express over 11,000 IL6 receptors per cell (Taga 1987). The cytotoxic activity was assessed by measuring the ability of various concentrations of IL6-PE40 to inhibit protein synthesis. We found that U266 cells were very sensitive to IL6-PE40 with a dose required to inhibit protein synthesis by 50% (ID_{50}) of 8 ng/ml (Siegall 1988; Fig. 2). We determined that this cytotoxicity was IL6 specific because excess IL6 competed for the toxic effects of IL6-PE40. The toxic effects of IL6-PE40 were also neutralized by the addition of an anti-IL6 antibody which binds and blocks IL6 binding to the IL6 receptor. Additionally, we produced a non-toxic form of IL6-PE40 termed IL6-PE40 [Asp^{553}] (Fig. 2). This molecule contains a single mutation at amino acid 553 in domain III of PE which inactivates its ADP-ribosylation activity. IL6-PE40 [Asp^{553}] was not toxic to U266 cells, demonstrating the killing of IL6-PE40 to be toxin based.

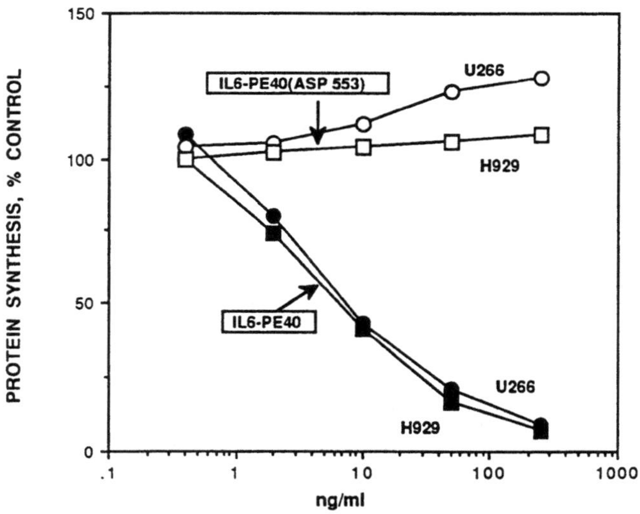

Fig. 2. Cytotoxic activity of IL6-PE40 and IL6-PE40(Asp553) on U266 and H929 myeloma cells. Chimeric toxins were added at various concentrations to the cells [5 x 10^5 cells/ml]. The cells were cultured for 22 hrs, pulsed with [^3H]-leucine for 1 hr, and the cellular protein was measured. [IL6-PE40]: (■, H929; ●, U266, [IL6-PE40(Asp553)]: (□, H929; ○, U266).

We tested a variety of other myeloma cells for their sensitivity to IL6-PE40 and found that H929 cells which have 16,500 IL6 receptors were also killed by IL6-PE40 (Siegall 1990; Fig. 2). In contrast, H1112, LB-84-1 and SKO-007 myeloma cells were not killed by IL6-PE40 suggesting that they have low levels of IL6 receptors. This was subsequently confirmed by binding assays using ^{125}I-IL6.

For a more effective treatment of IL6 receptor bearing myeloma cells, we have also developed reagents with higher cytotoxic activities. By altering the PE gene, we produced two new reagents termed IL6-PE66^{4Glu} and IL6-domain II-PE40 (Fig. 1, and below) which were significantly more toxic than IL6-PE40 to the myeloma cell lines U266 and H929 (Siegall submitted; Table 1). Additionally, the myeloma cell line H1112 which was not killed by IL6-PE40 was sensitive to IL6-PE66^{4Glu} indicating the presence of IL6 receptors on H1112 cells (Table 1). However, the myeloma lines SKO-007 and LB-84-1 were not sensitive to the new forms of IL6-Pseudomonas toxin.

Table 1.

Myeloma cell line	IL6-PE40 (ng/ml)	IL6-domain II-PE40 (ng/ml)	IL6-PE44(4Glu) (ng/ml)
U266	8	5	1.0
H929	8	5	1.5
H1112	>625	>625	250

ID_{50} is based on protein synthesis in a 24 hr. assay; experiments were done in duplicate or triplicate. Protein was measured by [^{3}H]-leucine incorporation.

IL6-PE66^{4Glu} is composed of IL6 fused to the entire PE molecule which contains mutations at position 57,246,247 and 249. These amino acids originally coded for Lys, His, Arg and His respectively and were all converted to Glu. These mutations decrease the cytotoxicity of PE by impairing the binding function found in domain I (Chaudhary, submitted). IL6-domain II-PE40 is composed of IL6 fused to domain II (amino acids 253-364) of PE followed by PE40. Domain II is responsible for processing and translocation of the toxin across cell membranes. The processes responsible for enhanced toxicity of these new IL6 toxin molecules are not fully known although indirect evidence points to increased translocation into the cytosol.

To determine the potential of these molecules in treating myeloma, their usefulness in animal models will be evaluated. Before performing animal experiments we studied the toxicity of the various IL6-Pseudomonas toxins in nude mice (Siegall, submitted). The amount of chimeric toxin that is lethal to

50% of the mice injected (LD_{50}) was 20 μg for both IL6-PE40 and IL6-domain II-PE40. IL6-PE66^{4Glu} was about twice as toxic with an LD_{50} of 10 μg (Table 2). We also measured serum levels in mice injected with IL6-toxin. When injected I.P., peak concentrations of all three forms were detected about 1 hr after injection and were detectable for up to 8 hrs in the blood (Table 2). We were able to detect more IL6-PE66^{4Glu} in the blood at 1 hr. (80% of injected amount) than both IL6-PE40 or IL6-domain II-PE40, (33-40% of the injected amount).

Table 2.

Molecule	LD_{50} (μg)	Peak Serum Level	Detection Limit	Amount Injected (μg/ml) / Maximum Detected (μg/ml)
IL6-PE40	20	1 hr	8 hr	15/5
IL6-domain II-PE40	20	1 hr	8 hr	15/6
IL6-PE66^{4Glu}	10	1 hr	8 hr	15/12

Toxicity of the molecules (LD_{50}) to nude mice was assessed by a single administration into the intraperitoneal (I.P.) cavity and observation for three days. Serum levels of the chimeric toxins were determined at various time points following I.P. administration by assaying cytotoxic activity on U266 cells.

REFERENCES

Allured V, Collier RJ, Carroll SF, McKay DB (1986) Structure of exotoxin A of Pseudomonas aeruginosa at 3.0-angstrom resolution. Proc Natl Acad Sci USA 83:1320-1324

Asaoku H, Kawano M, Iwato K, Tanabe O, Tanaka H, Hirano T, Kishimoto T, Kuramoto A (1988) Decrease in BSF-2/IL-6 response in advance cases of multiple myeloma. Blood 72:429-432

Chaudhary VK, Jinno Y, FitzGerald D, Pastan I (1990) Pseudomonas exotoxin contains a specific sequence at the carboxyl terminus that is required for cytotoxicity. Proc Natl Acad Sci USA 87:308-312

Chaudhary VK, Jinno Y, Gallo MG, FitzGerald D, Pastan I (1990) Sequences responsible for the animal toxicity of pseudomonas exotoxin. (submitted)

FitzGerald D, Pastan I (1989) Targeted toxin therapy for the treatment of cancer. J Natl Cancer Inst 81:1455-1463

Gauldie J, Richards C, Harnish D, Landsdorp P, Baumann H (1987) Interferon β2/B-cell stimulating factor type 2 shares identity with monocyte-derived

hepatocyte-stimulating factor and regulates the major acute phase protein responsive in liver cells. Proc Natl Acad Sci USA 84:7251-7255
Hirano T, Yasukawa K, Hirada H, Taga T, Watanabe Y, Matsuda T, Kashiwamura S, Nakajima K, Koyama K, Iwanatu A, Tsunaswa S, Sakiyama F, Matsui H, Takahara Y, Taniguchi T, Kishimoto T (1986) Complementary DNA for a novel human interleukin (BSF-2) that induces B lymphocytes to produce immunoglobulin. Nature 324:73-76
Hwang J, FitzGerald DJ, Adhya S, Pastan I (1987) Functional domains of Pseudomonas exotoxin identified by deletion analysis of the gene expressed in *E. coli*. Cell 48:129-136
Jinno Y, Chaudhary VK, Kondo T, Adhya S, FitzGerald DJ, Pastan I (1988) Mutational analysis of domain I of Pseudomonas exotoxin which reduce cell binding and animal toxicity. J Biol Chem 263:13203-13207
Jinno Y, Ogata M, Chaudhary VK, Willingham MC, Adhya S, FitzGerald D, Pastan I (1989) Domain II mutants of Pseudomonas exotoxin deficient in translocation. J Biol Chem 264:15953-15959
Kondo T, FitzGerald D, Chaudhary VK, Adhya S, Pastan I (1988) Activity of immunotoxins constructed with modified Pseudomonas exotoxin A lacking the cell recognition domain. J Biol Chem 263:9470-9475
Kawano M, Hirano T, Matsuda T, Taga T, Horii Y, Iwato K, Asaoku H, Tang B, Tanaba O, Tanaka H, Kuramoto A, Kishimoto T (1988) Autocrine generation and requirement of BSF-2/IL6 for human multiple myeloma. Nature 332:83-85
Nordan RP, Potter M (1986) A macrophage-derived factor required by plasmacytomas for survival and proliferation in vitro. Science 233:566-569
Pastan I, Willingham MC, FitzGerald DJ (1986) Immunotoxins. Cell 47:641-648
Pastan I, FitzGerald D (1989) Pseudomonas exotoxin: chimeric toxins. J Biol Chem 264:15157-15160
Schwab G, Siegall CB, Aarden LA, Neckers LM, Nordan RP (1990) Characterization of an interleukin-6 mediated autocrine growth loop in the human multiple myeloma cell line, U266. (submitted)
Siegall CB, Chaudhary VK, FitzGerald DJ, Pastan I (1988) Cytotoxic activity of an interleukin 6-Pseudomonas exotoxin fusion protein on human myeloma cells. Proc Natl Acad Sci USA 85:9738-9742
Siegall CB, Nordan RP, FitzGerald DJ, Pastan I (1990) Cell-specific toxicity of a chimeric protein composed of interleukin-6 and Pseudomonas exotoxin (IL6-PE40) on tumor cells. Mol Cell Biol (in press)
Siegall CB, FitzGerald DJ, Pastan I (1990) Cytotoxicity of IL6-PE40 and derivatives on tumor cells expressing a range of IL6 receptor levels (submitted)
Studier FW, Moffatt BA (1986) Use of bacteriophage T7 RNA polymerase to direct selective high level expression of cloned genes. J Mol Biol 189:113-130

Taga T, Kawanishi K, Hardy RR, Hirano T, Kishimoto T (1987) Receptors for B cell stimulatory factor 2: quantitation, specificity, distribution and regulation of their expression. J Exp Med 166:967-981

Van Snick JA, Vink S, Cayphas S, Uyttenhove C (1987) Interleukin-HP1, a T cell-derived hybridoma growth factor that supports the *in vitro* growth of murine plasmacytomas. J Exp Med 165:641-649

Vitteta ES, Fulton RJ, May RD, Till M, Uhr JW (1987) Redesigning natures poisons to create anti-tumor reagents. Science 238:1098-1104

In Vitro Culture of a Primary Plasmacytoma that has Retained Its Dependence on Pristane Conditioned Microenvironment for Growth

A. Degrassi, D. M. Hilbert, [2]A. O. Anderson, M. Potter, H. G. Coon

Laboratory of Genetics, National Cancer Institute, National Institutes of Health, Bethesda, MD, 20892 USA; [2]US Army Medical Research Institute of Infectious Diseases, Fort Detrick, Frederick, MD, 21701 USA

INTRODUCTION

In BALB/cAn mice, intraperitoneal implantation of plastics (Merwin and Redmon 1963), paraffin oils or the pure alkane pristane (2,6,10,14 tetremethylpentadecane), induces the formation of plasmacytomas (PCT) that arise in the peritoneal inflammatory tissues evoked by these agents (Potter and MacCardle 1964). The paraffin oils and pristane are known to induce the formation of a chronic oil granulomatous (OG) tissue on peritoneal surfaces (Potter and MacCardle 1964, Anderson et al 1985). This tissue contains predominantly macrophages and neutrophiles that have ingested or surrounded oil droplets and become fixed to peritoneal surfaces (Anderson et al 1985). The OG tissue is vascularized by angiogenesis from mesenteric blood vessels and covered by mesothelium. In addition, the OG tissue contains a rich supportive stroma of fibroblasts and reticular fibroblasts.

One of the hallmarks of primary plasmacytomas is the fact that free ascitic tumor cells in doses of 10^5 cannot be successfully transplanted when introduced into the normal peritoneal cavities of syngeneic BALB/cAn mice (Potter et al 1972). Such tumors can only be transplanted into pristane primed animals. However, the dependence of primary plasmacytomas on the oil-induced microenvironment is usually lost after several transplant generations. These findings suggest the growth requirements of primary plasmacytomas may differ from those of long term transplanted plasmacytomas. Consistent with this view has been the inability to grow primary plasmacytomas in vitro. In contrast, many serially transplanted plasmacytomas have been adapted to tissue culture and have proved invaluable in defining their growth requirements (Nordan and Potter 1986). In an effort to understand the growth requirements of granuloma-dependent tumors we have established granuloma-derived adherent stromal cell lines. We report here the in vitro growth of a primary plasmacytoma (5-25)on such lines. The resultant PCT cell line has retained its dependence on the oil-induced microenvironment for in vivo growth in spite of 8 months of in vitro passage.

RESULTS AND DISCUSSION

The in vivo dependence of primary plasmacytoma growth on

granulomatous tissue suggests that successful in vitro growth of primary plasmacytomas may require the presence of stromal/adherent cells from the primary site of plasmacytomagenesis, the granuloma. Accordingly, a primary plasmacytoma tumor (5-25) was removed from the peritoneal cavity of a BALB/cAn previously injected with pristane (0.5ml ip.) 150 days prior to sacrifice. The granulomatous tissue and plasmacytoma cells were disassociated for one hour at $37^{o}C$ in isotonic buffer containing collagenase (60U/ml) trypsin (0.1%) and chicken serum (2%) with mild agitation. Cells were washed twice in medium (RPMI 1640 containing 10% fetal calf serum (FCS), 2mM glutamine, 20mM HEPES, non-essential amino acids, and 50μg/ml gentamycin) and placed in 60mm dishes at a density of 10^{6} cells/ml. Cultures were maintained at $37^{o}C$ in a humidified atmosphere of 5% CO_2. The medium was changed every two days and when necessary the cells were passed to new dishes following treatment with collagenase and trypsin as described above. EM analyses of the resultant cultures revealed an adherent stromal layer (Figure 1, left) on which foci of semi-adherent plasmacytoma cells grew (Figure 1b). Fluorescent Activated Cell Sorter (FACS) analyses of the plasmacytoma cells indicated they were $CD\text{-}45^{+}$, FcR^{+}, IgA^{+}, ThB^{+}, $CD\text{-}45R^{-}$, $Thy\text{-}1^{-}$, and $Mac\text{-}1^{-}$, a phenotype typical of many plasmacytomas.

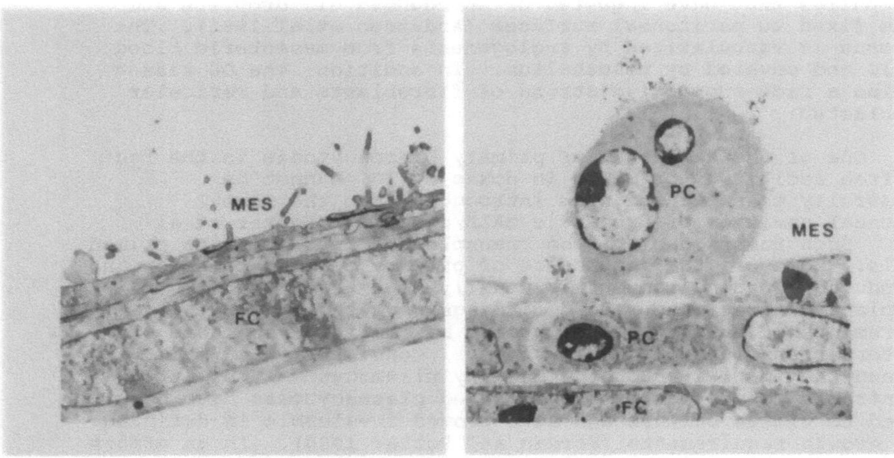

Fig. 1: Electronmicrographs of in vitro cultured granuloma-derived adherent cell layer (left) depicting mesothelial cell (MES) and fibroblastic cell (FC) types. The right micrograph contains 5-25 plasmacytoma cells (PC) which preferentially grow either on the microvillous surface of the mesothelial cells or in the intercellular space between the fibroblastic and mesothelial cells.

To date, two novel growth characterisitcs of in vitro cultured 5-25 PCT cells have been observed. First, establishment of the 5-25 PCT in vitro required an autologous, granuloma-derived adherent cell layer consisting of mesothelial, and fibroblastic cells. This requirement could not be replaced by rIL-6 or by conditioned medium from autologous feeder cultures. Second, EM analyses suggest the spacial relationship of the PCT and its stromal layer is critical since PCT growth is restricted to either the microvillous surface of the mesothelial cells or the intercellular space between the mesothelial and fibroblastic cells (Figure 1). It remains unclear whether the oriented growth of 5-25 PCT is dictated by local concentrations of diffusable factors or by the interactions of PCT- and stromal-specific adhesion molecules. These interactions are currently under investigation.

To determine if in vivo growth of the 5-25 cell line was characteristic of a primary plasmacytoma (i.e. granuloma dependent), 5-25 PCT cells were injected into a total of 17 BALB/cAn mice, 8 of which received 0.5ml of pristane ip. 3 days prior to tumor transplant. Four weeks post injection PCT growth was observed in 6/8 of the pristane primed mice. In contrast, 17 weeks post injection no tumor growth was observed in the 9 untreated mice indicating the in vitro passaged 5-25 PCT had retained its granuloma-dependence despite more than seven months of in vitro passage.

These results provide evidence for the critical role of adherent cell layers in the growth of early stage plasmacytomas and suggests the possibility that any such tumor can be grown on appropriate feeder layers. Furthermore, the ability to maintain such tumors in the "primary" state in vitro should provide a system for studying early stages of plasmacytoma development not previously accessible.

REFERENCES

Anderson AO, Wax JS, and Potter M (1985) Differences in the peritoneal response to pristane in BALB/cAnPt and BALB/cJ mice. Current Topics in Microbiology and Immunology 122:242.

Cancro M and Potter M (1976) The requirement of an adherent cell substratum for the growth of developing plasmacytoma cells in vivo. J. Exp. Med. 144:1554.

Merwin RM and Redmon, LW (1963) Induction of plasma cell tumors and sarcomas in BALB/c mice by diffusions chambers placed in the peritoneal cavity. J Natl Cancer Inst 31:998.

Nordan RP and Potter M (1986) A Macrophage-derived factor required by plasmacytomas for survival and proliferation in vitro. Science 233:566.

Potter M and MacCardle RC (1964) Histology of developing plasma cell neoplasia induced by mineral oil in BALB/c mice. J. Natl. Cancer Inst. 33:497-515.

Potter M, Pumphrey JG and Walters JL (1972) Growth of primary plasmacytomas in the mineral oil-conditioned peritoneal environment. J Natl Cancer Inst 49:305.

Experimental Plasmacytomas and B-Cell Tumors

Modulation of Growth and Differentiation of Murine Myeloma Cells by Immunoglobulin Binding Factors

Richard G. Hoover[*], Sheryl Roman, Jonni S. Moore, Christine Darby, and Susan Müller[+]

[*]Departments of Pathology, Medicine and Biochemistry, University of Arkansas for Medical Sciences, Arkansas Cancer Research Center, Little Rock, Arkansas 72205, and [+]Department of Pathology, University of Pennsylvania School of Medicine, Philadelphia, Pa. 19104

INTRODUCTION:

The role of immunoglobulin-binding factors (IBF) in the isotype-specific regulation of immunoglobulin production has been the subject of many studies (reviewed by Fridman 1987). A variety of soluble factors,that bind all classes of immunoglobulins in an isotype-specific manner, have been described and can augment or suppress B cell growth and differentiation in an isotype-specific fashion (Fridman 1987). Evidence has been presented that certain IBF have homology at the protein level with cell surface Fc receptors (Huff 1984; Peterson 1985). Thus, at least for some IBF, it is likely that the factor is a secreted or shed portion of the cell surface Fc receptor. IgA-binding factors, produced by cloned lines of T cells, that bear surface IgA-Fc receptors, or by T cell blasts induced with IgA, have been described and depending upon the circumstances, can mediate either IgA-specific help or suppression (Adachi 1984; Kiyono 1985; Yodoi 1983).

We have previously reported that mice and patients with IgA myeloma develop a marked expansion of Lyt 1-2+, T cells with surface membrane IgA-Fc receptors (Tα cells) (reviewed by Hoover 1981). This expansion of Tα cells was shown to be related to the high levels of serum IgA produced by the myeloma tumor cells. A direct relationship between IgA production and the induction of Tα cells was shown by the observation that affinity purified, polymeric IgA could interact directly with T cells in vitro to induce an expansion of Tα cells. In some instances, Tα cells from mice with IgA-producing myelomas could function as IgA-isotype-specific suppressor cells in normal immune responses (Hoover 1983). Finally, additional studies showed that IgA-induced T cells could suppress the growth and secretion of a IgA-producing myeloma tumor (Müller 1985). Collectively, these studies suggested that an exaggerated, IgA-isotype specific suppressor T cell circuit might be activated in a host bearing an IgA secreting tumor.

Numerous reports have shown that myeloma tumor cells can respond to a wide variety of immunoregulatory T cell signals and provide excellent models for the study of various aspects of B cell regulation (reviewed by Lynch 1979). A particular advantage of myeloma cells as a model of B cell regulation is the ability to exploit such a model to study regulatory events at the molecular level. In the present report we have studied the molecular events associated with the suppression of growth and secretion of MOPC-315 mediated by IgA-binding factor produced by IgA induced T cells.

Current Topics in Microbiology and Immunology, Vol. 166

RESULTS:

SUPPRESSION OF GROWTH AND SECRETION OF MOPC-315 BY IgA-BINDING FACTOR.

We had previously reported that co-culture of MOPC-315 tumor cells, with IgA induced T cells, resulted in suppression of growth of the tumor both in vivo and in vitro and suppression of secretion by the tumor cells (Müller 1985). To ascertain whether this observation was due to IgA-binding factor, we prepared IgA-binding factor by culturing nylon wool non-adherent T cells with affinity purified IgA for 24 hours. The IgA-induced T cells were then washed and recultured in IgA-free media for an additional 24 hours. IgA-binding factor was then purified from cell free supernatants by affinity chromatography on IgA-Sepharose columns. Control material was prepared from the supernatants of T cells cultured in media alone followed by affinity chromatography on IgA-Sepharose columns (Roman 1988). We treated MOPC-315 tumor cells with various dilutions of this material for 24 hours. As can be seen from Table 1, treatment of MOPC-315 cells with IgA-binding factor resulted in a dose dependent suppression of incorporation of ^{3}H-thymidine into tumor cells. Similarly, as shown in Table 2, when MOPC-315 was treated with IgA-binding factor for 24 hours there was inhibition of secretion by the tumor cells (mean suppression 83% ± 2%). Thus, IgA-binding factor prepared from the supernatants of IgA induced $T\alpha$ cells can mimic the effector function of the whole cells and is the likely mediator of the immunoregulatory function of IgA induced $T\alpha$ cells.

SUPPRESSION OF IMMUNOGLOBULIN PROTEIN SYNTHESIS IN MOPC-315 BY IgA-BINDING FACTOR.

Since IgA-binding factor could suppress the secretion of IgA by MOPC-315 tumor cells, we next examined whether this suppressor molecule could effect biosynthesis of IgA protein. As described above, MOPC-315 tumor cells were treated with IgA-binding factor or control material and then pulse labeled with ^{35}S-methionine. Cell lysates were then prepared and radiolabeled IgA was immunoprecipitated using a mono-specific rabbit anti-mouse IgA (heavy and light chain specific). Total cell lysates and IgA-immunoprecipitates were prepared and then subjected to analysis by SDS-PAGE under reducing conditions. As can be seen in Figure 1A, both heavy and light chain biosynthesis is suppressed in IgA-binding factor treated MOPC-315 as compared to controls. Scanning densitometry revealed 97% suppression of heavy chain synthesis and 95% suppression of light chain synthesis. When total cell lysates were subjected to SDS-PAGE and scanning densitometry, heavy chain synthesis was suppressed 74% and light chain synthesis was suppressed 73% (Figure 1B). However, there were no appreciable qualitative or quantitative differences in the non-IgA proteins synthesized in MOPC-315 (densitometry data not shown). Thus, IgA-binding factor selectively suppresses the biosynthesis of both heavy and light chain proteins.

SUPPRESSION OF HEAVY AND LIGHT CHAIN mRNA ACCUMULATION BY IgA-BINDING FACTOR.

Since IgA-binding factor suppressed the biosynthesis of heavy and light chain protein, we next sought to examine the effect of this suppressor factor on accumulation and half-life of heavy and light chain mRNA. As outlined above, MOPC-315 tumor cells were treated with IgA-binding factor or control material for 24 hours. Total cellular RNA was then isolated and subjected to gel electrophoresis and Northern blot analysis. As can be seen in Figure 2, the accumulation of alpha heavy chain mRNA was suppressed in IgA-binding factor treated MOPC-315 cells as compared to controls (mean suppression for four experiments,75% ± 5% as determined by scanning densitometry). Similar results were seen when accumulation of lambda 2 mRNA was measured. As can be seen in Figure 2, accumulation of lambda 2 mRNA was inhibited in IgA-binding factor treated MOPC-315 as compared to controls (mean suppression for three experiments 68% ± 6% as determined by scanning densitometry). To control for the specificity of this effect we probed the same blots for β-actin. As can be seen in Figure 2, IgA-binding factor has no detectable effect on accumulation of β-actin mRNA, thus this effect appears to be specific for heavy and light chain mRNA. In order to ascertain whether this suppression of mRNA accumulation might be mediated at the transcriptional level, we measured the effect of IgA-binding factor on the heavy and light chain mRNA half-life. MOPC-315 cells were treated with IgA-binding factor or control material for 24 hours. Actinomycin D was then added to the cultures. Aliquots of cells were harvested at various time points thereafter, and RNA was isolated and subjected to gel analysis as described above. As can be seen from Figure 3, IgA-binding factor did not significantly effect heavy or light chain mRNA half-life as compared to controls. Thus, the suppression of heavy and light chain mRNA accumulation mediated by IgA-binding factor likely occurs at the transcriptional level.

SUPPRESSION OF C-MYC mRNA ACCUMULATION BY IgA-BINDING FACTOR:

MOPC-315 and other murine plasmacytomas bear reciprocal chromosomal translocations involving chromosomes 15 and 12 (Cory 1986). Similar to what has been observed in human Burkitt lymphomas, the c-myc proto-oncogene is translocated from its normal position on chromosome 15 to chromosome 12 which bears the immunoglobulin heavy chain cluster. In many of these tumors, including MOPC-315, the c-myc gene is rearranged into the alpha switch region. C-myc is constituitively activated in these cells. Although it has not been definitively proven, the constitutive activation of c-myc in plasmacytoma cells has been implicated in the neoplastic transformation of these cells. Since IgA-binding factor suppresses the growth of MOPC-315 tumor cells and also suppressed expression of the alpha heavy chain gene, we sought to examine the effects of this factor on c-myc proto-oncogene expression. As can be seen in Figure 2, c-myc mRNA accumulation is decreased in MOPC-315 tumor cells treated with IgA-binding factor as compared to controls (mean suppression in three experiments, 76% ± 5% as determined by scanning densitometry).

DISCUSSION:

In many cases the regulation of isotype production mediated by FcR+ T cells is mediated by immunoglobulin binding factors produced by the effector cells (Fridman 1987). Although the biochemical nature of immunoglobulin-binding factors is as yet ill-defined the specificity and biological function of these molecules is well documented in the literature.
IgA-binding factors have been described that are capable of potentiating or suppressing IgA production by B cells in an isotype-specific fashion (Adachi 1984; Kiyono 1985; Yodoi 1983). Little information is available, however, with regard to the mechanism whereby IgA-binding factors modulate IgA production by B cells. In the present report we provide some insight into how IgA-binding factor might regulate IgA production.

The regulation of immunoglobulin gene expression appears to be a very complex process. Evidence has been presented in the literature for regulation at the level of transcription, mRNA processing, mRNA stability, translation and post-translational modification and protein stability (reviewed by Calame 1985). The data, presented in this report, suggests that IgA-binding factor regulates immunoglobulin genes at the level of transcription. The magnitude of suppression observed for heavy and light chain mRNA correlate well with the amount of suppression observed at the protein level, suggesting that other levels of regulation may play little, if any, role. It will be of interest to see whether other immunoglobulin-binding factors or other biologic response modifiers regulate immunogloublin gene expression in a similar fashion or whether our observations are unique to IgA-binding factor.

A number of DNA sequences have been identified that likely play significant roles in the regulation of immunoglobulin gene expression (reviewed by Calame 1985). It will be of interest to ascertain whether IgA-binding factor induces intracellular factors that can bind to DNA with subsequent inhibition of transcription and whether a common sequence adjacent to heavy and light chain genes is recognized by such molecules. The MOPC-315 model should provide a powerful tool for addressing such issues, particularly as they relate to IgA-specific regulatory factors.

The regulation of c-myc gene expression is also a complicated issue. It has been suggested by a variety of studies that both transcriptional and post-transcriptional mechanisms are important in a number of cell types including fibroblasts, B cells and T cells (reviewed by Cole 1986). As mentioned above, MOPC-315 bears a reciprocal 15:12 chromosomal translocation (Cory 1986). The c-myc gene, normally present on chromosome 15, is translocated and rearranged into the alpha switch region of the IgH gene. This translocation and rearrangement affects the nonproductively rearranged IgH chromosome (Cory 1985). The high constitutive levels of c-myc mRNA in plasmacytomas result from active transcription of the translocated allele of c-myc which may be due to exposure of cryptic, active promoters as a result of the chromosomal translocation. The studies of Adams (1985) have provided compelling evidence that constitutive activation of c-myc in B cell neoplasms is the pivotal oncogenic event. It is attractive to postulate that the suppression of growth mediated by

IgA-binding factor may relate to coordinate suppression of alpha heavy chain and c-myc gene expression. However, the activation of c-myc gene expression has been observed to accompany cell proliferation in a wide variety of systems (Kelley 1985). Therefore, it cannot be ruled out that our observation represents an epiphenomenen associated with IgA-binding factor mediated growth suppression which acts through another mechanism.

Table 1. Suppression of MOPC-315 Proliferation by IgABF

Treatment[a]	Dilution	cpm[b]
Control	1/2	106,736 ± 4,624
IgABF	1/2	360 ± 60
IgABF	1/10	20,777 ± 548
IgABF	1/100	54,626 ± 2,796
IgABF	1/1000	108,940 ± 8,621

a) A total of 5×10^5 MOPC-315 cells were treated with IgABF or control material at the dilutions indicated for 6 hours at 37^oC. The cells were then pulsed with 2uCi ^{3}H-TdR and cultured for an additional 24 hours.

b) MOPC-315 cells were harvested and the amount of radioactivity incorporated was assessed by scintillation counting. Data represent mean ± S.E.M.

Table 2. Suppression of MOPC-315 Secretion by IgABF

Treatment[a]	PFC/culture[b]
None	256 ± 6
Control	278 ± 7
IgABF	49 ± 2

a) Approximately 1800 MOPC-315 cells were cultured for 24 hours at 37^oC in microtiter wells alone or in the presence of a 1/10 dilution of control material or IgABF.

b) Each well was harvested and assayed individually for anti-TNP antibody secreting cells by an indirect Jerne plaque assay. Data represents mean ± S.E.M.

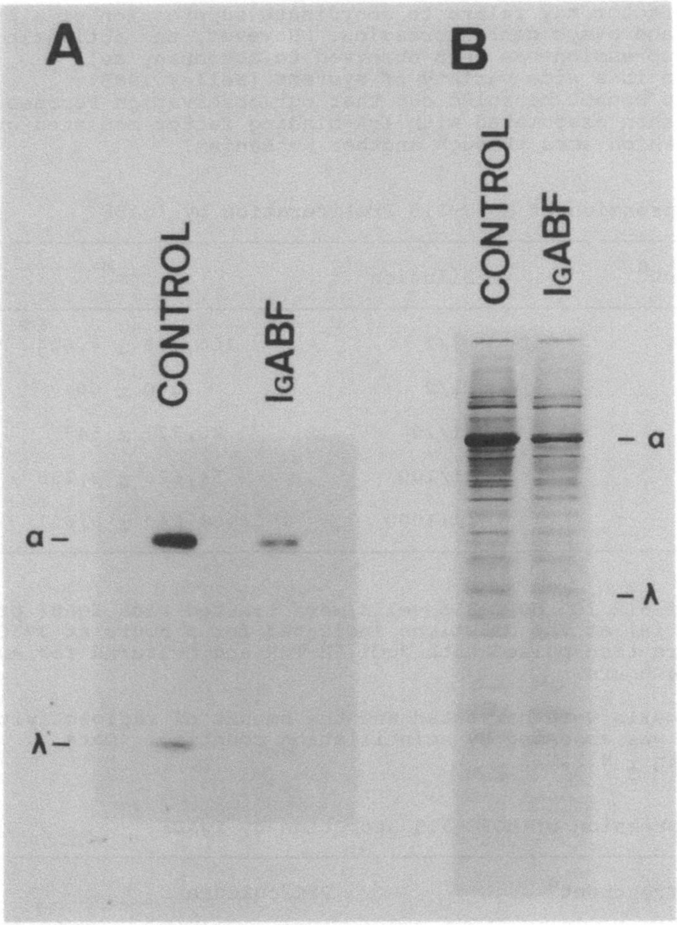

Figure 1: SDS-polyacrylamide gel electrophoresis analysis of immunoglobulin heavy and light chain proteins in suppressed MOPC-315 cells.

1 x 10^6 MOPC-315 cells were treated with a 1:100 dilution of either control material or IgABF for 24 hours and the proteins were radiolabeled with ^{35}S methionine. IgA immunoprecipitates (A) or total cell lysates (B) were then subjected to analysis by SDS-polyacrlyamide gel electrophoresis. Unlabeled, affinity purified RA315 was run in the gel as a molecular weight standard to determine the positions of the alpha heavy chain and lambda 2 light chain proteins. These positions are indicated in the margins. Data presented is one representative experiment of four separate experiments.

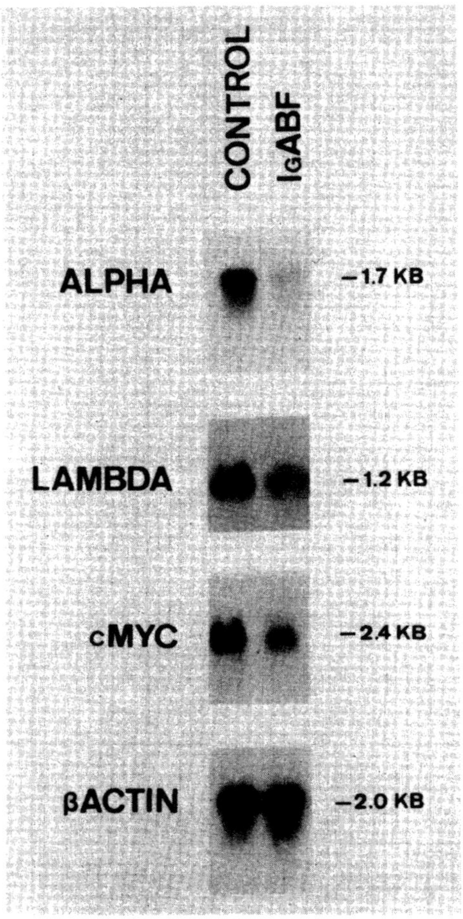

Figure 2: Northern blot analysis of mRNA expression in IgABF suppressed MOPC-315.

5×10^6 MOPC-315 cells were treated with a 1:100 dilution of either control material or IgABF for 24 hours. Total cellular RNA was then isolated and subjected to Northern blot analysis. The size of each transcript was determined relative to the positions of ribosomal RNA in ethidium stained gels and is indicated in the right hand margin. Data presented is one representative experiment of at least three separate experiments.

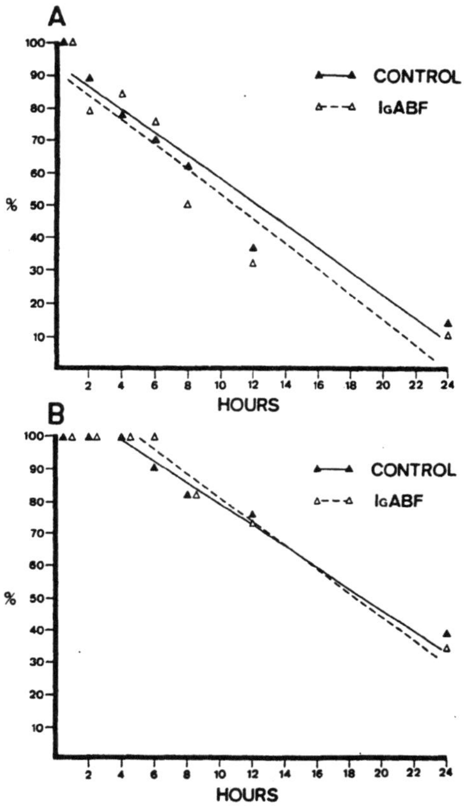

Figure 3: Half-life of mRNA encoding the alpha heavy chain and the lambda 2 light chain in IgABF suppressed MOPC-315.

MOPC-315 cells were treated with a 1:10 dilution of either control material or IgABF for 24 hours. The cells were then treated for an additional 2 hours with 5ug/ml Actinomycin D. Treated cells were then harvested at 0, 2, 4, 6, 8, 10, 12, and 24 hours after treatment with Actinomycin D. RNA was isolated and analyzed by Northern blotting. The quantity of each mRNA species was established by scanning laser densitometry. Each data point was normalized to the 0 time starting point and the data plotted relative to time. The data was analyzed by least squares regression and this was used to establish the decay time for the alpha heavy chain mRNA and the lambda 2 light chain mRNA. A) The half-life of the alpha heavy chain mRNA was: Control = 12.2 hours, IgABF = 11 hours. B) The half-life of the lambda 2 light chain mRNA was: Control = 19 hours, IgABF = 18.5 hours.

Adachi, M., J. Yodoi, N. Noro, T. Masuda, and H. Uchino. (1984). Murine IgA-binding factors produced by Fc R(+) T cells: Role of Fc R(+) T cells for the induction of Fc R and formation of IgA-binding factor in Con-A activated cells. J. Immunol. 133:65.
Adams, J.A., W. Harris, C.A. Pinkert, L.M. Corcoran, W.S. Alexander, S. Cory, R.D. Palmiter, and R. L. Brinster. (1985). The c-myc oncogene driven by immunoglobulin enhancers induces lymphoid malignancy in transgenic mice. Nature 318:533.
Calame, K.L. (1985). Mechanisms that regulate immunoglobulin gene expression. Annual Rev. Immunol. 3:159.
Cole, M. (1986). The myc oncogene: Its role in transformation and differentiation. Annual Rev. Genet. 20:361.
Cory, S. (1986). Activation of cellular oncogenes in hemopoietic cells by chromosomal translocation. Adv. Cancer Res. 47:189.
Fridman, W.H., J.-L. Teillaud, S. Amigorena, M. Daeron, U. Blank, and C. Neauport-Sautes. (1987). The isotype circuit: Immunoglobulins, Fc receptors and immunoglobulin-binding factors. International Rev. Immunol. 2:221.
Hoover, R.G., H.M. Gebel,, B.K. Dieckgraefe, S. Hickman, N.F. Rebbe N. Hirayama, Z. Ovary and R.G. Lynch. (1981). Occurrence and potential significance of increased numbers of T cells with Fc receptors in myeloma. Immunol. Rev. 56:115.
Hoover, R.G. and R.G. Lynch. (1983). Isotype specific regulation of IgA: Suppression of IgA responses in orally immunized mice by T cells. J. Immunol. 130:521.
Huff, T.F., J. Yodoi, T. Uede, and K. Ishizaka. (1984). Presence of an antigenic determinant common to rat IgE-potentiating factor, IgE-suppressive factor, and Fc receptors on T and B lymphocytes. J. Immunol. 132:406.
Kelley, K. and U. Siebenlist. (1985). The role of c-myc in the proliferation of normal and neoplastic cells. J. Clin. Immunol. 5:65.
Kiyono, H., L.M. Mosteller-Barnum, A.M. Pitts, S. I. Williamson, S. M. Michalek and J.R. McGhee. (1985). Isotype-specific immunoregulation: IgA-binding factors produced by Fc receptor-positive T cell hybridomas regulate IgA responses. J. Exp. Med. 161:731.
Lynch, R.G., J.W. Rohrer, B.F. Odermatt, H.M. Gebel, J.R. Autry, and R.G. Hoover. (1979). Immunoregulation of murine myeloma cell growth and differentiation: A monoclonal model of B cell differentiation. Immunol. Rev. 48:45.
Müller, S. and R.G. Hoover. (1985). T cells with Fc receptors in myeloma: Suppression of growth and secretion of MOPC-315 by T cells. J. Immunol. 134:644.
Peterson, L. and D. Conrad. (1985). Fine specificity, structure and proteolytic susceptibility of the human lymphocyte receptor for IgE. J. Immunol. 135:2654.
Roman, S., J.S. Moore, C. Darby, S. Müller, and R.G. Hoover. (1988). Modulation of Ig gene expression by Ig binding factors: Suppression of alpha-heavy chain and lambda 2-light chain mRNA accumulation in MOPC-315 by IgA-binding factor. J. Immunol. 140:3622.
Yodoi, J., M. Adachi, K, Teshigawara, M. Miyama-Inaba, T. Masuda, and W. H. Fridman. (1983). T cell hybridomas co-expressing Fc receptors (FcR) for different isotypes. II. IgA-induced formation of suppressive IgA-binding factor(s) by a murine hybridoma bearing Fc R and Fc R. J. Immunol. 131:303.

A New Cell Adhesion Mechanism Involving Hyaluronate and CD44

K. Miyake and P.W. Kincade

Immunobiology and Cancer Program
Oklahoma Medical Research Foundation
825 N.E. 13th Street, Okla. City, OK 73104

INTRODUCTION

Aside from the basic importance of understanding various cell adhesion/interaction mechanisms, these processes may also be informative with respect to malignancy and/or metastasis of tumor cells. During studies of adhesion molecules which may mediate attachment of developing lymphocytes to microenvironmental elements in the bone marrow, we discovered an apparently new cell adhesion mechanism. An adhesion model, comprised of a B lineage hybridoma and a cloned stromal cell line, was used to screen monoclonal antibodies prepared by immunizing with the stromal cells. Four antibodies were selected that inhibited the avid binding between these two cell types and, when purified and added to long term bone marrow cultures, they completely prevented lympho-hemopoiesis. Biochemical and other determinations revealed that these antibodies recognize epitopes on a well studied cell surface glycoprotein, Pgp-1/CD44 (Miyake et al., 1990). Another recently completed study revealed that adhesion between the hybridoma and stromal cells was almost completely dependent on hyaluronate (Miyale et al., submitted). This cell recognition/adhesion mechanism may have general importance and direct binding of cells to hyaluronate was particularly noteworthy with myeloma and hybridoma cells. Thus, hyaluronate recognition by CD44 on such tumor cells could have an influence on growth and metastatic ability.

RESULTS AND DISCUSSION

Long term bone marrow culture studies indicate that the adhesion of developing lymphocytes to stromal cells is an apparently complex process (Kincade, 1987). A variable fraction of lymphocytes is released from the adherent layer by treatment with phosphatidylinositol specific phospholipase C (PI-PLC) or EDTA and several known, or suspected, cell adhesion molecules are expressed in such cultures (Kincade et al., 1989; Gimble et al., 1989). The latter include N-CAM, N-cadherin, LFA-1, MEL 14, Thy-1, and Pgp-1/CD44. It proved difficult to determine if any of these were particularly important and a new cell adhesion assay was developed (Miyake et al., 1990). Cloned stromal cells were allowed to form adherent monolayers and radiolabeled lymphoid cells added 18 hours later. A survey revealed that the BM-2 hybridoma was especially useful and it quickly and firmly attached to the BMS2 stromal cell clone.

A panel of rat monoclonal antibodies was prepared by immunizing with the BMS2 stromal cell clone. One screening strategy involved selection of antibodies which inhibited BM-2 - BMS2 adhesion and four clones were further characterized (Miyake et al., 1990). All inhibited long term bone marrow cultures and all recognized Pgp-1/CD44. None of a series of antibodies to this glycoprotein was notably toxic for lymphoid or stromal cells and they did not affect cytokine dependent proliferation responses.

Several laboratories have prepared and sequenced cDNA clones corresponding to Pgp-1/CD44 from mice, man and other primates (Zhou et al., 1989; Idzerda et al., 1989; Stamenkovic et al., 1989; Goldstein et al., 1989; Wolffe et al., 1990). Of particular interest was the homology of its

Current Topics in Microbiology and Immunology, Vol. 166

N-terminal domain to proteins which are known to interact with extracellular matrix components. These include link protein and the core protein of large proteoglycans. A hyaluronate receptor had previously been studied which had many similarities to Pgp-1/CD44 (Underhill, 1989). Furthermore, other laboratories obtained evidence that hyaluronate can interact directly with hemopoietic cells (Lesley et al., 1989). We found that BM-2 cells attached to hyaluronate coated dishes, but not to wells pretreated with several other glycosaminoglycans (Miyake et al., submitted). This was completely inhibited by treatment of the dishes with a specific hyaluronidase, or by inclusion of the KM 201 monoclonal antibody. Similarly, adhesion of BM2 cells to BMS2 cells was hyaluronate and Pgp-1/CD44 dependent. Separate pretreatment of the two cell types with hyaluronidase or KM 201 suggested that lymphocyte bound Pgp-1/CD44 may interact with hyaluronate on the surface of stromal cells.

A survey indicated that lymphoid cells can bind to adherent cell lines derived from embryonic tissues and adult spleen as well as from bone marrow (Miyake et al., submitted). However, the degree to which Pgp-1/CD44 and hyaluronate were involved varied considerably. We now show that two myeloma lines also bind avidly to hyaluronate (Table 1). This interaction was completely blocked by the KM 201 antibody to Pgp-1/CD44. In contrast, neither the 70Z/3 pre-B lymphoma or WEHI-3 myelomonocytic leukemia cell lines showed affinity for hyaluronate coated dishes. The latter two lines can bind to certain stromal cell clones, but our observations indicate that completely independent adhesion mechanisms are involved (data not shown).

Table 1. Differential Adhesion to Hyaluronate of Established Cell Lines

		Percent Cells Bound		
Cell line		None	Rat IgG1	KM201
BM-2	Hybridoma	83.0 ± 0.9	82.4 ± 0.9	0.3 ± 0.1
XS63	Myeloma	86.2 ± 0.3	86.2 ± 1.4	0.4 ± 0.1
Sp2/0	Myeloma	88.0 ± 0.5	85.8 ± 0.7	0.4 ± 0.1
7OZ/3	PreB lymphoma	2.1 ± 0.1	2.3 ± 0.2	0.4 ± 0.1
WEHI-3	Myelomonocyte	2.0 ± 0.1	1.7 ± 0.1	0.6 ± 0.1

Hyaluronate (5 mg/ml) was coated on the bottom of plastic wells, and radiolabeled cell lines were added with the purified monoclonal antibodies (10 μg/ml). The values are presented as means ± SE for quadruplicate wells.

All five of these hemopoietic tumor cell lines expressed cell surface Pgp-1/CD44 (Figure 1). However, flow cytometric analysis suggested that the density was considerably higher on the cells which adhered to hyaluronate. These results are similar to ones recently obtained by Lesley and colleagues (Lesley et al., 1989). A thymoma line bound fluorescein labeled hyaluronate, but only after pretreatment of the cells with PMA. The mechanism of hyaluronate binding was Pgp-1/CD44 dependent and the density of this glycoprotein increased during stimulation.

Many questions remain about this new cell adhesion mechanism. Of particular interest is the fact that all Pgp-1/CD44 positive cells do not demonstrably bind to hyaluronate coated surfaces. This may result from a suboptimal density of this glycoprotein, alternative Pgp-1/CD44 isoforms, or a requirement for interaction with other cell adhesion molecules. All of these could be a reflection of differentiation and/or "activation" states (Dustin et al., 1989; Van Kooyk et al., 1989).

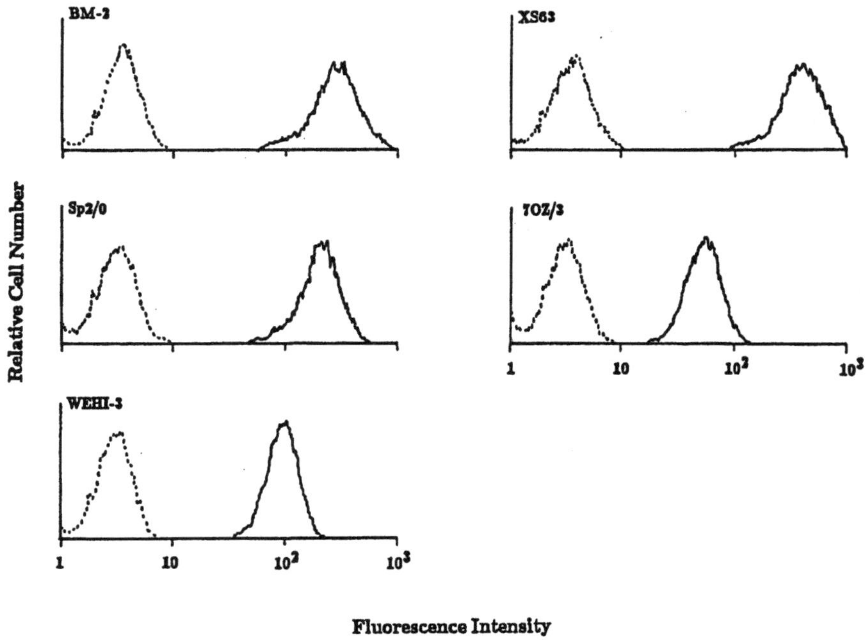

Figure 1. Flow cytometric analysis of Pgp-1/CD44 expression on established cell lines. Cells were labeled with purified, biotinylated IM7.8.1 antibody and then stained with FITC avidin (solid lines). Controls were stained with avidin alone (dotted lines).

The antibody secreting BM-2 line was initially selected on the basis of its ability to penetrate bone marrow when injected intravenously (Miyake, unpublished observations). Plasma cells in bone marrow can contribute significantly to antibodies made during secondary immune responses (Koch et al., 1982) and these might be the normal counterparts of BM-2 cells. It also seems possible that interaction with hyaluronate may be a common feature of plasma cells and myelomas. Human myelomas commonly spread to multiple medullary sites and it is important to learn if this cell adhesion mechanism is involved in metastasis. It has previously been speculated that modulation of CD44 may contribute to the spread of other types of malignancies (Stamenkovic et al., 1989).

REFERENCES

Dustin ML, Springer TA (1989) T-cell receptor cross-linking transiently stimulates adhesiveness through LFA-1. Nature 341:619-624

Gimble JM, Pietrangeli CE, Henley A, et al. (1989) Characterization of murine bone marrow and spleen derived stromal cells: Analysis of leukocyte marker and growth factor mRNA transcript levels. Blood 74:303-311

Goldstein LA, Zhou DFH, Picker LJ, et al. (1989) A human lymphocyte homing receptor, the Hermes antigen, is related to cartilage proteoglycan core and link proteins. Cell 56:1063-1072

Idzerda RL, Carter WG, Nottenburg C, Wayner EA, Gallatin WM, StJohn T (1989) Isolation and DNA sequence of a cDNA clone encoding a lymphocyte adhesion receptor for high endothelium. Proc Natl Acad Sci USA 86:4659-4663

Kincade PW (1987) Experimental models for understanding B lymphocyte formation. Adv Immunol 41:181-267

Kincade PW, Lee G, Pietrangeli CE, Hayashi S-I, Gimble JM (1989) Cells and molecules that regulate B lymphopoiesis in bone marrow. Ann Rev Immunol 7:111-143

Koch G, Lok BD, van Oudenaren A, Benner R (1982) The capacity and mechanism of bone marrow antibody formation by thymus-independent antigens. J Immunol 128:1497-1501

Lesley J, Schulte R, Hyman R (1989) Binding of hyaluronic acid to lymphoid cell lines. Exp Cell Res

Miyake K, Medina KL, Hayashi S-I, Ono S, Hamaoka T, Kincade PW (1990) Monoclonal antibodies to Pgp-1/CD44 block lympho-hemopoiesis in long-term bone marrow cultures. J Exp Med 171:477-488

Stamenkovic I, Amiot M, Pesando JM, Seed B (1989) A lymphocyte molecule implicated in lymph node homing is a member of the cartilage link protein family. Cell 56:1057-1062

Underhill, C.B. (1989) The interaction of hyaluronate with the cell surface: The hyaluronate receptor and the core protein. In: The Biology of Hyaluronan. Ciba Foundation Symposium, p 87-106

Van Kooyk Y, Van de Wiel-van Kemenade P, Weder P, Kuijpers TW, Figdor CG (1989) Enhancement of LFA-1-mediated cell adhesion by triggering through CD2 or CD3 on T lymphocytes. Nature 342:811-813

Wolffe EH, Gause WC, Pelfrey CM, Holland SM, Steinberg AD, August JT (1990) The cDNA sequence of mouse Pgp-1 and homology to human CD44 cell surface antigen and proteoglycan core/link protein. J Biol Chem 265:341-347

Zhou DFH, Ding JF, Picker LJ, Bargatze RF, Butcher EC, Goeddel DV (1989) Molecular cloning and expression of Pgp-1: The mouse homolog of the human H-CAM (Hermes) lymphocyte homing receptor. J Immunol 143:3390-3395

The Activity of an *ABL-MYC* Retrovirus in Fibroblast Cell Lines and in Lymphocytes

David Largaespada, Donal Kaehler, Evi Weissinger[1], Harold Mischak[1], Fred Mushinski[1], and Rex Risser

McArdle Laboratory for Cancer Research, University of Wisconsin, Madison, WI 53706

[1]Laboratory of Genetics, National Cancer Institute, NIH, Bethesda, MD 20892

INTRODUCTION

Evidence from many laboratories indicates that v-*abl* and c-*myc* oncogenes both can participate in the induction of B-cell neoplasia, and under appropriate conditions both may synergize in that process. The first indication of the interaction of these two oncogenes was the experiments of Potter et al (1973) in which it was found that infection of pristane-primed mice with Abelson murine leukemia virus (A-MuLV) greatly decreased the latent period for plasmacytoma development. Molecular analysis of those tumors indicated that most tumors expressed the v-*abl* oncogene product and contained a c-*myc* gene deregulated by chromosomal translocation (Ohno et al 1984). Although c-*myc* deregulation has clearly been implicated in plasmacytoma development (Shen-Ong et al 1982; Adams et al 1983; Mushinski et al 1983), retroviruses that express v-*myc* or c-*myc* appear to induce tumors of the myeloid series of cells (Bambaugh et al 1985; Vennstrom et al 1984). Moreover, infection with A-MuLV does not accelerate plasmacytoma development in Eμ-*myc* transgenic mice (Dyall-Smith et al 1988). On the other hand, spleen cells from such mice can develop into plasmacytomas following infection with A-MuLV (Sugiyama et al 1989). To investigate the interaction of v-*abl* and c-*myc* deregulated oncogenes under conditions where both are introduced simultaneously into cells we have constructed a virus that expresses both genes.

The ABL-MYC Retrovirus

Exons 2 and 3 of the c-*myc* gene were placed distal to the herpes simplex thymidine kinase promoter (*tk*). This element was inserted downstream of the v-*abl* coding sequences in a molecular clone of A-MuLV (Green et al 1987a). The *myc* gene in ABL-MYC virus is thus transcribed as part of a viral genome and from the internal *tk* promoter. A control virus, ABL-CYM, was constructed such that the c-*myc* gene was in opposite transcriptional orientation, thus potentially encoding an anti-sense c-*myc* RNA.

Current Topics in Microbiology and Immunology, Vol. 166

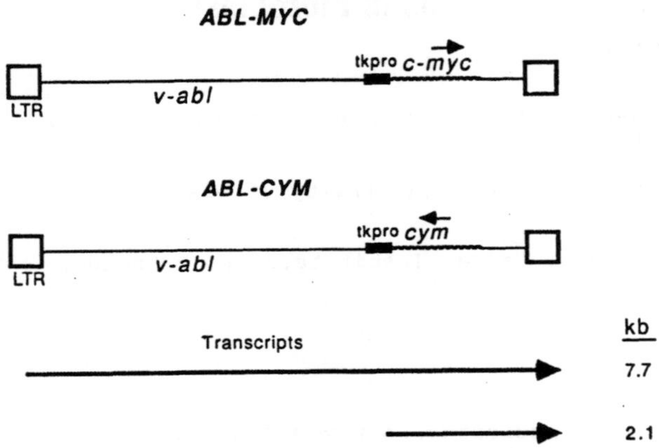

Fig. 1. A-MuLV and ABL-MYC Retroviruses.

Helper-containing ABL-MYC virus was prepared by co-transfection of NIH3T3 with ABL-MYC DNA and a molecular clone of Moloney MuLV. Helper-free ABL-MYC was prepared by transfection of the ψ-2 packaging cells and recovery of transformed cell clones as previously described (Green et al 1987b). Viruses were titered on NIH3T3 cell by focus formation and colony-formation in agarose suspension (Table 1). The focus-forming titers of A-MuLV and ABL-MYC were comparable, however ABL-MYC was 10-fold more efficient at inducing agarose colonies. These results suggest that ABL-MYC virus is more efficient at some aspects of transformation than A-MuLV. ABL-MYC transformed cells were examined for the presence of proviral sequences and for expression of viral RNA. The predicted proviral structure was found and no rearrangement of the endogenous c-*myc* locus was detected. *myc* RNA that co-migrates with viral U3 sequences was detected.

Table 1: Titers of ABL Retroviruses in NIH3T3 cells

Virus	FFU/ml[a]	CFU/ml[b]	CFU/FFU
ABL-MYC(ψ2)	8.7×10^4	1.3×10^4	0.15
ABL-CYM(ψ2)	2.8×10^3	5.5×10^1	0.020
A-MuLV(ψ2)	1.3×10^5	2.0×10^3	0.015

[a] Measured by foci of morphologically transformed cells.
[b] Measured by colony formation in semi-solid medium.

Activity of ABL-MYC Virus in Mice

To determine the activity of the ABL-MYC *in vivo* we inoculated mice intravenously with helper-containing or helper-free virus pools. ABL-MYC virus induced a high incidence of lymphomas in mice normally susceptible to A-MuLV disease (Table 2). The latent period for tumor development was approximately 30-40 days, a result typical of A-MuLV disease. Gross and microscopic pathology of diseased mice was usually indistinguishable from that of A-MuLV. As found with A-MuLV tumors (Green et al. 1987a), ABL-MYC tumors were oligoclonal in origin because they showed only a few proviral integration sites. Examination of the Ig locus in these tumors indicated that J_K rearrangement had taken place, however no rearrangement of the kappa light chain locus was detected. These results suggest that these ABL-MYC tumors are of pre-B cell origin.

Table 2: Tumor Induction by A-MuLV or ABL-MYC Retroviruses

MOUSE STRAIN	AGE (D)	VIRUS DOSE	VIRUS A-MuLV (ψ2)	VIRUS ABL-MYC (ψ2)
BALB/c	20-40	2×10^5	25/27[a]	17/17[c]
	20-40	2×10^3	4/7[a]	3/8[b]
	80-90	2×10^4	8/44[a]	9/9[b]
SWR	65	2×10^4	0/20	16/20[b]
B6	24	2×10^5	0/26	7/17[b]
			A-MuLV (M-MuLV)	A-Myc (M-MuLV)
BALB/c	30	10^4	5/5[a]	10/10[c]
B6	35	10^4	0/13	6/6[b]

[a] pre-B lymphoma
[b] plasmacytoma
[c] pre-B and plasmacytoma

In mice resistant to A-MuLV lymphomagenesis, a different pattern of pathology was seen with ABL-MYC virus. A high percentage of the mice developed plasmacytomas with a median latent period of 80-90 days. The tumors appeared as masses in the mesenteries of the gut and were frequently found growing out of the intestinal wall. Microscopic analysis confirmed that the major cell type in these tumors was a plasma cell. Cells showed large amounts of cytoplasm, acentric nuclei and clear perinuclear spaces (M. Potter, personal communication). Frequently, mice also developed ascites cell growth. Examination of the sera or ascites fluid of the mice indicated that each contained an unusually high titer of a single class of immunoglobulin heavy chain. Of 18 samples examined, 12 contained high levels of IgA, 4 contained high levels of IgM, 1 contained IgG, and 1 contained IgA and IgM. Therefore, if ABL-MYC-infected mice do not develop pre-B cell tumors, they usually go on to develop plasmacytomas, even in the absence of mineral oil priming.

Attempts to culture plasmacytomas *in vitro* uniformly failed unless supernatant from P388 tumor cells was included in the culture medium. Examination of the DNA of the plasmacytomas indicated that they contained the expected ABL-MYC provirus and did not show any rearrangement of the endogenous c-*myc* locus (Fig. 2). Plasmacytomas were of clonal origin as judged by the pattern of kappa gene rearrangement.

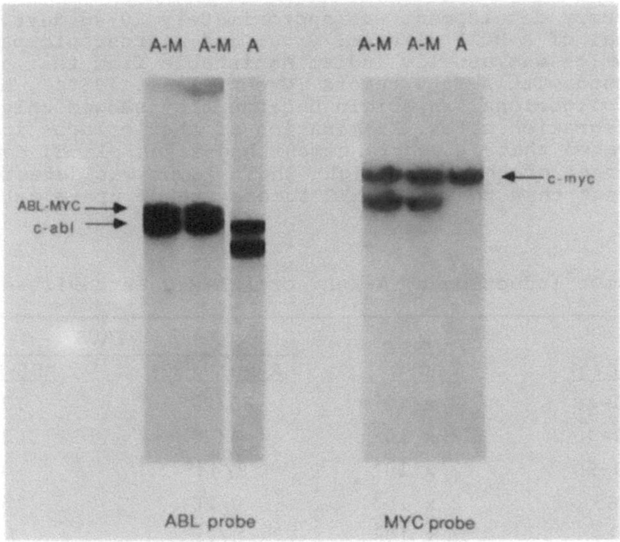

Fig. 2. Southern blot of DNA from ABL-MYC-induced **pre-B cell** lymphoma (AM left lane) or plasmacytoma (A-M middle lane) or A-MuLV-induced pre-B cell lymphoma (A). DNA was digested with Xba I which cleaves in viral LTR and hybridized with *abl* and *myc* probe.

A Plasmacytoma that Developed from a Single Infected Cell

One tumor arose following reconstitution of a lethally irradiated mice with bone marrow infected with ABL-MYC(ψ2) virus. This tumor, which initially appeared at day 54 post-reconstitution, presented as an enlarged brachial lymph node. On autopsy at day 77 the mouse was found also to contain a mesenteric tumor similar to the plasmacytomas described above. Cells were isolated from both tumors and examined microscopically and molecularly. The mesenteric tumor was clearly a plasmacytoma with the major cell type having the clear perinuclear space and cart-wheel nucleus features characteristic of plasma cells. The brachial tumor was lymphoblast in appearance. Examination of the DNA of these two tumors indicated that both descended from the same infected cell because both had the same pattern of proviral integration (Fig. 3). In contrast, the pattern of kappa gene rearrangement in the two samples differed (Fig. 3). In order for the same infected cell to give rise to two different patterns of kappa rearrangement it is likely that the cell was infected before kappa rearrangement was initiated. If that was the case, then the plasmacytoma target for ABL-MYC must have been an immature B-cell.

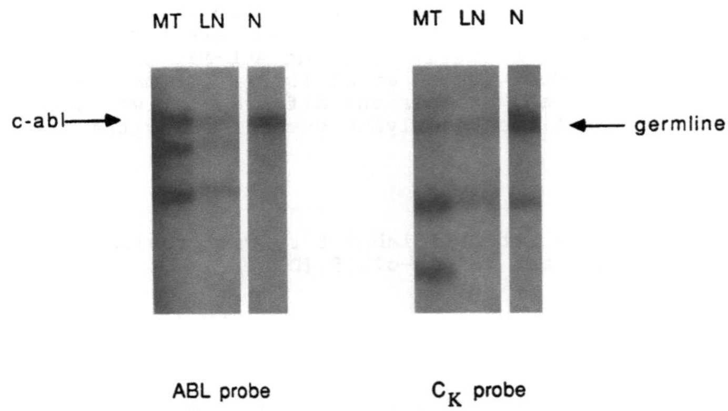

Fig. 3. Southern blot of DNA from mesenteric tumor (MT) or brachial lymph node (LN) tumor of a mouse which received bone marrow cells infected with ABL-MYC virus. DNA was cleaved with EcoRI (*abl* probe) or with BamHI (C_K probe).

DISCUSSION

Evidence has been presented that demonstrates the co-expression of c-*myc* with v-*abl* alters the apparent tissue tropism and host range of A-MuLV. Thus, B6 mice and adult SWR mice, which are resistant to A-MuLV lymphomagenesis, develop plasmacytomas when inoculated with ABL-MYC virus. The latent period for tumor development is much longer than that seen for pre-B cell lymphomas. This increased latent period may reflect the fact that these plasmacytomas develop in the absence of an oil granuloma. When pristane-primed mice are inoculated with ABL-MYC virus, plasmacytomas develop as rapidly as pre-B tumors (Mischak et al., this volume). Furthermore, plasmacytomas are strictly dependent on additional growth supplements when placed in tissue culture. This observation reinforces the idea that the long latent period for plasmacytoma development may reflect nutritional requirements.

The observation that plasmacytomas develop in the mice might suggest that a different target cell is being transformed by the combination of v-*abl* and c-*myc*. While this is entirely possible, evidence from the one tumor relevant to that point would argue that the transformation event by ABL-MYC took place prior to light chain gene rearrangement. This stage may be the same as or just slightly past the A-MuLV target. *In vitro* experiments clearly indicate that ABL-MYC can transform pre-B cells. The differing response of different strains to the ABL-MYC virus and to A-MuLV may reflect different numbers of target cells. When pre-B cell targets are present in sufficient numbers, pre-B cell tumors appear. Only when such targets are not present in sufficient numbers, can other types of tumors be recovered.

Do v-*abl* and c-*myc* oncogenes co-operate in transformation of lymphoid cells? From the evidence presented here, an unambiguous answer cannot be given. Clonally dominant ABL-MYC tumors appear, as is the case with A-MuLV (Green et al 1989). It may be that the addition of c-myc to v-abl extends the differentiative capacity of the target cell, but the underlying event remains the same.

ACKNOWLEDGEMENTS

Support for work in the authors' laboratory was provided by R01-CA22443 and training grant T32-CA-o7215 (D.L.).

REFERENCES

Bambaugh WR, Keath EJ, Cole MD (1986) A mouse c-*myc* retrovirus transforms established fibroblast lines *in vitro* and induces monocyte-macrophage tumors *in vivo*. J Virol 59: 276-283

Dyall-Smith D, Cory S (1988) Transformation of bone marrow cells from Eμ-*myc* transgenic mice by Abelson murine leukemia virus and Harvey murine sarcoma virus. Oncogene Res 2: 403-409

Green PL, Kaehler DA, Bennett LM, Risser R (1989) Multiple steps are required for the induction of tumors by Abelson murine leukemia virus. J Virol 63: 1989-1994

Green PL, Kaehler DA, Risser R (1987a) Clonal dominance and progression in Abelson murine leukemia virus lymphomagenesis. J Virol 61: 2192-2197

Green PL, Kaehler DA, Risser R (1987b) Cell transformation and tumor induction by Abelson murine leukemia virus in the absence of helper virus in the absence of helper virus. Proc Natl Acad Sci 84: 5932-5936

Mushinski JF, Bauer SR, Potter M, Reddy EP (1983) Increased expression of *myc*-related oncogene mRNA characterizes most BALB/c plasmacytomas induced by pristane or Abelson murine leukemia virus. Proc Natl Acad Sci 80: 1073-1077

Ohno S, Migita S, Wierner F, Babonits M, Klein G, Mushinski EJF, Potter M (1984) Chromosomal translocations activating *myc* sequences and transduction of v-*abl* are critical events in the rapid induction of plasmacytomas by pristane and Abelson virus. J Exp Med 159: 1762-1777

Potter M, Sklar MD, Rowe WP (1973) Rapid viral induction of plasmacytomas in pristane-primed BALB/c mice. Science 182: 592-594

Shen-Ong GLC, Keath EJ, Piccoli SP, Cole MD (1982) Novel *myc* oncogene RNA from abortive immunoglobulin-gene recombination in mouse plasmacytomas. Cell 31: 443-452

Sugiyama H, Weber G, Silva S, Babonitis M, Wiener F, Klein G (1989) The accelerating role of Abelson murine leukemia virus in murine plasmacytoma development: *in vitro* infection of spleen cells generates donor-type tumors after transfer to pristane-treated BALB/c mice. Int. J. Cancer 44:

Vennstrom B, Kahn P, Adkins B, Enrietto P, Hayman MJ, Graf T, Luciev P (1984) Transformation of mammalian fibroblasts and macrophages *in vitro* by a murine retrovirus encoding an avian v-*myc* oncogene. EMBO J 3: 3233-3229

Plasmacytoma Induction in BALB/c6;15 → DBA/2 Chimeras

S. Silva, H. Sugiyama, M. Babonits, F. Wiener and G. Klein.

Department of Tumorbiology, Karolinska Institutet, 10401 Stockholm, Sweden

INTRODUCTION

Murine plasmacytomas (MPC) can be induced by intraperitoneal injection of 2,6,10,14 tetramethylpentadecane (pristane) into inbred BALB/c mice. Susceptibility to MPC development under similar regimen of pristane, range between 10 % in BALB/cJ (Potter and Wax, 1981), and 61 % in BALB/cAnPt with a mean latent period of 210-220 days (Potter and Wax, 1983). Other inbred strains such as A/He,AKR, C3H, C57Bl and DBA/2 develop very rarely MPC and had been considered as resistant to MPC induction, (Potter, 1984). The F_1 hybrids between BALB/c and resistant parental strains, are also resistant to MPC induction, suggesting that resistance is under the control of genes that express dominancy, and susceptibility may represent the recesive allele, (Potter, 1984). The genes controlling susceptibility vs resistance have not been defined yet.

EXPERIMENTAL STRATEGY

One of the question risen is whether the susceptibility to MPC lies on the cellular level or lies on other level. To answer this question we have developed the following experimental model (Fig 1): We have produced non-radiation chimeras between the susceptible BALB/c and the resistant DBA/2 strains. We have used BALB/c mice carrying a Robertsonian 6;15 chromosome that serves as a marker to distinguish host from donor cells. The Rb6;15 fusion chromosome arose spontaneously in our BALB/cJ colony and was introduced into BALB/cAnPt background. Homozygous mice for the Rb6;15 marker were used as bone marrow (BM) donor. DBA/2 newborn mice were used as recipients. Our DBA/2 mice were obtained from Jackson Laboratories and bred in our department. The BM cell suspentions were injected via periorbital (p.o.) at days 1, 2 and 3 after birth, at a rate of 2 x 10^4 nucleated cells in a volume of 30 μl/day. The mice were assayed for chimerism after weaning, by karyotyping the peritoneal lymphocytes that were stimulated with newborn calf serum (NBCS). Chimeras carrying at least 10 % donor cells were divided into two groups. The first received 0,5 ml pristane i.p. three times at monthly intervals. The second group was injected with either 2x or 3x 0,5 ml of pristane at monthly intervals, followed by A-MuLV (helper free) injection via i.p. 2 - 3 weeks later. The Ψ-2pAB$_4$ cell line (kindly provided by Dr. R. Risser) was used to obtain Abelson virus helper free. The primed chimeras were injected with 0,5 ml supernatant. Cytosmears and chromosomal preparations were made from cells present in the peritoneal fluid. In parallel, the cytoplasmic Ig class was determined on cytosmears by indirect immunofluorescence methods using affinity purified goat anti-mouse antibodies.

Current Topics in Microbiology and Immunology, Vol. 166

The MPC diagnosis was confirmed both by the presence of MPC-associated translocations and by transplantation into primed syngeneic mice.

RESULTS

Eightynine DBA/2 newborns were p.o. injected with BALB/c6;15 BM cell suspentions and 50 of them survived and were further assayed for chimerism (Table 1) Thirtyfive mice showed a chimerism ≥ 10 % (Table 2) Six MPC of donor origin developed between the 10 chimeras treated only with pristane (Incidence = 60 %) (Table 3) after latency periods ranging between 165 and 244 days ($\bar{X}$= 221 days). Between the 25 chimeras that were treated with pristane and infected with Abelson virus, 8 of them (Incidence 32 %) developed donor type MPC after post virus latencies ranging from 21 to 160 days ($\bar{X}$= 84 days) (Fig 2 and 3,Table 4).

The typical (12;15) translocation was present in 5 of 6 TEPCs and in 4 of 8 (ABPCs) (Table 4). The short arm of one 6;15 chromosome has entered in a reciprocal combination with one copy of chromosome 12 resulting in a typical 12;15 (IgH/myc) translocation. The tumors TEPC-Ch-144-0 and ABPC-Ch-129-0 carried variant t(6;15) translocations, in which both arms of a Rb6;15 element went in a pericentrical invertion, (Fig 4). The other three ABPCs (119-3, 163-1 and 163-10) showed a new variant translocation in which the chromosome 15 of one Rb6;15 chromosome, was the receptor of the distal segment (bands B and C) of one chromosome 16, (Fig 4).

All MPC but one were analyzed for Ig class and the majority of them were either IgA_k or IgG_k producers. Two primary ABPCs (152-0 and 152-3) were non-producers.

As a test of malignancy and tranplantability, primary MPC were injected into primed and unprimed BALB/c,CDF_1 and DBA/2 recipients, (Table 5). With the exception of ABPC-Ch-163-1, all plasmacytomas grew in primed BALB/c and CDF_1 as ascites tumors. No MPC grew in primed DBA/2. None of the tumors grew in unprimed mice. The ABPC-Ch-163-1 was overgrowth by lymphosarcoma cells originating also from donor cells. Table 6 shows the non-plasmacytoma tumors, their latencies and cytogenetics. Non-plasmacytomas tumors originated either from donor cells or from host cells as well.

CONCLUSIONS

Our results show that neither the incidence nor the latency time of MPC development was affected by the DBA/2 background, suggesting that BALB/c susceptibility to MPC-genesis reside on the MPC-precursor level. Furthermore, we assume that the faculty to develop a functional oil granuloma (OG), essential for MPC-genesis and for transplantation of primary MPC, is independent of the susceptibility or resistance to adquire MPC-translocations.

REFERENCES

Potter M. (1984) Genetics of susceptibility to plasmacytoma development in BALB/c mice. Cancer Surveys 3:247-264

Potter M. and Wax J. (1981) Genetics of susceptibility to pristane-induced plasmacytoma in BALB/cAn: reduced susceptibility in BALB/cJ with a brief description of pristane induced arthritis. The Journal of Immunology 127:1591-1597

Potter M. and Wax J. (1983) Peritoneal plasmacytomagenesis in mice: Comparison of different pristane dose regims. J. Natl. Cancer Inst. 71:391-395

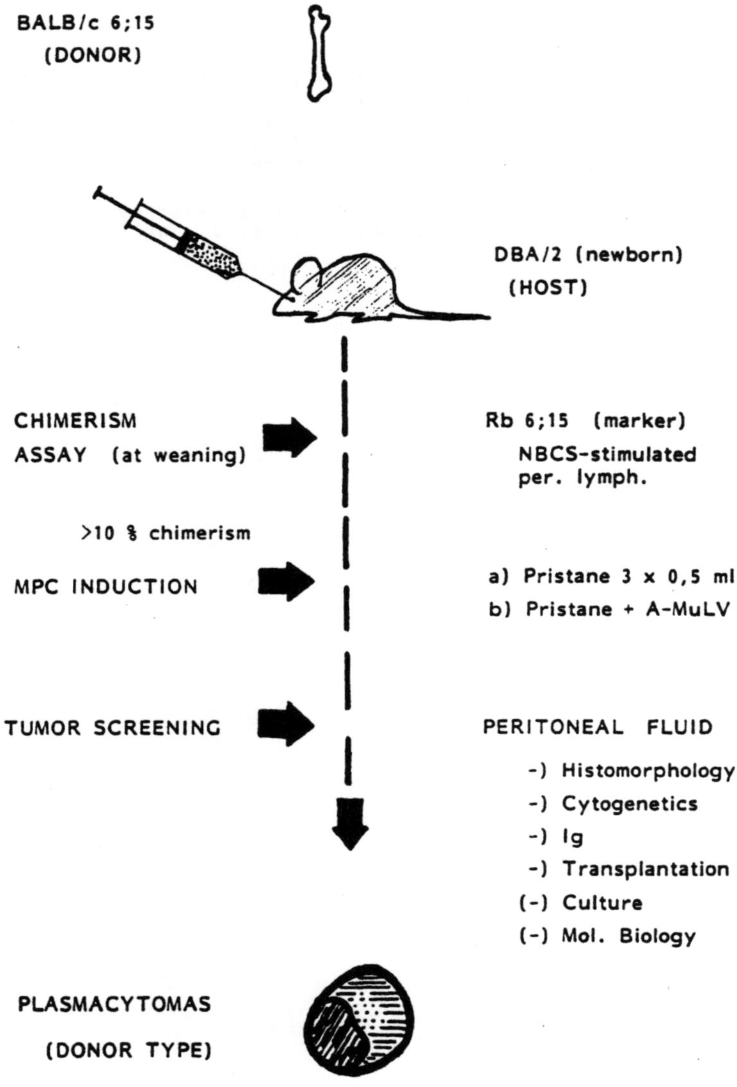

Fig. 1 Experimental design.

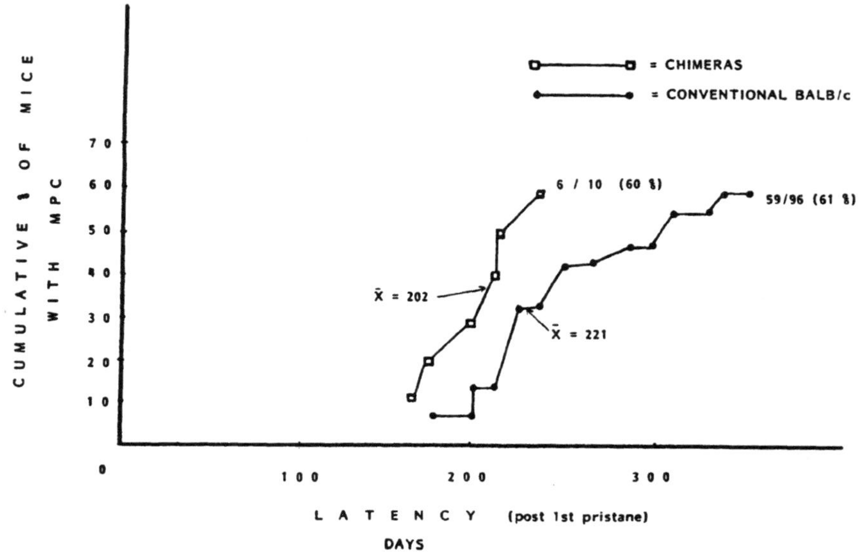

Fig. 2 Cumulative plasmacytoma incidence in pristane treated Chimeras. Comparison with MPC incidence in unmanipulated mice

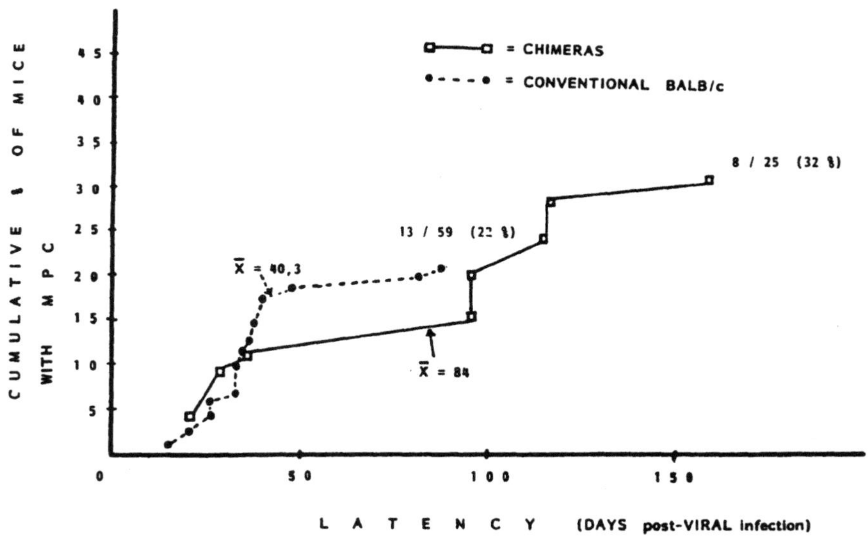

Fig. 3 Cumulative MPC incidence in pristane + A-MuLV treated chimeras. Comparison with MPC incidence in conventional mice

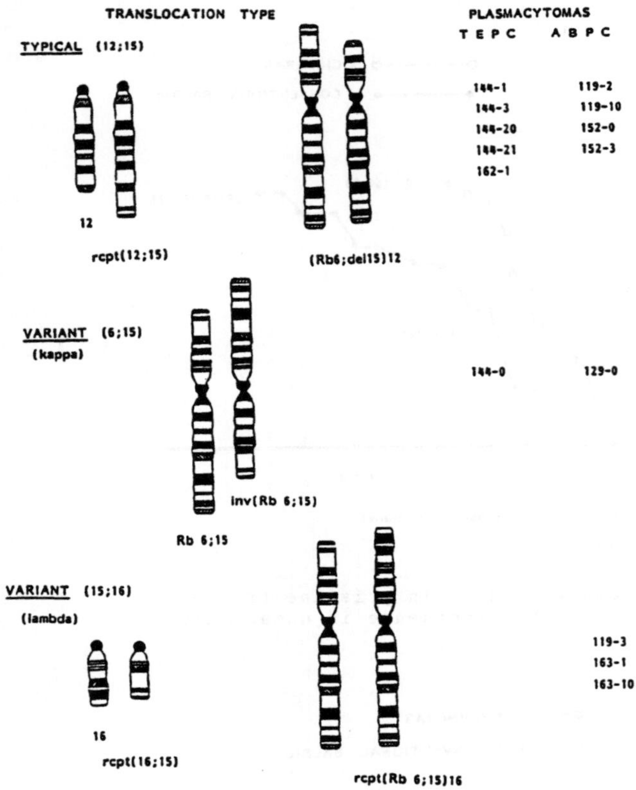

Fig. 4 TRANSLOCATIONS OBSERVED IN MPC INDUCED IN BALB/c6;15-->DBA/2 CHIMERAS

Table 1. BALB/c6;15-->DBA/2 chimeras and their treatment

TOTAL p.o. INJECTED NEWBORN	TOTAL WEANED MICE	DONOR/HOST CHIMERISM AT WEANING		MPC INDUCTION TREATMENT	
		<10%	>10%	PRISTANE 3x0,5 ml	PRISTANE + A-MuLV
89	50	15	35	10	20 * 5 **

* = 2x0,5 ml Pristane ** = 3x0,5 ml Pristane

Table 2. BALB/c6;15->DBA/2 CHIMERISM AT WEANING

EXP. MOUSE Nº	0	5	10	15	20	25	30	35	40	45	50	55	60	65
119-0	●													
119-1		●												
119-2			●											
119-3				●										
119-10				●										
119-20				●										
120-0	●													
120-1									●					
129-0								●						
129-1				●										
129-2					●									
129-3				●										
129-10		●												
129-20	●													
130-1		●												
130-2				●										
131-0		●												
131-1		●												
131-2						●								
132-0				●										
132-1		●												
132-2		●												
132-3				●										
144-0												●		
144-1														●
144-2										●				
144-3						●								
144-10											●			
144-11	●													
144-20						●								
144-21										●				
144-30						●								
151-0	●													
151-1		●												
151-2					●									
152-0							●							
152-1														●
152-2							●							
152-3									●					
152-10	●													
152-11					●									
152-20								●						
162-0		●												
162-1				●										
162-2								●						
163-0				●										
163-1													●	
163-2														●
163-3						●								
163-10								●						

Table 3. TUMOR INCIDENCE IN PRISTANE-TREATED AND PRISTANE+ A-MuLV TREATED CHIMERAS

INDUCTION AGENTS	Nº of MICE	TUMOR TYPE AND ORIGIN				MICE WITHOUT TUMORS
		M P C	L S	Mφ/ H	M M L	
PRISTANE	1 0	6 (d)	-	1 (d)		3
		6 0 %		1 0 %		3 0 %
PRISTANE + A-MuLV	2 5	8 (d)[§ **]	1 (h) 2 (d)** 3	2 (h) 1 (h/d) 3	2 (h)[§]	1 1
		3 2 %	1 2 %	1 6 %	8 %	4 4 %

§ = Two types of tumors (MPC + MML) originating from (d) respectively (h) cells.
** = Two types of tumors (MPC + LS) originating from (d) cells.

Table 4. Latency periods, Ig and Cytogenetics of MPC developed in chimeric mice

TUMOR NAME	INDUCTION AGENTS	MOUSE CHIMERISM AT WEANING (%)												LATENCY PERIOD (DAYS)		Ig	TRANSLOCATION TYPE		
		10	15	20	25	30	35	40	45	50	55	60	65	PP	PV		TYPICAL (12;15)	VARIANT (6;15)	VARIANT (15;16)
TEPC-Ch-144-21	ppp								★					165	-	A_kG_k	★		
" 144-20	ppp				★									176	-	A_k	★		
" 162-1	ppp		★											198	-	G_k	★		
" 144-0	ppp										•			214	-	A_k		•	
" 144-1	ppp												★	214	-	A_k M_k	★		
" 144-3	ppp				★									244	-	A_k	★		
ABPC-Ch-163-10	pp v						⊕							106	21	A_k			⊕
" 163-1	pp v											⊕		114	29	N.D.			⊕
" 119-10	ppp v		★											242	36	G_k	★		
" 119-2	ppp v	★												306	96	A_k	★		
" 119-3	ppp v		⊕											306	96	A_k			⊕
" 152-3	pp v							★						215	115	N. P.	★		
" 129-0	pp v						•							178	117	G_k		•	
" 152-0	pp v					★								260	160	N. P.	★		

p= pristane
v= A-MuLV
PP= post pristane
PV= post virus
N.D. = not done
N.P. = non producer

TABLE 5. MPC in vivo TRANSPLANTATION

TUMOR NAME		RECIPIENT MICE					
		BALB/c pristane		CDF_1 pristane		DBA/2 pristane	
		+	-	+	-	+	-
TEPC-Ch-	144-0	3/3	0/2	2/2	ND	0/2	0/2
"	144-1	6/6	0/2	2/4	ND	0/2	0/2
"	144-3	4/4	0/2	2/2	0/2	0/3	0/2
"	144-20	2/2	0/2	2/2	0/2	(‡) 2/2	0/2
"	144-21	6/6	0/2	3/3	0/2	0/2	ND
"	162-1	4/4	0/2	2/2	0/2	0/2	0/2
ABPC-Ch-	119-2	3/3	0/2	3/3	0/2	0/2	ND
"	119-3	2/2	0/5	2/2	ND	0/2	0/2
"	119-10	3/3	0/2	7/8	0/2	0/2	0/2
"	129-0	4/4	0/2	3/4	0/2	0/2	ND
"	152-0	3/4	0/2	3/3	0/2	0/2	0/2
"	163-1 (*)	9/9	1/7	2/2	ND	(‡) 2/2	ND
"	163-10	7/7	0/7	3/3	0/2	(‡) 2/3	0/2
"	152-3	3/4	0/2	2/2	0/2	0/2	0/2

(*) = LS overgrew the MPC

(‡) = LYMPHOMA originating from intermediary host cells.

TABLE 6 LATENCY AND CYTOGENETIC FEATURES OF 9 NON-PC TUMORS DEVELOPED AFTER PRISTANE OR AFTER PRISTANE + A-MuLV TREATMENT IN BALB/c 6;15-->DBA/2 CHIMERAS.

TUMOR TYPE	TUMOR DESIGNATION	INDUCTION AGENTS	MOUSE CHIMERISM AT WEANING	LATENCY PERIODS (DAYS) PP*	PV	ORIGIN	CHROMOSOME ABERRATIONS
LYMPHOSARCOMA	ABLS-Ch-129-1	p p V	15 %	264	206	(d)	DIPLOID
	ABLS-Ch-151-2	p p V	20 %	277	177	(h)	DIPLOID
	ABPC/LS-Ch 163-1	p p V	60 %	113	28	(d/d)	DIPLOID / DIPLOID t(15;16)
MACROPHAGE - HISTIOCYTIC	ABMφ-Ch-129-3	p p V	15 %	116	65	(h)	DIPLOID
	ABMφ-Ch-132-0	p p V	15 %	267	213	(h/d)	near DIPLOID / TETRAPLOID
	ABMφ-Ch-131-2	p p V	25 %	105	51	(h)	DIPLOID
	TEMφ-Ch-144-30	p p p	25 %	344	-	(d)	near DIPLOID +7, $+M_1$
MYELOID	ABML/ABPC-Ch-119-10	p p p V	15 %	242	32	(h/d)	HYPERDIPLOID / TETRAPLOID +6, +7, +15, +18 / t(15;16)
	ABMML-Ch-129-2	p p V	20 %	116	55	(h)	DIPLOID

* = After 1st pristane injection (h) = host (d) = donor

Mouse Plasmacytoma Associated (MPC) T(15;16) Translocation Occurs Repeatedly in New MPC Induction System

F. Wiener, S. Silva, H. Sugiyama, M. Babonits, H. Axelson, P. Chinmay and G. Klein

Department of Tumorbiology, Karolinska Institutet, 10401 Stockholm, Sweden

I would like to introduce my presentation with two slides published 11 years ago (Ohno et al. 1979). They illustrate the MPC associated 12:15 and 6:15 translocations. I do not intend to describe how successfully the whole story developed in the hands of molecular biologists. From the cytogenetic point of view, however, one intriguing question has remained unsolved (illustrated in Fig. 1).

While the rat immunocytomas carried only typical (IgH/myc) translocations in the relatively limited number of cases investigated (Wiener et al. 1982), the Burkitt lymphoma system gave evidence of three alternative translocations (Lenoir and Bornkamm 1987). The typical IgH/myc was the most frequent in this system as well (approximately 80%). The variant translocation that juxtaposed the kappa or the lambda constant region genes to myc occurred less frequently. The frequency of the myc/lambda translocation (8;22) was twice higher than of the myc/kappa (2;8) translocation. It was therefore puzzling that no translocation between myc carrying chr. 15 and the lambda light chain gene carrying chr. 16 (Francke et al. 1982) has been identified in more than 160 MPCs karyotyped in our laboratory. Now, this last puzzle is also clarified.

Before presenting the results of the cytogenetic analyses, a few words about the MPC induction system. The conventional way to induce MPCs is to inoculate i.p. into BALB/c mice either pristane alone or pristane followed by Abelson virus infection. Except for a few deletion-type (Wiener et al. 1984) and translocation-negative MPCs (Ohno et al. 1990), 95% of the tumors carried either the 12;15 or the 6;15 translocations. However, in newly developed unconventional induction systems (Fig. 2) the "missing" variant 15;16 MPCs have been identified in a relatively high percentage of cases (approximately 25%).

The results of the cytogenetic analysis are summarized in Table 1. The reciprocal translocation between chromosomes 15 and 16 was the single constant chromosomal aberration present in all metaphase plates. In ABPC-Ch-163-1, 163-10 and SRBC-3 the chromosomal element arising from the joining of the Rb6;15 chr. with the distal part of 16 chr. was present in two copies (Fig. 3). In the ABPC-Ch-119-3 the non-translocated RB6;15 chr. was duplicated, however (Fig. 4). Interestingly, in this latter tumor the t(15;16) chr. was absent.

The ABPC(4;12xCB.20)-128 PC was unique, since it contained both typical 12;15 and variant 15;16 translocations in the same metaphase plate (Fig. 5). This type of MPC had never been encountered in our previous cytogenetic studies. It has to be emphasized that this was the single 15;16 tumor that arose in the conventional induction system.

Current Topics in Microbiology and Immunology, Vol. 166

High resolution banding studies on selected metaphase plates of the ABPC-Ch-163-10 and ABPC-CD-3 in vitro lines confirmed the reciprocal translocation between chrs. 15 and 16. The breakpoint (BP) on chr. 15 was on the interface of sub-bands D2/3. On chr. 16 it was mapped to band B1 (Fig. 6).

The detection of cytogenetically identical 15;16 translocations in 7 MPCs suggests that it may represent a myc/lambda juxtoposition. The following facts lend support to this assumption:

i) The location of the BPs on chromosome 15 and 16. The BP on chr. 15 is identical with the myc-pvt-1 associated BP in the 12;15 and 15;16 translocations studied earlier. The lambda gene has been genetically mapped 15 cM from the centromere (Epstein et al. 1986) and chromosomally previously located to the proximal part of chr. 16, between the centromere and sub-band B5 (Francke et al. 1982). This is consistent with the B1 chromosomal segment that contains the translocation BP. We suggest that this is probably the chromosomal site of the Ig-lambda gene.

ii) Currently ongoing PFGE analysis of the 15;16 translocation carrying in vitro lines detected rearrangement of the pvt-1 but not of the myc locus (Fig. 7). This further increases the likelihood that the translocation may represent the "lambda counterpart" of kappa/myc variant translocation.

These findings raise the intriguing question: why did we not encounter any 15;16 translocations in our previous studies on more than 160 MPCs? And why did we find them in as many as seven tumors in our recent experimental series? To answer these questions I am proposing two alternative but interrelated scenarios that may explain the emergence of 15;16 translocation PCs in the unconventional induction systems.

First, it has to be noted that all seven 15;16 variant MPCs arose in pristane + Abelson virus exposed mice. It is noteworthy that the relative proportion of variant vs typical translocations was much higher in pristane + Abelson treated than in mice treated only with pristane (Table 2). This suggests that the combined effect of pristane + Abelson virus is more favorable to the outgrowth of variant translocation carrying cells than pristane alone.

Second, six of the seven 15;16 translocations carrying tumors have arisen in unconventional induction systems, based on the transfer of the in vitro Abelson infected spleen cells into pristane treated BALB/c mice or on partial or total chimerization of the recipient by previous i.v. bone marrow injections.

These latter cell populations may differ in their susceptibility to Abelson virus infection or transformation as compared to in situ cells of the conventional mice. The Abelson infection of this rapidly expanding B-cell population may recruit an additional category of precursor cells that become translocation prone at the time of lambda rearrangement. Thus, the development of 15;16 PCs in the unconventional system may be understood.

The alternative scenario is based on the assumption that the PC precursor cell population is composed of cells carrying all three types of translocations in different proportions (Fig. 8). In the case of long latency - when pristane alone is used as inducing agent - mainly the 12;15 translocation carrying cells had the chance to grow out as clonal tumors. When the latency was drastically reduced by Abelson virus infection that followed the pristane treatment, variant 6;15 PCs grew out in a higher number as it was shown in Table 2. In the unconventional induction systems that have even shorter latencies of PC development that in the conventional induction system, some of the 15;16 translocation carrying precursor cells may be able to develop into clonal MPCs.

REFERENCES

Epstein R, Davisson M, Lehmann K, Akeson EC, Cohn M (1986) Position of Igl-1, md, Bst loci on chromosome 16 of the mouse. Immunogenet 23:78-83

Francke U, Martenville B, D'Eustachio P, Ruddle FH (1982) Comparative gene mapping: Murine lambda light genes located in region cen-B5 of mouse chromosome 16 are non homologous to human chromosome 21. Cytogenet Cell Genet 33:267-277

Lenoir GM, Bornkamm GW (1987) Burkitt's lymphoma, a human cancer model for the study of the multistep develoment of cancer: proposal for a new scenario. Adv Viral Oncology 7:173-204

Ohno S, Babonits M, Wiener F, Spira J, Klein G, Potter M (1979) Non-random chromosome changes involving Ig-gene chromosomes (Nos 12 and 6) in pristane induced mouse plasmacytomas. Cell 18:1001-1007

Ohno S, Migita S, Murakami S (1990) C-myc-S_u rearrangement in a murine plasmacytoma without visible chromosomal translocation. Oncogene in press

Silva S, Babonits M, Wiener F, Klein G (1989) Plasmacytoma induction in radiochimeras. Int J Cancer 44:337-342

Sugiyama H, Weber G, Silva S, Babonits M, Wiener F, Klein G (1989) The accelerating role of the Abelson murine leukemia virus in murine plasmacytoma development: in vitro infection of spleen cells generates donor type tumors after transfer to pristane treated BALB/c mice. Int J Cancer 44:348-352

Wiener F, Babonits M, Spira J, Klein G, Bazin F (1982) Non-random chromosomal changes involving chromosomes 6 and 7 in rat immucytomas. Banding and translocation homologies between T(12;15) of mouse plasmacytoma and T(6;7) of rat immunocytoma. Int J Cancer 29:431-437

Wiener F, Ohno S, Babonits M, Sumegi J, Wirschubsky Z, Klein G, Mushinski JF, Potter M (1984) Hemizygous interstitial deletion of chromosome 15 (Band D) in three translocation negative plasmacytomas. Proc Natl Acad Sci USA 81:1159-1163

TABLE 1 Experimental system used for the induction of (15;16) translocation carrying plasmacytomas, latency periods, Ig production and translocation type

GROUP	TUMOR	PASSAGE GENERATION	EXPERIMENTAL SYSTEM	MOUSE STRAIN[4] DONOR	MOUSE STRAIN[4] HOST	LATENCY PERIOD IN DAYS PP	LATENCY PERIOD IN DAYS PV	Ig CLASS	TRANSLOCATION TYPE
I	ABPC-Ch-163-1[7]	0	Chimera[1]	BALB/c6;15	DBA/2	113	28	ND	rcpt(15:16)
I	ABPC-Ch-163-10[7]	0	" "	BALB/c6;15	DBA/2	106	21	A_k	rcpt(15;16)
I	ABPC-Ch-119-3[8]	1	" "	BALB/c6;15	DBA/2	306	96	A_k	rcpt(15;16)
II	ABPC-SRBC-3	0	Transfer[2]	BALB/c6;15	BALB/c	-	54	k	rcpt(15:16)
II	ABPC-Hi-12	0	" "	BALB/c6;15	BALB/c	-	40	NP	rcpt(15;16)
III	ABPC-RCh-CD-3	0	Radiochimera[3]	BALB/c6;15	DBA/2	191	101	ND	rcpt(15:16)
IV	ABPC(C4;12xCB.20)-128	0	Conventional	Not applicable[5]		124	39	NP	rcpt(15;16) + rcpt(12;15)

[7] 0,5 ml pristane 2 times at monthly intervals followed by A-MuLV infection

[8] 0,5 ml pristane 3 times at monthly intervals followed by A-MuLV infection

1=as described in Fig.2
2=as in (Sugiyama et al.1989)
3=as in (Silva et al.1989)
4=All tumors were of donor type
5=The tumor has developed in a (BALB/c4;12xCB.20)F_1 mouse

PP=post pristane
PV=post virus
NP=non producer
ND=not done

TABLE 2 Comparison of the latency and the number of variant plasmacytomas in the pristane only (TEPC) and pristane + Abelson (ABPC) series

EXPERIMENTAL SYSTEM	PC INDUCTION	LATENCY IN DAYS	TRANSLOCATION TYPE (12;15)	(6;15)	(15;16)
Conventional	TEPC	230[1]	65	7	0 [3]
Conventional	ABPC	110[1]	54	30	1
Chimera	ABPC	71[2]	4	1	3
Chimera	TEPC	194[1]	4	1	
Transfer	ABPC	52[2]	4	3	2
Radiochimera	ABPC	60[2]	-	1	1

[1]Post pristane

[2]Post virus

[3]The proportion of variant plasmacytomas in the ABPC series is significantly higher compared to TEPC series

$X^2 = 14,477 > X^2(0,001) = 10,827$ $p<0,001$

Fig.1

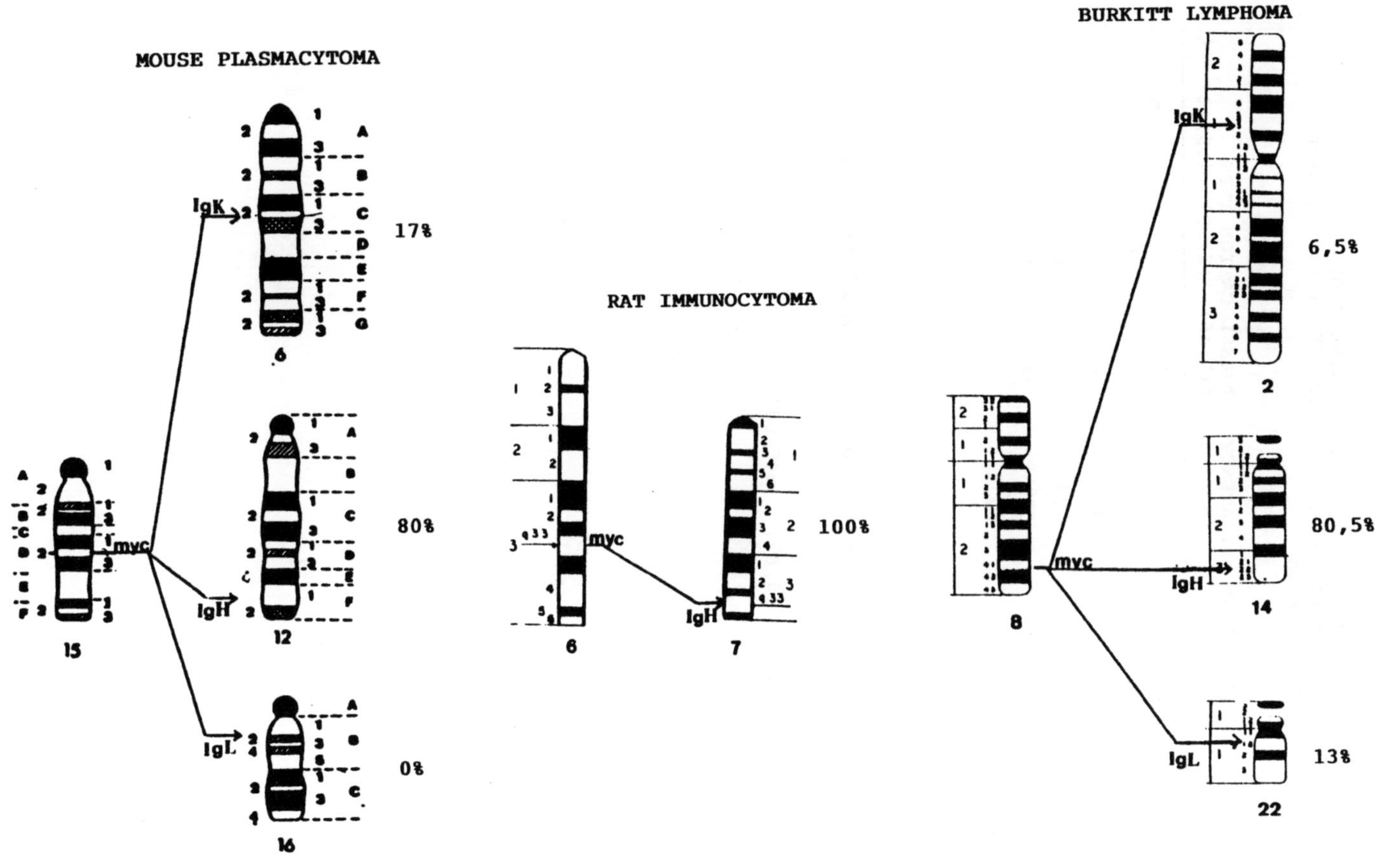
MOUSE PLASMACYTOMA
RAT IMMUNOCYTOMA
BURKITT LYMPHOMA
myc
IgK
IgH
IgL
15
6
12
16
17%
80%
0%
6
7
100%
8
2
14
22
6,5%
80,5%
13%

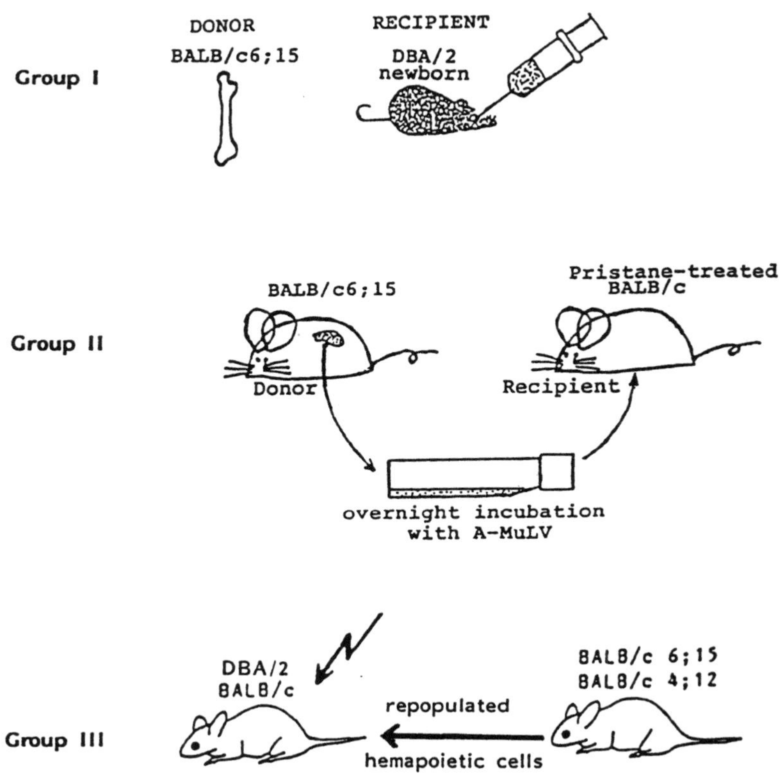

Fig.2. Group I. ABPC-Ch-163-1,163-109 and 119-3 were induced in BALB/c6;15-→DBA/2 chimeric mice.Newborn DBA/2 were ioculated with 10^5 BALB/c6;15 bone marrow cells by periorbital injection (3 times at daily intervals).Four to six weeks after inoculation the mice were treated with pristane + Abelson virus.All three tumors contained two Rb6;15 chrs, showing that they were of donor origin. Group II. ABPC-SRBC-3 and ABPC-Hi-12 MPCs were induced by transferring in vitro Abelson infected spleen cells of BALB/c6;15 origin into pristane treated BALB/c recipients.Both carried the Rb6;15 marker and were thus of donor origin. Group III. ABPC-RCh-CD-3 arose in a group of DBA/2 radiochimeras reconstituted with BALB/c6;15 bone marrow cells.Four to six weeks after reconstitution the radiochimeras were treated with pristane followed by Abelson virus infection.The presence of the RB6;15 chr. pair proved the donor origin of the tumor.

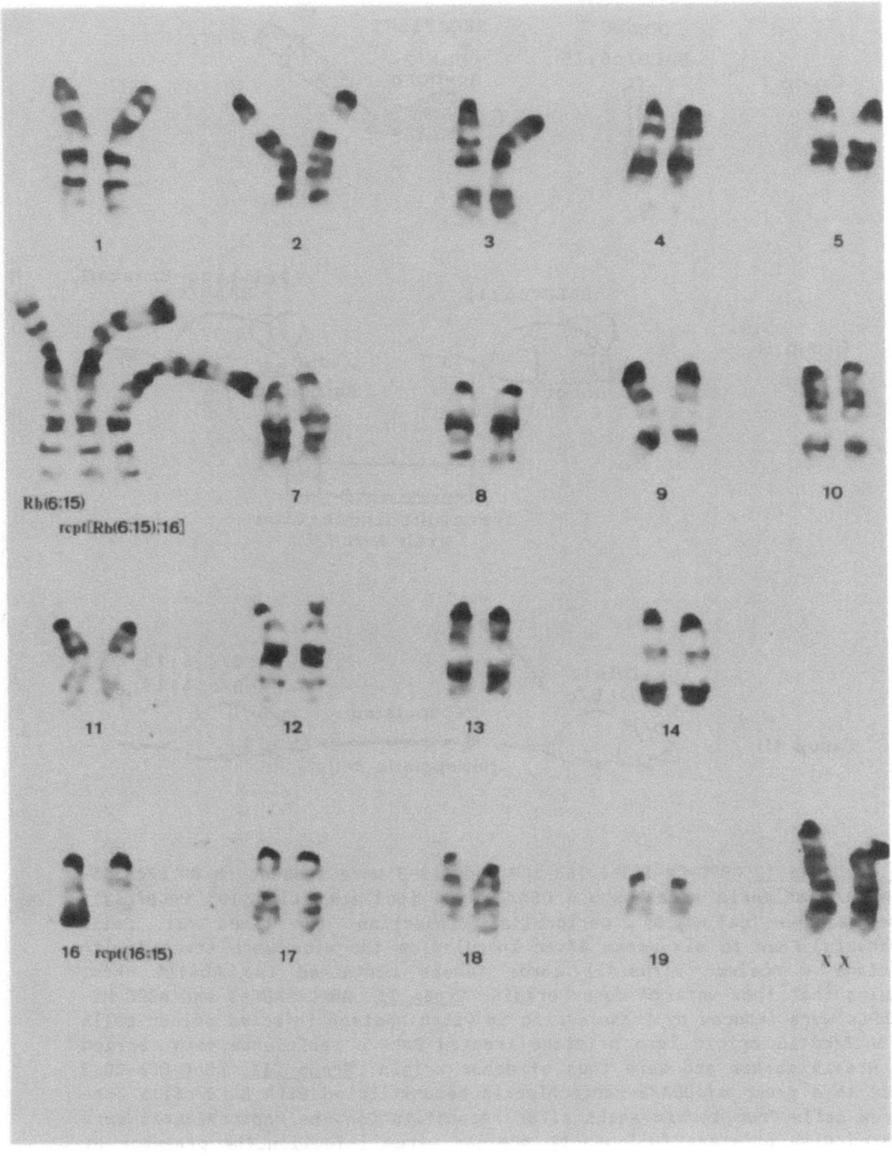
1
2
3
4
5
Rb(6;15)
rcpt[Rb(6;15);16]
7
8
9
10
11
12
13
14
16
rcpt(16;15)
17
18
19
X X

Fig. 3

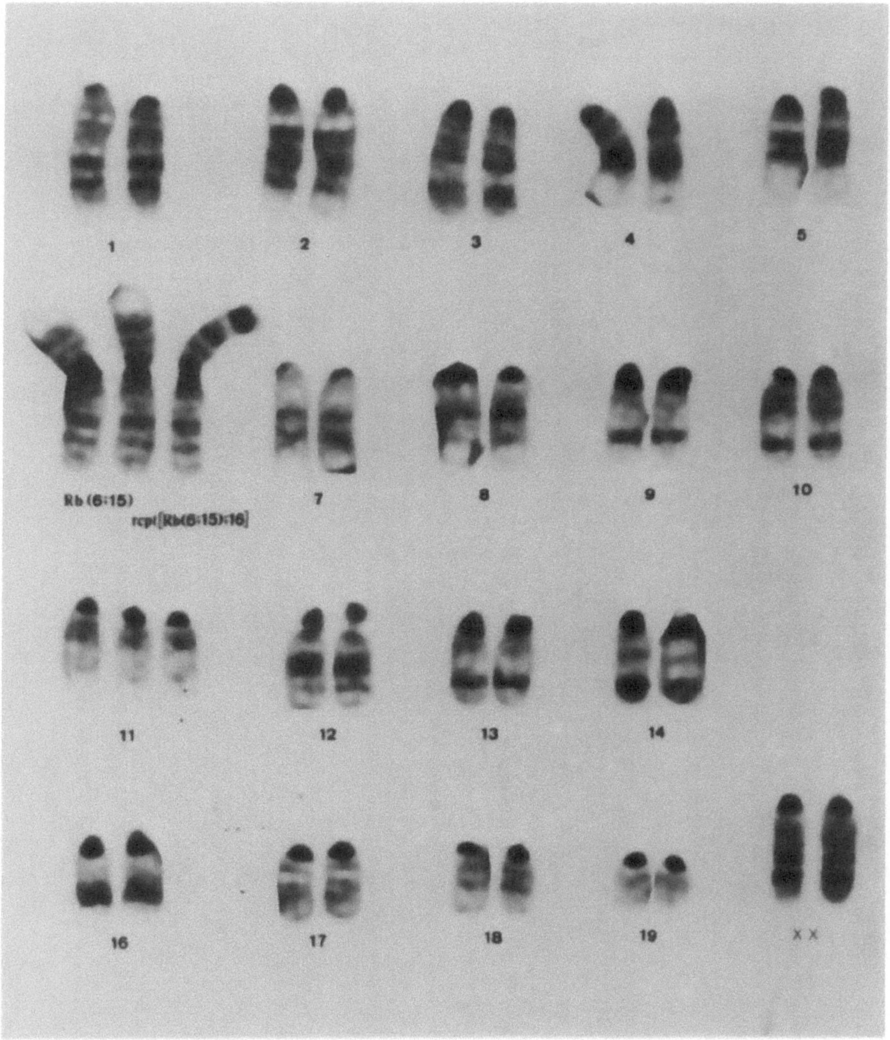

Fig. 4

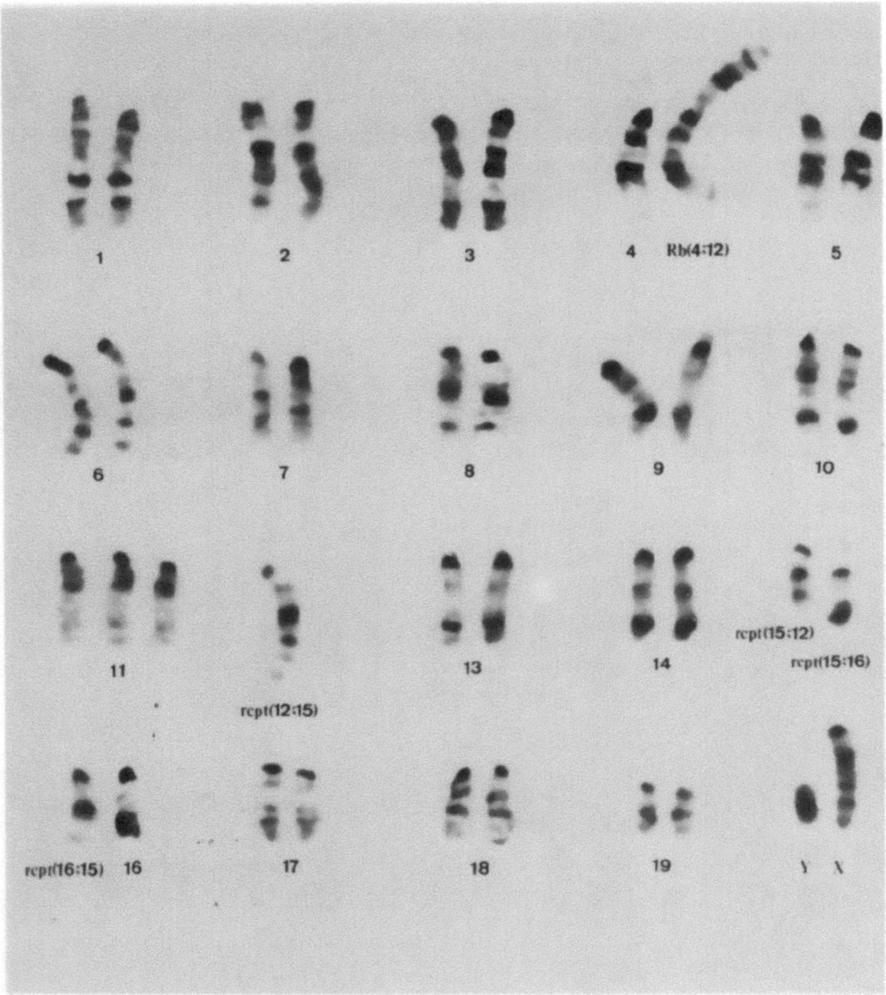
1
2
3
4
Rb(4:12)
5
6
7
8
9
10
11
rcpt(12:15)
13
14
rcpt(15:12)
rcpt(15:16)
rcpt(16:15)
16
17
18
19
Y
X

Fig. 5

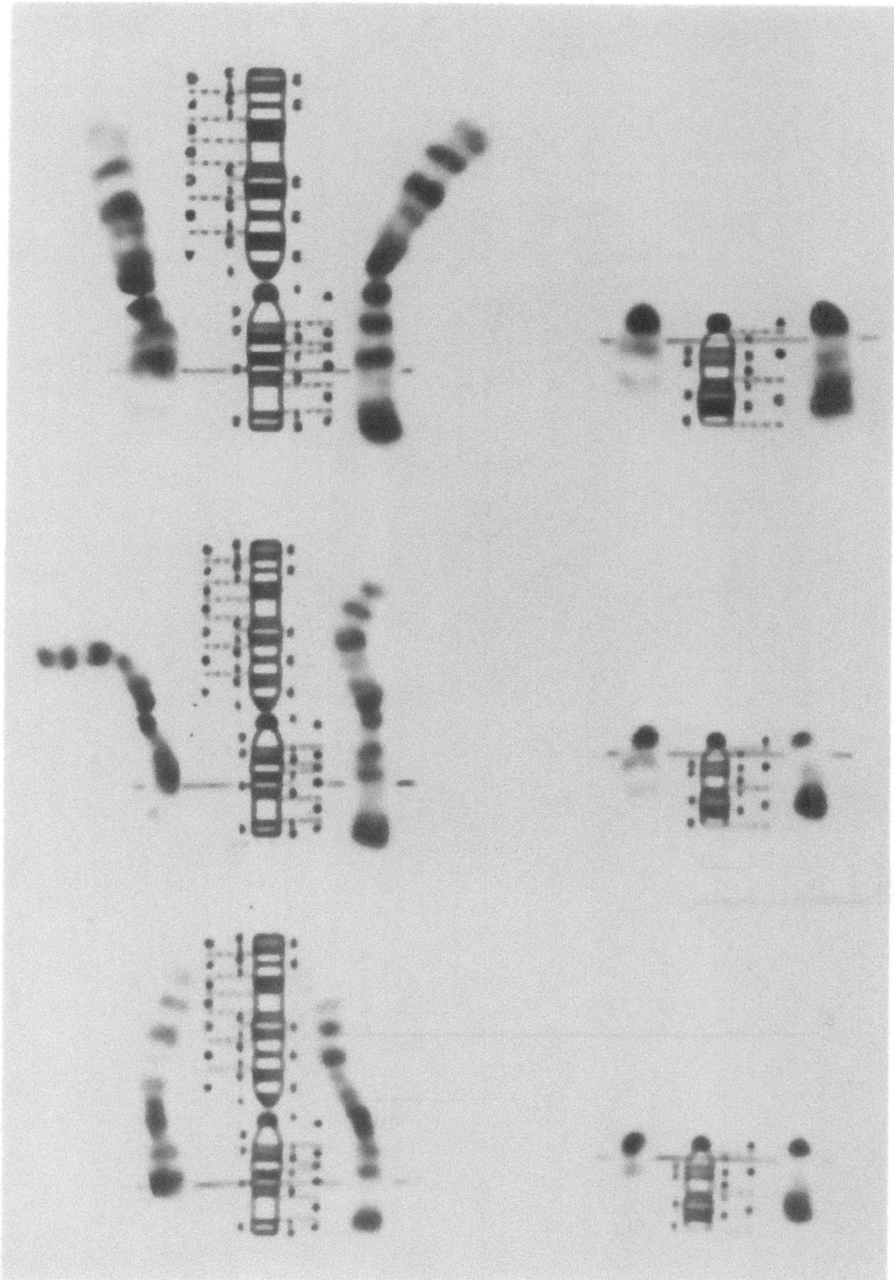

Fig. 6

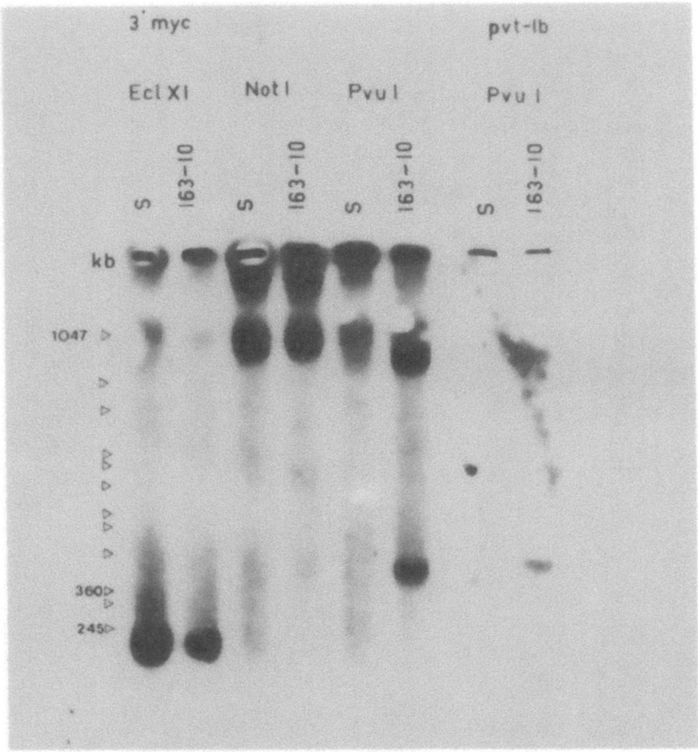
3' myc
pvt-Ib
Ecl XI
Not I
Pvu I
Pvu I
S
163-10
S
163-10
S
163-10
S
163-10
kb
1047
360
245

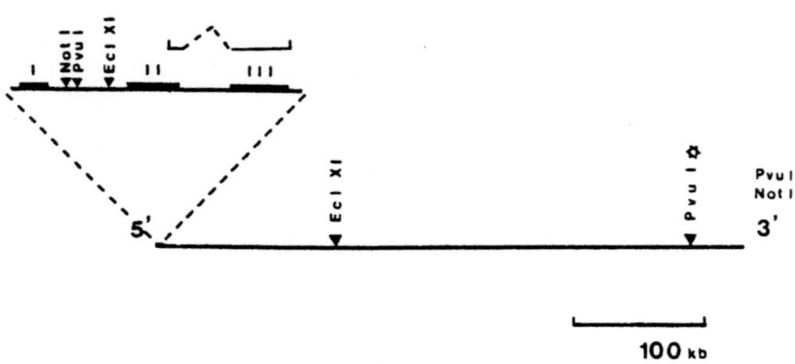
I
Not I
Pvu I
Ecl XI
II
III
5'
Ecl XI
Pvu I
Pvu I
Not I
3'
100 kb

Fig. 7

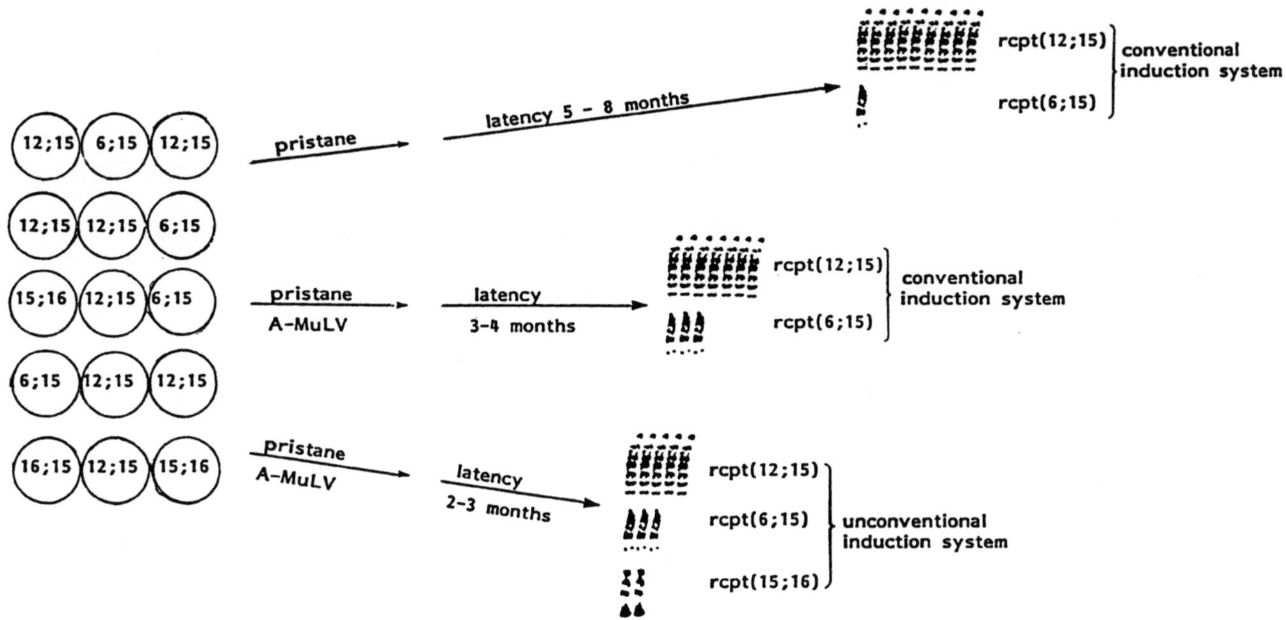

12;15
6;15
12;15
12;15
12;15
6;15
15;16
12;15
6;15
6;15
12;15
12;15
16;15
12;15
15;16
pristane
latency 5 - 8 months
rcpt(12;15)
rcpt(6;15)
conventional induction system
pristane
A-MuLV
latency
3-4 months
rcpt(12;15)
rcpt(6;15)
conventional induction system
pristane
A-MuLV
latency
2-3 months
rcpt(12;15)
rcpt(6;15)
rcpt(15;16)
unconventional induction system

Fig.8

A Retrovirus Expressing v-*abl* and c-*myc* Induces Plasmacytomas in 100% of Adult Pristane-Primed BALB/c Mice

E.M. Weissinger*, D. Largaespada#, S.J. Smith-Gill*, R. Risser#, J.F. Mushinski*, H. Mischak*

* Laboratory of Genetics, National Cancer Institute, NIH, Bethesda, MD 20892
# McArdle Institute for Cancer Research, University of Wisconsin, Madison, WI 53706

Intraperitoneal mineral oils such as pristane induce plasmacytomas in up to 60% of inbred BALB/c mice over the course of the period of two years (Potter et al. 1984). The appearance of the tumors can be speeded up dramatically by intraperitoneal injection of Abelson murine leukemia virus (A-MuLV) (Potter et al. 1973). However, the incidence of plasmacytomas does not increase, in part because other tumors, pre-B cell lymphomas and myeloid tumors also arise in this setting. Molecular genetic studies of plasmacytomas have revealed that 100% of these tumors show deregulated expression of the c-myc proto-oncogene (Mushinski 1988). The universal deregulation of c-myc may be necessary, but it appears to be only one step in a multistep scenario leading to plasmacytomagenesis. Recombinant retroviral vectors offer the possibility of studying the cooperation of myc with other oncogenes. So far two oncogenes, in combination with myc, have accelerated plasmacytoma induction as compared to pristane alone. These oncogene combinations, myc+raf (Troppmair et al. 1989, Kurie et al. 1990) or myc+ras (Clynes et al. 1988), in form of retroviruses, accelerated plasmacytoma onset, but did not increase the tumor incidence. Since v-abl apparently cooperated well with deregulated c-myc in A-MuLV-induced tumors in pristane-primed mice, we predicted that a retrovirus which coexpressed v-abl and c-myc would also induce plasmacytomas expeditiously. Thus the protein-encoding portion of c-myc cDNA, under the control of the widely active thymidine kinase promotor, was inserted in the genome of A-MuLV (Largaespada et al. 1990) and injected intraperitonially, either helper-free or with Moloney Murine Leukemia Virus (MoMuLV) helper virus, into immunocompetent inbred BALB/c-AnPt mice which had previously recieved a single injection of 0.5 ml pristane. When ascites accumulated and contained tumor-like cells, the mice were sacrificed and their tumors excised for transplantation, histopathological studies and for DNA and RNA extraction.

PLASMACYTOMA INDUCTION

Two to three month old, pristane-primed BALB/c mice were given 2×10^5 focus forming units (ffu) of the ABL-MYC virus, either helper-free or with MoMuLV helper virus by intraperitoneal injection.

Current Topics in Microbiology and Immunology, Vol. 166

After a latency of 25 to 55 days all animals developed tumors and ascites fluid (Fig 1). The tumor cells showed the histological characteristics typical of plasmacytomas. Solid tumors (between 1 and 20) occured as nodules in the mesentery as well as in the peritoneum, and some of them seemed to grow out of the gut. Spleen and liver of the animals were slightly enlarged, probably due to increased blood flow, but no enlarged lymph nodes could be detected.

Injection of pristane-primed BALB/c with helper-free ABL-MYC virus also resulted in the development of plasmacytomas in all animals. These ABL-MYC plasmacytomas were morphologically indistinguishable from those induced with helper virus and arose with the same mean latency (Fig.1).

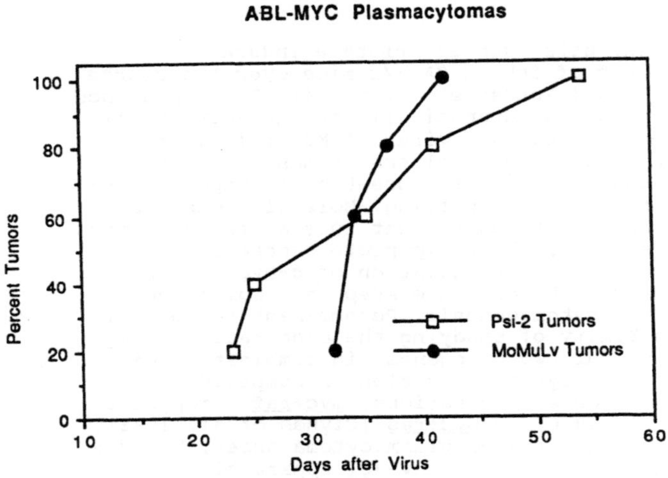

Fig.1. Time-course of plasmacytomas induced in pristane-primed BALB/c with the ABL-MYC virus with MoMuLV (MoMuLv tumors) and helper-free (Psi 2 tumors).

Immunoglobulin heavy and light chain classes in the ascites fluid were determined by a particle concentration fluorescence immunoassay (Jolley et al. 1984, Rousseau et al.1989), which was read in a Pandex Screen Machine. All samples were reacted with goat anti-mouse sera, which were specific for IgM, IgA or IgG heavy chains and for kappa or lambda light chains. The typing showed that the majority of the tumors secreted IgM with kappa light chains. Only one IgA-producing tumor could be detected. Upon transplantation of the tumors into the peritoneum of pristane-primed recipients, all animals developed plasma cell tumors after a period of 10 to 15 days. The transplanted tumors were shown to be derived from the primary tumor cells by Southern blot analysis, which detected the same virus integration sites and the same immunoglobulin J_H rearrangements in the primary and transplanted tumor.

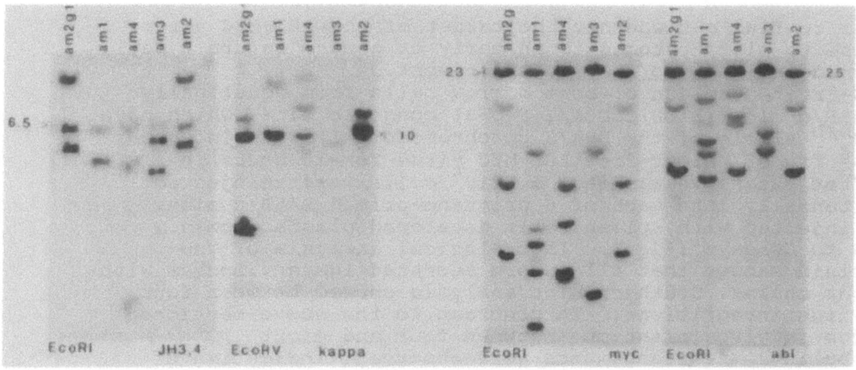

Fig.2. Southern blots of DNA from ABL-MYC-induced plasmacytomas. am-1 to am-4 represent primary tumors, while am-2g1 represents the first generation transplant of am-2. EcoRI digests were hybridized with probes specific for c-myc and v-abl. While the 23 and 25 kB bands represent the respective germ line bands, the other bands represent different viral integration sites (only one EcoRI restriction site is present in the ABL-MYC virus, located in the tk-promotor). To show immunoglobulin heavy chain rearrangement, a probe specific for $J_H3,4$ was used on an EcoRI digest. Kappa rearrangements were shown by probing an EcoRV digest with a probe specific for J_{kappa}.

Southern blots of DNA from independent tumors hybridized with JH probes revealed between one and four rearranged immunoglobulin heavy chain loci in each DNA, two to three rearranged kappa light chain bands, but no lambda rearrangement. Probing of the blots with probes specific for v-abl and c-myc showed a small number of rearranged bands, indicating between two and seven viral integration sites. Correct expression of the viral messages was proven by northern blotting (data not shown). These data suggest that each tumor consists of only one or a small number of clones.

In order to evaluate the contribution of pristane to plasmacytoma induction, we injected BALB/c mice with ABL-MYC virus plus helper without pristane priming. Again, all mice developed plasmacytomas, but without developing ascites-fluid. Solid tumors developed in the mesentery and in the thymus area 60 - 70 days after injection of the virus.

IN VITRO TRANSFORMATION OF SPLEEN AND BONE MARROW CELLS

In order to find out whether the target of the ABL-MYC virus giving rise to plasmacytomas is an early or a more mature B cell, we performed an *in vitro/in vivo* experiment.
$1x10^7$ LPS-treated spleen or bone marrow cells from genetically marked BALB/c mice (BALB/c.DBA/2-Pep-3) congenic mice, which have the Pep-$3^{b/b}$ allele of the DBA/2 on chromosome 1) were infected with $3x10^6$ ffu of helper-free ABL-MYC virus for 18 hours *in vitro*. After extensive washing, $1.5x10^6$ cells were reinjected intraperitoneally into each of 6 pristane-primed BALB/c mice. All mice injected with spleen cells developed plasmacytomas within 30 to 60 days (Fig.3). Immunological analysis of the ascites fluid showed that all tumors secreted IgM antibodies with kappa light chains. Southern blot analysis showed between four and six virus integrations. In contrast to the above-mentioned tumors from *in vivo* infection, between four and eight immunoglobulin J_H rearrangements were observed, indicative of the presence of several individually transformed B cell clones. This increased number of clones in each transplanted mouse could be due to the fact that 15 times the amount of the ABL-MYC virus that has been injected directly into each mouse had been used to transform the cells of one spleen *in vitro*. Therefore, more cells could be immediately infected with the virus *in vitro* than *in vivo*.

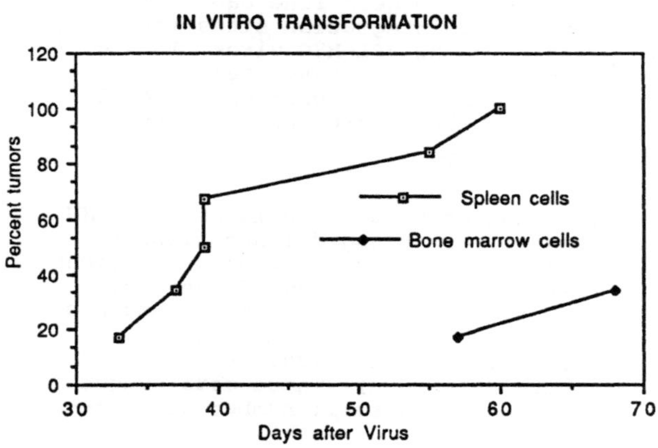

Fig.3. Time-course of tumor development in BALB/c inoculated with *in vitro*-transformed spleen and bone marrow cells.

Probing the Southern blots of the tumors for the genetic marker, Pep-$3^{b/b}$ (data not shown), showed that all tumors had arisen from the donor cells.

On the other hand, the _in vitro_-transformed bone marrow cells led to pre-B lymphomas in 2 out of 6 mice after a long latency. The first tumor appeared after 57 days. Massive enlargments occured in the paravertebral and inguinal lymph nodes. Livers and spleens of the sacrificed animals were also extremly enlarged. Histological examination of the tumors indicated that these tumors did not contain plasma cells. Southern blot analysis revealed one or two virus integration sites, a single immunoglobulin heavy chain rearrangement, but no light chain rearrangements. These data are consistent with the diagnosis of these tumors as pre-B lymphomas.

One mouse developed ascites fluid after a latency of 76 days and was sacrificed after plasma cells were detected in the ascites. This animal had tumors in the mesentery and the peritoneum, but no lymph node involvment could be detected. We also were able to show the presence of the genetic marker in the tumors derived from the bone marrow transplants.

DISCUSSION

The ABL-MYC retrovirus coexpresses v-_abl_ and c-_myc_ and is capable of inducing plasmacytomas in BALB/c mice with 100% incidence. This virus is unique in his ability to induce plasmacytomas without pristane priming (Largaespada et al. 1990), but pristane priming reduces the average latent period of plasmacytomagenesis in BALB/c mice from 70 days to 35 days and induces ascites formation. Furthermore, there is a remarkably greater percentage of tumors and a shorter latent period of ABL-MYC-induced plasmacytomas compared to those induced by other viruses in the same system. Every BALB/c mouse injected with the ABL-MYC virus developed a plasmacytoma, whereas in a comparable group of mice (2-4 month old pristane-primed BALB/c) treated with the RIM virus, only 28% of the mice developed plasmacytomas, and only after a much longer latency (60 -118 days, Clynes et al. 1988). J3V1 treatment of pristane-primed BALB/c leads to plasmacytoma development in 50% of the mice with a latency of 60 to 120 days (Troppmair et al. 1989) (Table 1). The results obtained with the helper-free ABL-MYC virus clearly indicate that viremia is not necessary for plasmacytoma induction. What is more, the rapid induction kinetics along with the 100% incidence suggest that the combination of v-_abl_ and c-_myc_ obviates the need for multiple secondary events in pristane conducted mice.

Although it has yet to be proven, it is our working hypothesis that a small subset of B cells is susceptible to single-hit transformation by ABL-MYC, requiring no further genetic alteration. The results obtained by implantation of in vitro infected spleen cells are in keeping with this hypothesis.

Table 1. Comparison of Plasmacytoma induction in pristane-primed BALB/c mice with ABL-MYC virus, RIM virus and J3V1 virus

VIRUS	PRISTANE ml	PLASMACYTOMA incidence %	LATENCY in days	REFERENCES
J3V1	0.2	50	60-120	Troppmair et al. 1989
RIM	0.5	28	60-118	Clynes et al. 1988
ABL-MYC	0.5	100	33-42	this paper
NONE	3x0.5	60	125-365	Potter et al., 1984

The transplantation experiments also helped to delineate the target-cell population of the ABL-MYC virus. There seem to be at least two B cell differentiation stages susceptible to transformation by this virus. In the spleen, a population of late-stage B cells seems to be preferred. In the bone marrow predominantly immature B cells are immortalized. The fact that we could also detect occasional plasmacytomas in mice with transformed bone marrow cells, suggests two possible explanations. An infected pre-B cell could be transformed and then further differentiate into a plasma cell. Since in vitro transformation of bone marrow with v-abl alone results exclusively in pre-B-cell tumors when reimplanted in mice, the coexpression of c-myc along with v-abl might foster differentiation. This hypothesis is consistent with the findings presented by Largaespada et al. (this volume). On the other hand, it cannot be ruled out that the virus infected a mature B cell normally found in the bone marrow.

In summary, the ABL-MYC virus provides a powerful tool to perform futher studies of mature plasmacytomas as well as to investigate B cell differentiation.

REFERENCES:

Clynes R, Wax J, Stanton LW, Smith-Gill S, Potter M, Marcu KB (1988) Rapid induction of IgM-secreting murine plasmacytomas by pristane and an immunoglobulin heavy-chain promotor/enhancer-driven c-myc/v-Ha-ras rertovirus. Proc Natl Acad Sci USA 85: 6067-6071

Largaespada D, Kaehler D, Weissinger EM, Mischak H, Mushinski JF, Risser R (1990) The activity of an ABL-MYC retrovirus in fibroblast cell lines and lymphocytes. Curr.Top.Microbiol. Immunol.(this volume)

Jolley ME, Wang CH, Eckenberg SJ, Zuelke MS, Kelso DM (1984) Particle concentration fluorescence immunoassay (PCFIA): a new, rapid immunoassay technique with high sensitivity. J.Immunol. Methods 67: 21-25

Kurie JM, Morse HC III, Principato MA, Wax JS, Troppmair J, Rapp UR, Potter M, and Mushinski JF (1990). V-myc and v-raf act synergistically to induce B-cell tumors in pristane-primed adult Balb/c mice. Oncogene 5: 577-582

Potter M, Wiener F, Mushinski JF (1984) Recent developments in plasmacytomagenesis in mice. Adv Viral Oncol 4: 139-162

Potter M, Sklar MD, Rowe WP (1973). Rapid induction of plasmacytomas in pristane primed BALB/c mice. Science 182:592-594

Mushinski JF (1988) C-myc Oncogene Activation and Chromosomal Translocation in BALB/c Plasmacytomas. In: Klein G (ed) Cellular Oncogene Activation. Marcel Dekker Inc., New York pp 181-221

Rousseau PG, Mallet CP, Smith-Gill SJ (1989) A substantial proportion of the adult BALB/c available B cell repertoire consists of multireactive B cells. Molecular Immunology 26, 10: 993-1006

Troppmair J, Potter M, Wax JS, and Rapp UR (1989) An altered v-raf is required in addition to v-myc in J3V1 virus for acceleration of murine plasmacytomagenesis. Proc Natl Acad Sci USA 86: 9941-9945

Role of *raf*-1 Protein Kinase in IL-3 and GM-CSF-Mediated Signal Transduction

U.R. Rapp,[+] J. Troppmair,[+] M. Carroll,[*] and S. May[*]

[+]National Cancer Institute, Laboratory of Viral Carcinogenesis, Frederick, MD 21701 [*]The Johns Hopkins Oncology Center, Baltimore, MD 21231

INTRODUCTION

Raf-1 is a cytosolic serine/threonine-specific protein kinase that is ubiquitously expressed (Storm et al., 1990). The *raf*-1 gene can be converted into an oncogene by N-terminal fusion, truncation, or by site-specific mutation (Heidecker et al., 1990). When activated forms of the *raf*-1 gene were expressed in E. coli, the purified protein had kinase activity *in vitro* and induced DNA synthesis as well as morphological transformation upon microinjection into quiescent NIH 3T3 cells (Smith et al., 1990). Thus, activated *raf*-1 is an intracellular mitogen.

Raf-1 kinase is a candidate downstream effector of mitogen signal transduction, since *raf* oncogenes overcome growth arrest resulting from a block of cellular *ras* activity due either to a cellular mutation (*ras* revertant cells) or microinjection of anti-*ras* antibodies (Rapp et al., 1988, Smith et al., 1986). Moreover, mitogen treatment of a variety of fibroblastic (Morrison et al., 1988, Morrison et al., 1989, App et al., 1990) epithelial (Blackshear et al., 1990) and lymphoid cells (Siegel et al., 1990, Thompson et al., in preparation, Turner et al., submitted) activates *raf*-1 kinase activity and induces a transient translocation of the activated enzyme to the perinuclear area and the nucleus as demonstrated for NIH 3T3 cells and brain tissue (Rapp et al., 1988, Olah et al., 1990). Cells containing activated *raf* have an altered transcription pattern (Heidecker et al., 1990), and *raf* oncogenes behave as transcriptional activators for Ap-1/PEA 3-dependent promoters in transient transfection assays (Wasylyk et al., 1989, Kaibuchi et al., 1989, Jamal et al., 1990).

There are at least two independent pathways for *raf*-1 activation by extracellular mitogens: one involving PKC and a second initiated by intrinsic or associated receptor tyrosine kinases (Morrison et al., 1988; App et al., 1990; Baccarini et al., 1990; Blackshear et al., 1990; Thompson et al., in preparation; Carroll et al., 1990; Turner et al., submitted). In either case, activation involves *raf*-1 protein phosphorylation. This may be a consequence of a direct phosphorylation in a kinase cascade amplified by autophosphorylation or may be caused entirely by autophosphorylation initiated by binding of an activating ligand to the *raf*-1 regulatory site, analogues to PKC activation (Nishizuka et al., 1986). In the latter scheme, production of such a ligand would be under the control of PKC or the receptor tyrosine kinase. Support for a kinase-cascade mechanism of

Current Topics in Microbiology and Immunology, Vol. 166

activation of *raf*-1 comes from experiments with the PDGF-β receptor. High levels of tyrosine phosphorylation of *raf*-1 were observed in SF-9 insect cells doubly-infected with baculoviruses expressing the PDGF receptor and *raf*-1. Moreover, direct tyrosine phosphorylation, correlating with *raf*-1 kinase activation, could be achieved with partially purified enzymes *in vitro* (Morrison et al., 1989). However, the fraction of *raf*-1 molecules that were found tyrosine-phosphorylated in NIH 3T3 or BALB 3T3 cells in a ligand-dependent (PDGF) manner is very low ($\leq$ 1%), and its role in *raf*-1 activation remains to be established.

We have decided to extend the studies on mitogen regulation of *raf*-1 kinase to include the IL-3-dependent murine myeloid cell lines FDC-P1, 32D, and DA3, in which we previously evaluated the ability of a variety of oncogenes to abrogate growth factor dependence (Rapp et al., 1985, Cleveland et al., 1986, Dean et al., 1987, Cleveland et al., 1989).

RESULTS

Combination of v-*raf* and v-*myc* Efficiently Abrogates IL-3 Dependence

Earlier studies have demonstrated that the most efficient inducers of factor independence were exogenous tyrosine kinase oncogenes, irrespective of their normal expression in myeloid cells, followed by *myc* oncogenes, which showed dose-dependent abrogation (Cleveland et al., 1986). *Myc* transcription is normally regulated by IL-3 in these cells (Dean et al., 1987, Cleveland et al., 1989). *Ras*, *mos*, or *raf* oncogenes were either ineffective or had very low activity (Cleveland et al., 1986, Rapp et al., 1988). Since *raf* and *myc* oncogenes synergize or cooperate for transformation, depending on the target tissue (Rapp et al., 1985, Blasi et al., 1985, Troppmair et al., 1989), we examined this oncogene combination for the efficiency of IL-3 abrogation. FDC-P1 and 32D cells were infected with equal doses of v-*raf* and v-*myc* carrying viruses. Cultures were carried in IL-3 and challenged weekly for IL-3-independent growth (Fig. 1-A&B). v-*raf* virus alone caused emergence of factor-independent cells only after a long lag period in FDC-P1 and not at all in 32D cells. In contrast, combination with v-*myc* in J2 virus resulted in efficient growth factor abrogation of both FDC-P1 and 32D cells, comparable to *abl* MuLV. None of the factor-independent FDC-P1 and 32D cells were induced for IL-3 or GM-CSF message. We conclude, based on the effects of v-*raf*, that activated endogenous *raf*-1 protein kinase is a candidate mediator for IL-3-induced mitogenesis.

IL-3 Induces *raf*-1 Phosphorylation and Kinase Activation of FDC-P1 Cells

To examine the effect of IL-3 on endogenous *raf*-1 activity, the ability of IL-3 to induce *raf*-1 protein phosphorylation and kinase activation was tested. Addition of IL-3 to the growth factor-deprived cultures of the myeloid cell line FDC-P1 induces concentration-dependent phosphorylation of *raf*-1, as seen in a shift of the apparent molecular weight from 72 to 74 kDa in DSD-PAGE. This is paralleled by an increase in the *in vitro* kinase activity of *raf*-1 immunoprecipitates using histone (H1) as substrate (data not shown).

IL-3 and GM-CSF Induce *raf*-1 Phosphorylation and Kinase Activation in DA-3 Cells

In order to determine whether *raf*-1 is more commonly involved in hemopoietic growth signal transduction, we carried out similar experiments with DA-3 cells whose growth can be supported by either IL-3 or GM-CSF. The results obtained demonstrate that both factors induce rapid, time-dependent phosphorylation of *raf*-1 (Fig. 2A). *In vitro* histone phosphorylation by *raf*-1 immunoprecipitates obtained with the *raf*-specific polyclonal peptide antiserum sp63 was increased after stimulation with either mitogen (Fig. 2B). Almost no kinase activity is detectable when the immunoprecipitation is carried out in the presence of excess competing peptide, confirming that the *in vitro* kinase activity is associated with *raf*-1.

IL-3 and GM-CSF-Induced *raf*-1 Phosphorylation Occurs on Serine and Tyrosine

To characterize the amino acid sites on *raf*-1 that are phosphorylated upon IL-3 or GM-CSF stimulation, phosphoamino acid analysis of the ^{32}P labeled *raf*-1 protein from DA-3 cells was carried out. The results demonstrate a marked increase in both phosphoserine and phosphotyrosine content of *raf*-1 in IL-3 or GM-CSF-stimulated cells compared to untreated cells (Fig. 3). This suggests similar mechanisms for IL-3 and GM-CSF-induced *raf*-1 phosphorylation.

Time Course of IL-3-Induced Tyrosine and Serine Phosphorylation of *raf*-1

Further studies compared the kinetics of IL-3-induced tyrosine phosphorylation of *raf*-1 with the known rapid tyrosine phosphorylation of a 140 kd component of the IL-3 receptor (Isfort et al., 1988). Immunoprecipitates of IL-3-stimulated DA-3 cells with antiphosphotyrosine antibodies show three dominant phosphoproteins: a 74 kd, a 55 kd, as well as the 140 kd putative IL-3 receptor component. Tyrosine phosphorylation is apparent one minute after IL-3 addition and peaks after five minutes. The identity of the 74 kd protein with *raf*-1 was established with anti-*raf*-1 immunoblots. Whereas tyrosine phosphorylation of *raf*-1 is transient overall *raf*-1 phosphorylation, reflecting serine and tyrosine phosphorylation continues to increase up to 60 minutes (data not shown).

FDC-P1 Cells Abrogated for IL-3 Dependence by ts v-*abl* Show Temperature Sensitive *raf*-1 Protein Phosphorylation

The effects of a temperature-sensitive (ts) mutant of v-*abl*, DP, on IL-3 requirement, and *raf*-1 phosphorylation in IL-3 dependent cells were examined. DP contains an amino acid insertion in the tyrosine kinase domain of v-*abl* which renders it temperature-sensitive with respect to its tyrosine kinase activity and transforming efficiency (Kipreos et al., 1987). The FDDP-2 cell line was derived by introducing a retroviral construct expressing DP into FDC-P1 cells (Cleveland et al., 1989). At the non-permissive temperature (39°C), ts v-*abl* has low tyrosine activity, and the FDDP-2 cells are IL-3-dependent. Addition of IL-3 to the cells induces rapid *raf*-1 phosphorylation (Fig. 5). At the permissive temperature (32°C), ts v-*abl* kinase activity is increased 11-fold (Kipreos et al., 1987), and the cells become independent of IL-3 for growth (Cleveland et al., 1989). Under these conditions, *raf*-1 is constitutively phosphorylated, and addition of IL-3 has no further effect. Phosphoamino acid analysis of *raf*-1 immunoprecipitates from ^{32}P-labeled FDDP-2 cells at a permissive temperature shows high levels of phosphorylation on tyrosine consistent with increased protein tyrosine kinase activity of ts v-*abl* at 32°C (data not shown). Additionally, serine phosphorylation of *raf*-1 is increased.

DISCUSSION

Infection with 3611MSV facilitates emergence of IL-3-independent FDC-P1, but not 32D cells. However, infection of both FDC-P1 and 32D cells with J2 virus that carries v-*raf* and v-*myc* efficiently abrogates IL-3 dependence. We therefore speculate that two pathways (one which leads to induction of c-*myc* and a second involving *raf*-1 kinase activation) are being activated by IL-3 and are jointly required for mitogenesis.

The ability of IL-3 and GM-CSF to activate *raf*-1 kinase was demonstrated for two cell lines: FDC-P1 and DA-3. Activation occurred at mitogenic doses and correlated with rapid phosphorylation of *raf*-1 on both serine and tyrosine. Moreover, FDC-P1 cells conditionally abrogated for IL-3 dependence by introduction of a temperature-sensitive *abl* tyrosine kinase show temperature-dependent *raf*-1 phosphorylation and kinase activation. IL-3 regulation of *raf*-1 kinase activity presumably involves a non-receptor tyrosine kinase as an intermediate since the known component of the IL-3 receptor does not have a kinase domain (Itoh et al., 1990).

Comparison of *raf*-1 phosphorylation induced by a variety of receptors revealed phosphorylation predominantly on serine in case of receptors with intrinsic kinase activity and to equal degrees on serine and tyrosine when activation was triggered through non-kinase receptors IL-2 and CD4. The intracellular tyrosine kinase that is activated by CD4 crosslinking is *lck* (Veillette et al., 1988), whereas IL-2 stimulation activates an as yet unknown intracellular tyrosine kinase (Turner et al., submitted). It is possible that direct tyrosine phosphorylation of *raf*-1 by both classes of receptors, those with intrinsic and those with associated tyrosine kinase activity, is involved in *raf*-1 kinase activation under physiological conditions. However, it is also evident from related work (Morrison et al., 1988, Siegel et al., 1990) that at least one other serine kinase, protein kinase C, which is known to be activated by several receptor systems including the PDGF receptor and the T cell receptor, may act as an intermediate in *raf*-1 activation. While PKC is an essential intermediate in T cell receptor triggering of *raf*-1 in certain T cells (Siegel et al., 1990), it is not required for PDGF- and EGF-mediated *raf*-1 activation (Morrison, et al., 1988; App et al., 1990). Thus there appear to be redundant pathways for receptor-mediated *raf*-1 activation.

It is likely that *raf*-1 kinase is also a key component of proliferation control in B cells. First, *raf*-1 RNA is expressed in B cells during various stages of differentiation (Storm et al., 1990). Second, 3611 MSV, which expresses an activated version of *raf*-1, v-*raf* immortalizes B lineage cells *in vitro* at the pre-B cell stage (Principato et al., 1988, Principato et al., 1990, Morse et al., 1988). Finally, 3611 MSV accelerates pristane-induced plasmacytoma development in BALB/cAn mice (Shaughnessy et al., unpublished observations). Future work will examine *raf*-1 regulation by B cell-specific growth modulators.

REFERENCES

App H, Hazan R, Zilberstein A, Schlessinger J, Rapp UR (1990) *Raf*-1 protein kinase activation by the EGF receptor, submitted

Baccarini M, Sabatini DM, App H, Rapp UR, Stanley ER (1990) Colony stimulating factor-1 (CSF-1) stimulates temperature-dependent phosphorylation and activation of the *raf*-1 proto oncogene product. EMBO J, in press

Blackshear PJ, Haupt DM, App H, Rapp UR (1990) Insulin activates the *raf*-1 protein kinase. J Biol Chem, in press

Blasi E, Mathieson B, Varesio L, Cleveland JL, Borchert PA, Rapp UR (1985) Selective immortalization of murine macrophages from fresh bone marrow by a *raf/myc* recombinant murine retrovirus. Nature 318:667-670

Carroll MP, Clark-Lewis I, Rapp UR, May WS (1990) Interleukin-3 and grandulocyte-macrophage colony-stimulating factor mediate rapid phosphorylation and activation of cytosolic c-*raf*. J Biol Chem, in press

Cleveland JL, Dean M, Rosenberg N, Wang JYJ, Rapp UR (1989) Tyrosine kinase oncogenes abrogate interleukin-3 dependence of murine myeolid cells through signaling pathways involving c-*myc*; conditional regulation of c-*myc* transcription by temperature-sensitive v-*abl*. Mol Cell Biol 9:5685-5695

Cleveland JL, Jansen HW, Bister K, Frederickson TN, Morse HC III, Ihle JN, Rapp UR (1986) Interaction beteen *raf* and *myc* oncogenes in transformation *in vivo* and *in vitro*. J Cell Biochem 30:195-218

Dean M, Cleveland JL, Rapp, UR, Ihle JN (1987) Role of *myc* in the abrogation of IL-3 dependence of myeloid FDC-P1 cells. Oncogene Res 1:279-296

Heidecker G, Cleveland JL, Beck T, Huleihel M, Kolch W, Storm SM, Rapp, UR (1989) Role of *raf* and *myc* oncogenes in signal transduction. In: Colburn NH (ed) Genes and signal transduction in multistage carcinogenesis. Marcel Dekker, New York, p 379-374

Heidecker G, Huleihel M, Cleveland JL, Beck TW, Lloyd P, Pawson T, Rapp UR (1990) Mutational activation of c-raf-1 and definition of the minimal transforming sequence. Mol Cell Bioll, in press

Isfort RA, Huhn RD, Frackelton AR, Ihle JN (1988) Stimulation of factor-dependent myeloid cell lines with interleukin-3 induces tyrosine phosphorylation of several cellular substrates. J Biol Chem 263:19203-19209

Isfort RA, Stevens DA, May WS, Ihle JN (1988) Interleukin-3 binds to a 140 kDA phophotyrosine-containing cell surface protein. Proc Natl Acad Sci USA 75:7982-7986

Itoh N, Yonehara S, Schreurs J, Gorman DM, Maruvama K, Ishii A, Yahara I, Miyaiima A (1990) Cloning of an interleukin-3 receptor gene: a member of distinct receptor gene family. Science 247:324-327

Jamal S, Ziff E (1990) Transactivation of c-*fos* and β-actin genes by *raf* as a step in early response to transmembrane signals. Science 344:463-466

Kaibuchi K, Fukumoto Y, Oku N, Hori Y, Yamaoto T, Toyoshima K, Takai Y (1989) Activation of the serum response element and 12-O-tetradecanoylphorbol-13-acetate response element by the activated c-*raf*-1 protein in a manner independent of protein kinase C. J Biol Chem 264:20855-20858

Kipreos ET, Lee GJ, Wang JYJ (1987) Isolation of temperature-sensitive tyrosine kinase mutants of v-*abl* oncogene by screening with antibodies for phosphotyrosine. Proc Natl Acad Sci USA 84:1345-1349

Morrison DK, Kaplan DR, Escobedo JA, Rapp UR, Robert TM, Williams TL (1989) Direct activation of serine/threonine kinase activity of *raf*-1 through tyrosine phosphorylation by the PDGF-β receptor. Cell 58:649-657
Morrison DK, Kaplan DR, Rapp UR, Roberts TM (1988) Signal transduction from membrane to cytoplasm: Growth factors and membrane-bound oncogene products increase *raf*-1 phosphorylation and associated protein kinase activity. Proc Natl Acad Sci USA 85:8855-8859
Morse HC, Rapp UR (1988) Tumorigenic activity of artificially activated c-*onc* genes. In: Klein G (ed) Cellular oncogene activation. Marcel Dekker, Inc., New York, p 335.
Nishizuka Y (1986) Studies and perspectives of protein kinase C. Science 233:305
Olah Z, Komoly S, Nagashima N, Joo F, Rapp UR, Anderson WB (1990) Cerebral ischemia induces transient intracellular redistribution resulting in intranuclear translocation of the *raf* proto oncogene product in hippocampal pyramidal cells. Exp Brain Res, Submitted
Principato M, Klinken SP, Cleveland JL, Rapp UR, Holmes KL, Pierce JH, Morse HC III (1988) *In vitro* transformation of murine bone marrow cells with a v-*raf*/v-*myc* retrovirus yields clonally-related mature B cells and macrophages. Curr Top Microbiol Immunol 141:31-41
Rapp UR, Cleveland JL, Bonner TI, Storm SM (1988) The *raf* oncogenes. In: Curran T, Reddy EP, Skalka A (eds) The oncogene handbook, vol 14. Elsevier Science Publishers, The Netherlands, p 213-253
Rapp UR, Cleveland JL, Frederickson TN, Holmes KL, Morse HC III, Jansen HW, Patschinsky T, Bister K (1985) Rapid induction of hemopoietic neoplasms in newborn mice by a *raf*/*myc* recombinant murine retrovirus. J Virol 55:23-33
Rapp UR, Heidecker G, Huleihel M, Cleveland JL, Choi WC, Pawson T, Ihle JN, Anderson WB (1988) *Raf* family serine/threonine protein kinases in mitogen signal transduction. Cold Spring Harbor Symposia Quant Biol vol 53, Cold Spring Harbor Laboratory, 173-184
Siegel JN, Klausner RD, Rapp UR, Samelson LE (1990) T-cell antigen receptor stimulation activates the c-*raf* kinase via a protein kinase C-dependent pathway. J Biol Chem, in press
Smith MR, DeGubicibus SJ, Stacey DW (1986) Requirement for c-*ras* proteins during viral oncogene transformation. Nature 320:540-543
Smith MR, Heidecker G, Rapp UR, Kung H (1990) Induction of transformation and DNA synthesis after microinjection of *raf* proteins. Mol Cell Biol, in press
Storm SM, Cleveland JL, Rapp UR (1990) Expression of *raf* family proto oncogenes in normal mouse tissue. Oncogene 5:345-351
Troppmair J, Potter M, Wax JS, Rapp UR (1989) An altered v-*raf* is required in addition to v-*myc* in J3V1 virus for acceleration of murine plasmacytomagenesis. Proc Natl Acad Sci USA 86:9941-9945
Turner BC, Rapp UR, App H, Greene MI, Dobashi K, Reed JC (1990) Interleukin 2 (IL-2) induces phosphorylation and activation of p72-74 *raf*-1 kinase in T lymphocytes. Submitted
Veillette A, Bookman MA, Horak EM, Bolen JB (1988) The CD4 and CD8 T-cell surface antigens are associated with the internal membrane tyrosine-protein kinase p56. Cell 55:301-308
Wasylyk C, Wasylyk B, Heidecker G, Huleihel M, Rapp UR (1989) Expression of *raf* oncogene activates the PEA1 transcription factor motif. Mol Cell Biol 9:2247-2250

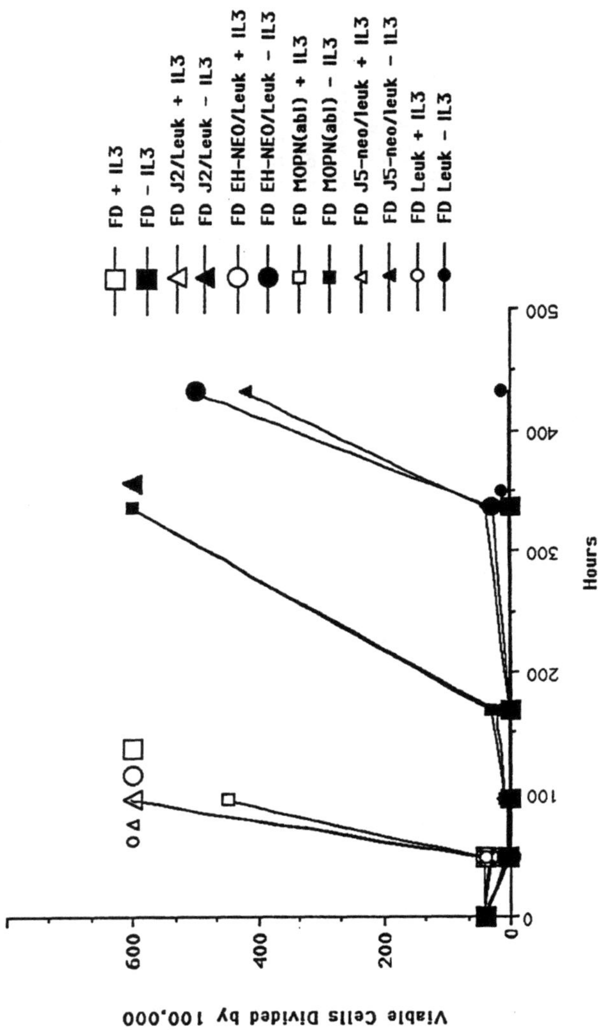

Fig. 1A

Figures 1A & 1B: IL-3 dependent growth of FDC-P1 (A) and 32D cells (B) infected with v-*raf*, v-*myc*, v-*raf*/v-*myc*, and v-*abl*-expressing retroviruses. Cells were infected at a multiplicity of infection of 0.5 with leuk pseudotypes of recombinant retroviruses expressing v-*raf* (EH-neo), v-*myc* (J5), v-*myc*, and v-*raf* (J2) or v-*abl* (MOPN) and cultured for 2 weeks in the presence of IL-3 (10 U/ml). To assay for growth in the absence of IL-3, $4x10^6$ cells were seeded into 75 cm^2 tissue culture flasks, and cell numbers were determined over a period of 20 days.

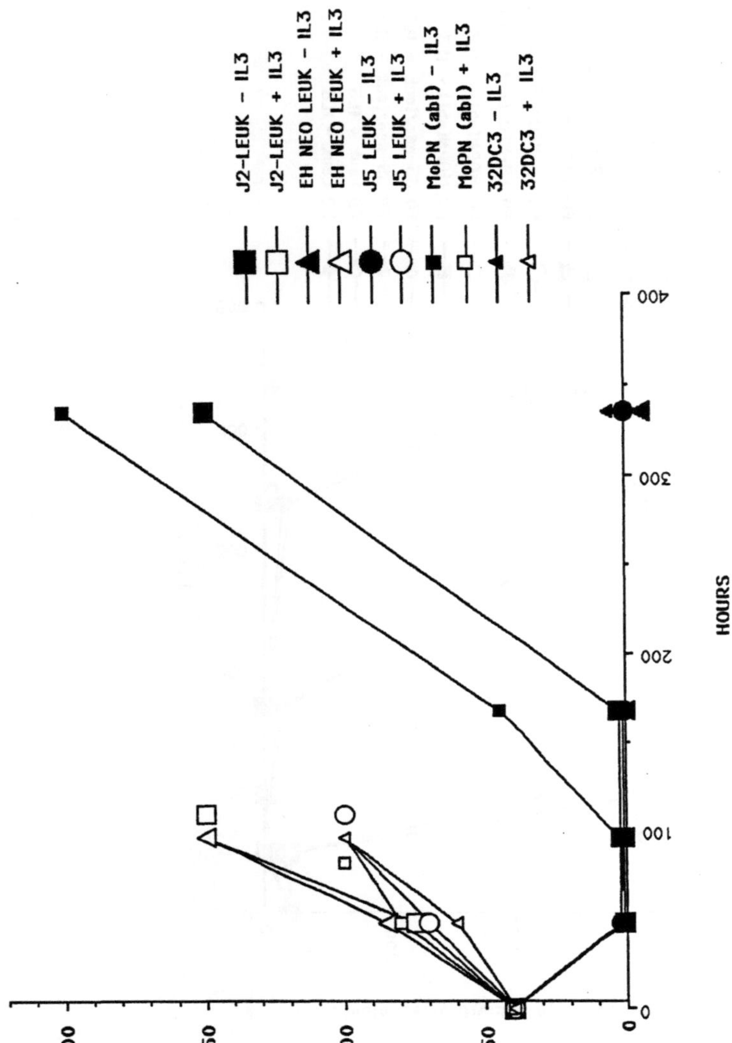

J2-LEUK - IL3
J2-LEUK + IL3
EH NEO LEUK - IL3
EH NEO LEUK + IL3
J5 LEUK - IL3
J5 LEUK + IL3
MoPN (abl) - IL3
MoPN (abl) + IL3
32DC3 - IL3
32DC3 + IL3
Viable Cells Divided by 100,000
HOURS
0
50
100
150
200
0
100
200
300
400

Fig. 1B

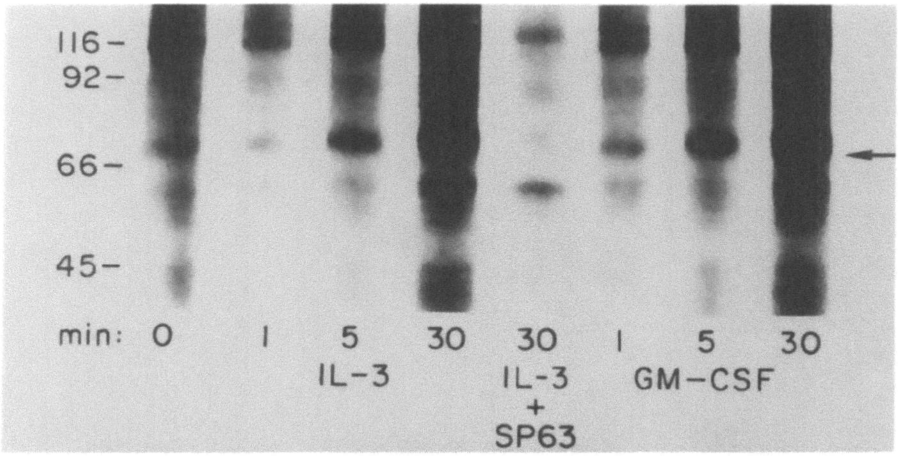

Figure 2A: Time course of growth factor-induced *raf*-1 phosphorylation in DA-3 cells. Cells were growth factor deprived for three hours, washed, and resuspended at $1x10^7$ cells/ml. Labeling was carried out with 100 µCi/ml of ^{32}P orthophosphoric acid for 60 minutes. Radiolabeled cells were stimulated with 10U/ml of IL-3 of GM-CSF for the time indicated. *Raf*-1 protein was immunoprecipitated with *raf*-1 specific peptide antiserum sp63 in the presence or absence of competing peptide

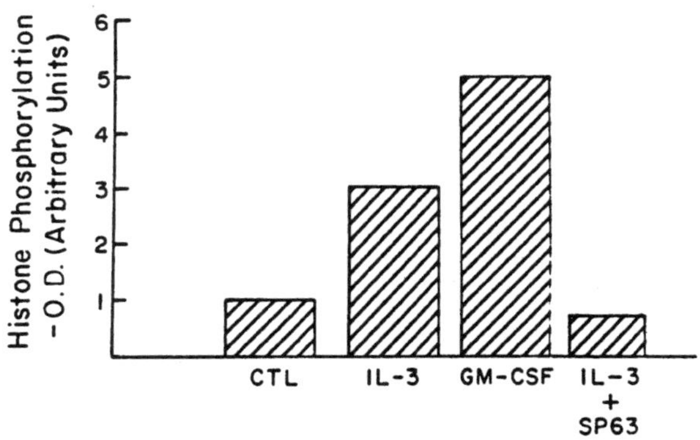

Figure 2B: DA-3 cells were treated with 10U/ml IL-3 or GM-CSF for 15 minutes. *Raf*-1 immunoprecipitates obtained with the *raf*-1 specific peptide antiserum sp63 in the presence or absence of competing peptide were assayed for *in vitro* kinase activity in the presence of [γ-^{32}P] ATP and histone H5 as substrate. Reaction mixtures were resolved by 15% SDS PAGE and analyzed by autoradiography. Histone phosphorylation was quantitated using optical densiometry.

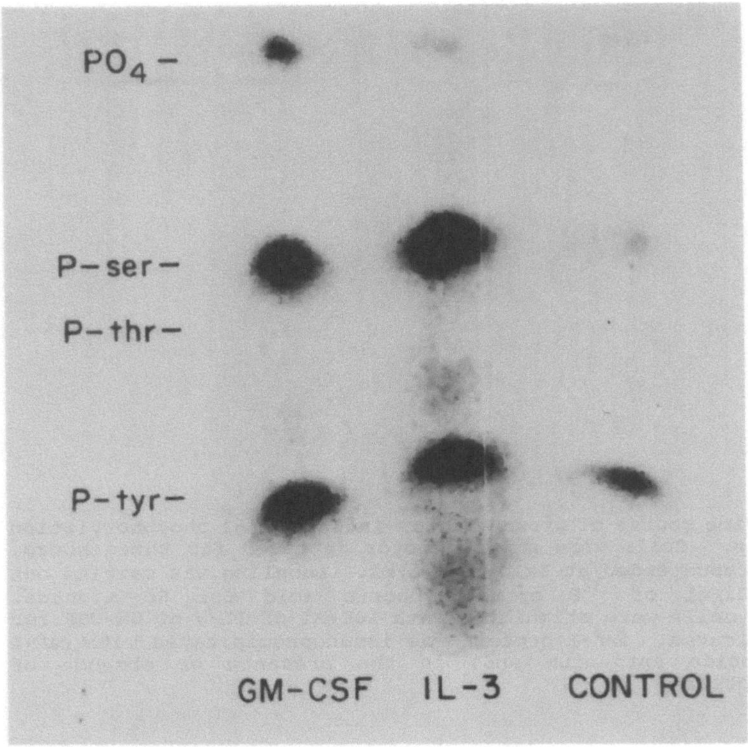

Figure 3: IL-3 and GM-CSF induces *raf*-1 phosphorylation on serine and tyrosine residues. DA-3 cells were growth factor-deprived for 3 hours, metabolically labeled with 100 μCi/ml of ^{32}P orthophosphoric acid for 60 minutes and stimulated with 10U/ml of IL-3 or GM-CSF. *Raf*-1 immunoprecipitates prepared with *raf*-1-specific peptide antiserum sp63 were resolved by 10% SDS-PAGE. Labeled *raf*-1 was excised and subjected to phosphoamino analysis.

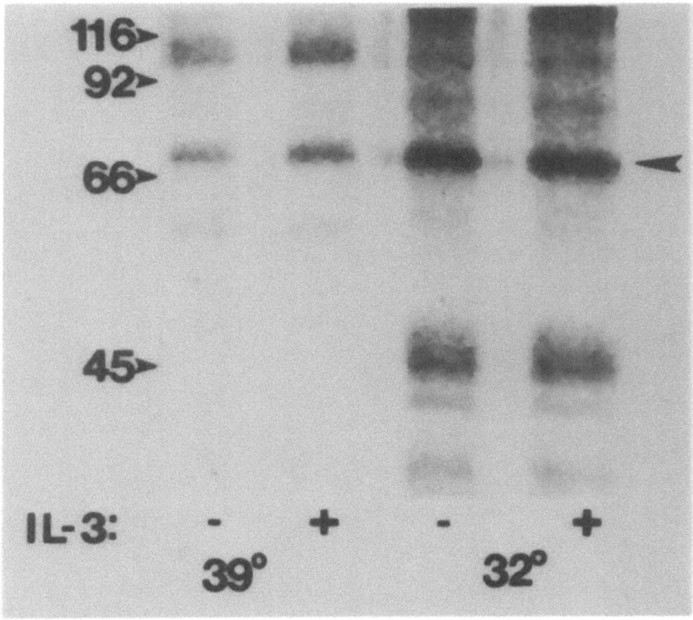

Figure 4: Effect of IL-3 on *raf*-1 phosphorylation in FDDP-2 cells carrying a ts v-*abl* mutant retrovirus at nonpermissive (39°C) or permissive temperature (32°C) cells were labeled with ^{32}P orthophosphoric acid, stimulated with 10U/ml of IL-3 for 15 minutes at the temperature indicated, and lysates were immunoprecipitated with *raf*-1-specific peptide sp63. Immunoprecipitates were resolved by 10% SDS-PAGE.

Identification of Consensus Genes Expressed in Plasmacytomas but Not B Lymphomas

C. Timblin, P.L. Bergsagel, and W.M. Kuehl

NCI-Navy Medical Oncology Branch, National Cancer Institute, National Institutes of Health and Naval Hospital, Bethesda, MD 20814-5015

INTRODUCTION

Mouse plasmacytoma cell lines differ from pre-B and B lymphoma cell lines by a number of notable features: 1) the distinctive plasmacytoid morphology including extensive and well-developed rough endoplasmic reticulum and Golgi apparatus (Matter et al. 1972, Potter 1987); 2) the lack of expression of some B cell markers (e.g. Ia antigens) (Boothby et al. 1988); 3) the lack of significant c-myc mRNA expression from the normal allele, which correlates with the presence of plasmacytoma repressor factor (PRF) gel-shift activity (Kakkis et al. 1989); 4) quantitatively increased expression of immunoglobulin; and 5) the consistent dominance of the plasmacytoma phenotype in hybrids with all B cells, including normal B cells and tumor B cell lines (Wall and Kuehl 1983). Our goal is to determine if there are consensus genes which are expressed in murine plasmacytomas, but not in B lymphomas. We have combined subtractive cDNA and PCR technologies to construct and analyze a plasmacytoma minus highly differentiated B lymphoma subtractive cDNA library.

RESULTS

The 4N2.1 variant of the MPC 11 plasmacytoma cell line was used as a source of mRNA for first strand cDNA synthesis (Choi et al. 1980). The wild type MPC 11 plasmacytoma cell line synthesizes and secretes approximately 10-15% of its protein as an IgG_{2b}, kappa immunoglobulin. The 4N2.1 variant does not express detectable immunoglobulin heavy chain mRNA and has deleted the functional kappa light chain gene. However, 4N2.1 continues to express the kappa constant region fragment, which corresponds to about 1% of the mRNA present in these cells. As a source of mRNA for subtractive hybridization, we used a highly differentiated B lymphoma cell line called A20.2J. The A20.2J B lymphoma line expresses and secretes IgG_{2a} and kappa light chain, the latter corresponding to about 0.1% of the mRNA in this cell line (Word and Kuehl 1981). In contrast to the 4N2.1 plasmacytoma cell line, A20.2J cells have a lymphoid morphology, express surface Ia antigens, express c-myc mRNA from an apparently normal c-myc allele, and have no detectable PRF activity (C. Timblin and M Kuehl, unpublished). In addition, fusion of A20.2J cells to 70Z/3B pre-B lymphoma cells results in somatic cell hybrids with a phenotype like that of the pre-B lymphoma partner (Bender and Kuehl 1987). By Northern blot analysis, we have found that 4N2.1 and A20.2J cells express similar amounts of J chain, intracisternal A particle, and PC.1 antigen mRNAs (C. Timblin and M. Kuehl, unpublished). When we began this study we knew of no genes that are expressed in MPC 11 and other mouse plasmacytoma cells, but not in A20.2J and other B cell lines.

Current Topics in Microbiology and Immunology, Vol. 166

The construction and initial analysis of the subtractive library have been described previously (Timblin et al. 1990). Briefly, first strand cDNA was made from cytoplasmic mRNA from the plasmacytoma cell line 4N2.1 using random hexamer primers and MMLV reverse transcriptase. The cDNA was subtracted twice and positively selected once, in each case with a 2.5 fold excess of the appropriate cytoplasmic mRNA, i.e. from A20.2J and 4N2.1 cells, respectively. The positively selected cDNA/RNA heteroduplex was converted to double-stranded cDNA, blunt-end ligated to a PCR linker, and PCR amplified for 30 cycles. Amplified cDNA species larger than 200 bp were then cloned into lambda gt10. To facilitate screening, we picked 1290 plaques onto duplicate grids.

No highly or moderately abundant subtractive clones were detected by screening with differential first strand cDNA probes, or by using the subtractive cDNA per se as a probe. To screen the library for low abundance subtractive clones, we randomly selected plaques (excluding incompletely subtracted C_k clones), amplified the insert by PCR using primers to lambda gt10 left and right arms, and labelled the insert using random primers. We then probed Northern blots containing cytoplasmic RNAs from the parental plasmacytoma cell line 4N2.1, an unrelated plasmacytoma cell line 653FB, and the subtractive partner A20.2J. Based on this screening procedure of 115 clones, the clones from the library can be divided into five categories: 1) no signal (23%) for any of the three cell lines on Northern blots derived from gels loaded with 5 ug of cytoplasmic poly (A)+ RNA per lane; 2) nonsubtractive (17%), i.e. expressed at approximately equivalent levels in all three cell lines; 3) tumor specific (12%), i.e. expressed only in the 4N2.1 parental plasmacytoma cell line; 4) quantitatively subtractive (14%), i.e. expressed at a level approximately threefold or more higher in the plasmacytoma cell lines compared to the B lymphoma cell line; and 5) qualitatively subtractive (34%), i.e. expressed in both plasmacytoma cell lines and not detected in the B lymphoma cell line.

We are interested primarily in the quantitatively and qualitatively subtractive clones that may represent consensus genes expressed by all (or most) plasmacytomas, but expressed at a lower level or not at all in B lymphomas. To extend our analysis of these candidate clones, we have probed Northern blots that contain RNA from additional cell lines.

By examining only a small fraction of the genes whose expression is quantitatively increased in plasmacytomas compared to the A20.2J B lymphoma, we have identified a potentially interesting gene, i.e. corresponding to clone 315, that is expressed as 2.4 and 7 kb mRNAs in 12 out of 13 plasmacytoma cell lines, 5 out of 8 B lymphoma cell lines, and 2 out of 10 pre-B lymphoma cell lines examined (Fig. 1). In two of the plasmacytoma cell lines, 4N2.1 and S107.3.4, there is an aberrant form of the mRNA at about 2.6 kb (the 2.4 and 2.6 kb mRNA species are both present in 4N2.1 cells but only the latter species is present in S107.3.4 cells). The mRNAs detected by clone 315 are expressed at a 5-10 fold higher level in the plasmacytoma cell lines than in the B and pre-B lymphoma cell lines. It is also interesting that the expression of this mRNA appears to be inducible by IL-6 in an IL-6 dependent hybridoma (Matsuda et al. 1988, C. Timblin and R. Wall, data not shown).

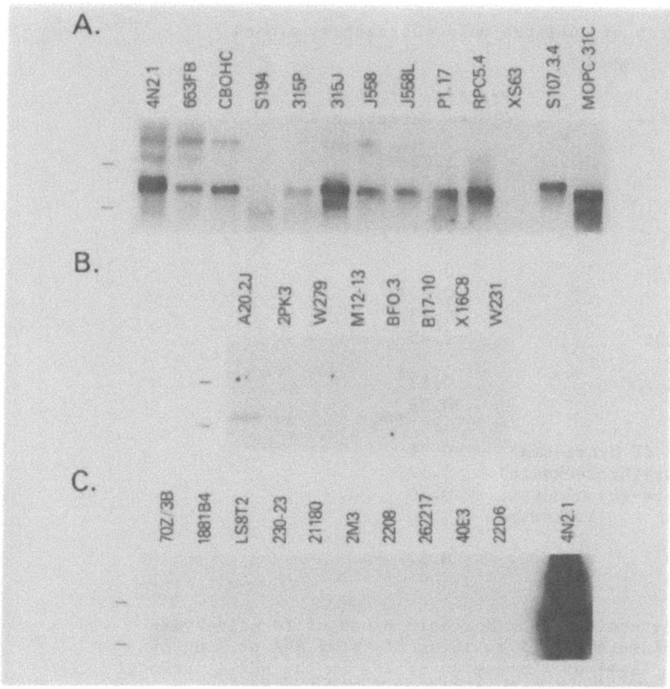

Figure 1. Expression of clone 315 in plasmacytoma, B lymphoma, and pre-B lymphoma cell lines. Northern blot analysis was done with 5 ug of cytoplasmic poly(A)+ RNA loaded in each lane of a formaldehyde-agarose gel. The blots were hybridized with random primer labelled insert from clone 315. A, plasmacytoma cell lines (93 hr exposure); B, B lymphoma cell lines (14 day exposure); C, pre-B lymphoma cell lines and 4N2.1 plasmacytoma control (14 day exposure). Indicated above each lane is the cell line used (the cell lines are referenced in Bender and Kuehl 1987, Timblin et al. 1990 except for the last 7 pre-B lymphoma cell lines which were generously provided by Naomi Rosenberg). The positions of 5.0 and 1.9 kb rRNAs are indicated by horizontal marks.

The mRNA expression patterns determined with 31 unrelated qualitatively subtractive clones in a variety of hematopoietic and non-hematopoietic cell lines are summarized in Table 1. Qualitatively subtractive clones were initially classified on the basis that mRNA was detected in 2 plasmacytoma cell lines (4N2.1 and 653FB), but not in the subtractive B lymphoma partner (A20.2J). Nonetheless, it is apparent that these clones have a high probability of detecting mRNA in other B and pre-B lymphoma cell lines, as well as in non-B hematopoietic and fibroblast cell lines. However, this group of clones is more likely to detect mRNAs in two unrelated plasmacytoma cell lines (i.e. CBOHC or S194) than in any other cell line examined.

Table 1. RNA expression patterns of qualitatively subtractive clones [1]

Cell type	Cell line	Fraction of cDNA Clones Detecting RNA [2]
Plasmacytomas	4N2.1	1.00 [3]
	653FB	1.00
	CBOHC	0.74
	S194	0.71
B cell lymphomas	A20.2J	0
	2PK3	0.35
	M12.13	0.13
	WEHI 231	0.45
Pre-B lymphomas	70Z/3B	0.52
	18.81	0.36
Hematopoietic, non-B	HTB157 (T hybridoma)	0.36
	C19 (erythroleukemia)	0.57
	32D (pre-granulocytic leukemia)	0.50
Non-hematopoietic	NIH3T3	0.57

1. The expression of RNA was determined from Northern blots of formaldehyde-agarose gels loaded with cytoplasmic RNA (i.e. 10 ug of total RNA or 5 ug of poly(A)+ RNA, depending on the level of expression of RNA in the 4N2.1 parental plasmacytoma).
2. The four plasmacytomas, four B lymphomas, and the 70Z/3B pre-B lymphoma were tested with 31 qualitatively subtractive cDNA clones; the other cell lines were tested with only the first 14 of the 31 qualitatively subtractive clones identified.
3. By definition, all of the qualitatively subtractive cDNA clones detected mRNA in the 4N2.1 (parental) and 653FB plasmacytomas, but not in the A20.2J B lymphoma subtractive partner (see text).

We have analyzed a number of the qualitatively subtractive clones on extended Northern blot panels of 13 plasmacytoma cell lines, 8 B lymphoma cell lines and 10 pre-B lymphoma cell lines (Table 2). Based on this analysis, we present data on 6 different qualitatively subtractive cDNA clones, each of which identifies mRNA in most plasmacytomas but in one or none of 8 B lymphoma cell lines examined. These 6 clones actually identify 7 different genes since clone 289 is a hybrid cDNA, i.e. 289A and 289B, as noted in Table 2. There appear to be two kinds of expression patterns of these 7 genes: 1) expression in plasmacytomas and pre-B lymphomas; and 2) expression in plasmacytomas only.

Table 2. RNA expression of qualitatively subtractive clones [1]

CLONE	70	260	251*	289A*	289B	326	291
mRNAs(kb)	3.3	10	2.7 4.4	1.8	1.3	4.4 4.9	4.7
PLASMACYTOMA							
CBOHC	+	+		+	+	+	+
S194	+	+	+		+		+
315J(P)	+	+	+(0)	+	+	+	+
J558	+	+	+	+		+	+
P1.17	+	+	+	+	+	+	+
RPC5.4	+	+	+	+	+	+	+
XS63			+	+	+	+	+
S107.3.4		+	+	+	+	+	+
MOPC-31C		+	+		+	+	+
B LYMPHOMA							
2PK3							+
WEHI 279	+	+					
M12.13							
BFO3							
Bal 17							
X16C.5							
WEHI 231							
PRE-B LYMPHOMA							
70Z/3B	+	+					
LS8.T2	+	+					
1881.B4	+	+					
230.23	+	+					
211.80	+	+					
2M3	+	+					
220.8	+	+					
2.6.2.217	+	+					
40E3	+	+					
22D6	+	+					

1. The results summarized in Table 2 are derived from Northern blot analysis of formaldehyde/agarose gels loaded with 5 ug of cytoplasmic poly(A)+ RNA per lane. The data for clones 251, 289, and 326 have been shown elsewhere (11); however, preliminary sequence data (L. Bergsagel, unpublished) indicates that clone 289 is actually a hybrid cDNA (i.e. 289A and 289B) derived from two unrelated mRNAs. The data for clone 70 are shown in Fig. 2. Clones 260 and 291 are described in the text, but the data is not shown.
2. An (*) is used to indicate microsomal localization of the mRNA.

Two genes are expressed in most (or all) plasmacytomas and pre-B lymphomas, but rarely in B lymphomas. Clone 70 detects a 3.3 kb mRNA that is expressed in 10 of 13 plasmacytoma cell lines and all of the pre-B lymphoma cell lines examined (Fig. 2 and Table 2). Clone 70 does not detect this mRNA in any of the B lymphoma cell lines examined, with the single exception of WEHI 279. The second gene in this category is identified by clone 260, which detects a 10 kb mRNA (data not shown). This second gene is expressed in 12 of 13 plasmacytoma cell lines, all 10 pre-B lymphoma cell lines examined, and in one of eight B lymphoma cell lines, i.e. WEHI 279 again (Table 2). It is intriguing that we have been able to identify two genes that are expressed only in cell lines corresponding to the very early and terminal stages of B lymphocyte differentiation.

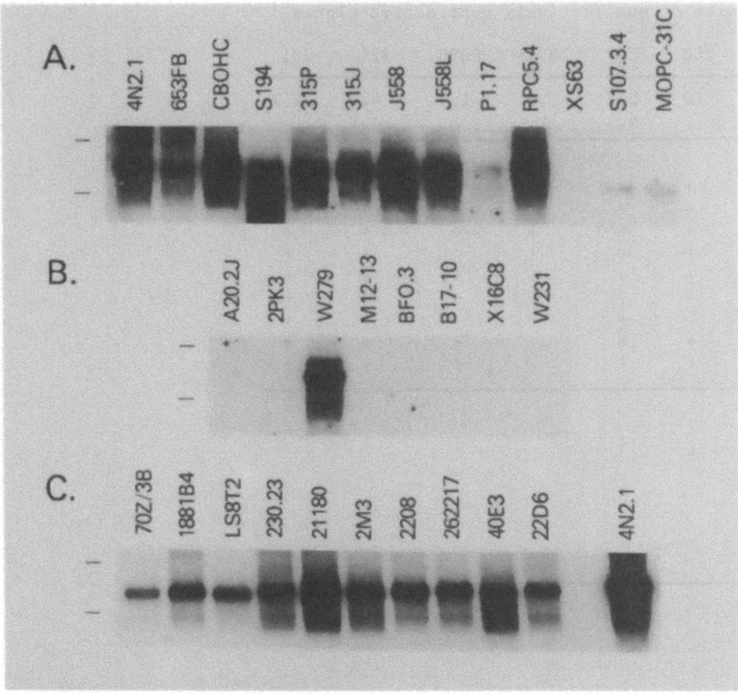

Figure 2. Expression of Clone 70 in plasmacytoma, B lymphoma, and pre-B lymphoma cell lines. Northern blot analysis was done as described in Fig. 1 except the cDNA insert from clone 70 was used as a probe. All blots were exposed for 13 days. A, plasmacytoma cell lines; B, B lymphoma cell lines; C, pre-B lymphoma cell lines. Horizontal marks indicate the positions of the 5.0 and 1.9 kb rRNAs.

In addition to the plasmacytoma/pre-B lymphoma specific genes, we have identified four plasmacytoma specific genes that are expressed in most plasmacytomas, but not in any pre-B or B lymphomas examined (Timblin et al. 1990, and Table 2). Clone 251 detects 2.7 and 4.4 kb microsomal mRNAs that are expressed in 10 of 11 independent plasmacytoma cell lines examined. It is noteworthy that clone 251 detects mRNA in plasmacytoma 315J, but not in 315P, a nonsecreting variant of 315J (Gebel et al. 1979). Clone 289A, which encodes a 1.8 kb microsomal mRNA, is expressed in 10 of 13 plasmacytoma cell lines examined. Clone 289B, which encodes a 1.3 kb non-microsomal mRNA, and clone 326, which encodes 4.6 and 4.9 kb non-microsomal mRNAs, are each expressed in 10 of 11 independent plasmacytoma cell lines examined. We have also identified clone 291 (Table 2), which is the only plasmacytoma specific clone identified thus far that detects mRNA (4.7 kb) in every plasmacytoma cell line examined. The mRNA identified by clone 291 is not detected in any of the pre-B lymphomas, but is detected in a single B lymphoma cell line, i.e. in the highly differentiated 2PK3 B lymphoma cell line.

SUMMARY

We have combined subtractive cDNA and PCR technologies to construct and analyze a plasmacytoma minus a highly differentiated B lymphoma subtractive cDNA library. We detected no plasmacytoma-specific clones by hybridization with differential cDNA probes or the subtractive insert. However, random selection of 115 clones has identified 16 quantitatively subtractive and 39 qualitatively subtractive clones. From these clones we have identified 8 potentially interesting genes. One quantitatively subtractive clone (clone 315) identifies an mRNA that is expressed in most plasmacytoma cell lines, but is expressed at an approximately 10-fold lower level in B and pre-B lymphoma cell lines; preliminary evidence suggests that the expression of this gene is increased by IL-6. From the 31 unrelated qualitatively subtractive clones, we have identified two classes of genes that are expressed in one or none of 8 B lymphomas examined: 1) those expressed in most plasmacytoma and pre-B lymphoma cell lines (clone 70 and clone 260); and 2) those expressed in most plasmacytoma cell lines, but not in any of the ten pre-B lymphomas examined (clones 251, 289A, 289B, 326, 291).

REFERENCES

Bender TP, Kuehl WM (1987) Differential expression of the c-myb proto-oncogene marks the pre-B cell/B cell junction in murine B lymphoid tumors. J Immunol 139:3822-3827

Boothby M, Gravallese E, Liou H-C, Glimcher LH (1988) A DNA binding protein regulated by IL-4 and by differentiation in B cells. Science 242:1559-1562

Choi E, Kuehl WM, Wall R (1980) RNA splicing generates a variant light chain from an aberrantly rearranged k gene. Nature 286:776-779

Gebel HM, Autry JR, Rohrer JW, Lynch RG (1979) In vitro and in vivo studies of a 2,4,6-trinitrophenyl (TNP)-binding immunoglobulin A (IgA) lymphoma isolated from MOPC-315, a TNP-binding IgA plasmacytoma of BALB/c mice. J Natl Cancer Inst 62:201-211

Kakkis E, Riggs KJ, Gillespie W, Calame K,(1989) A transcriptional repressor of c-myc. Nature 339:718-721

Matsuda T, Hirano T, Kishimoto T (1988) Establishment of interleukin 6 (IL-6) B cell stimulatory factor 2 (BSF-2)-dependent cell line and preparation of anti-IL-6/BSF-2 monoclonal antibodies. Eur J Immunol 18:951-956

Matter A, Lisowska-Bernstein B, Ryser JE, Lamelin J-P, Vassalli P (1972) Mouse thymus-independent and thymus-derived lymphoid cells. II Ultrastructural studies. J Exp Med 136:1008-1030

Potter M (1987) Plasmacytoma development in BALB/c mice. Adv Viral Oncol 7:99-122

Timblin C, Battey J, Kuehl WM (1990) Application for PCR technology to subtractive cDNA cloning: identification of genes expressed specifically in murine plasmacytoma cells. Nucleic Acids Research 18:1587-1593

Wall R, Kuehl M (1983) Biosynthesis and regulation of immunoglobulins. Ann Rev Immunol 1:393-422

Word CJ, Kuehl WM (1981) Expression of surface and secreted IgG_{2a} by a murine B lymphoma before and after hybridization to myeloma cells. Molec Immunol 18:311-322

Proliferation of B Cell Precursors in Bone Marrow of Pristane-Conditioned and Malaria-Infected Mice: Implications for B Cell Oncogenesis

D.G. Osmond, S. Priddle and S. Rico-Vargas

Department of Anatomy, McGill University,
3640 University Street, Montreal, Quebec, Canada H3A 2B2

INTRODUCTION

Precursor B cells in the bone marrow undergo a sequence of proliferative and differentiative steps in giving rise to primary B lymphocytes destined for the peripheral lymphoid tissues (Osmond 1986, Osmond & Park 1987). In recent years, we have defined discrete populations of B cell precursors in mouse bone marrow, based on immunofluorescence labeling of the nuclear enzyme, terminal deoxynucleotidyl transferase (TdT), the cell surface B lineage-associated B220 glycoprotein and cytoplasmic μ heavy chains of IgM (Opstelten & Osmond 1985, Park & Osmond 1987, 1989a). As a working model of B cell genesis, we have proposed that 6 phenotypically distinct cell populations form a differentiation sequence extending from the stage of V_H gene rearrangement, during which TdT is expressed, to the mature B lymphocyte expressing surface IgM (sIgM). Three populations comprise Pro-B cells before the expression of μ chains, while two populations represent Pre-B cells which express cytoplasmic μ chains (cμ) and which finally mature into sIgM$^+$ B lymphocytes (Table 1).

Table 1. Phenotype of B lineage cells in mouse bone marrow[a]

Cell population		Phenotype				Mitoses[b]
		TdT	B220	cμ	sIgM	
1. Pro-B cells:	early	+				1
2.	intermediate	+	+			1
3.	late		+			3
4. Pre-B cells:	large		+	+		1
5.	small		+	+		0
6. B lymphocytes			+		+	0

[a] In 8-10 wk C3H/HeJ mice
[b] Estimated number of mitoses occurring at each differentiation stage (Park & Osmond 1989a).

Current Topics in Microbiology and Immunology, Vol. 166

Studies of cell population dynamics have revealed both proliferative expansion and cell loss during the differentiation of precursor B cells (Opstelten & Osmond 1985, Deenen, Hunt & Opstelten 1987, Park & Osmond 1987, 1989a). The accumulation of cells in metaphase after inducing mitotic arrest in vivo has indicated that each pro-B and large pre-B cell stage includes at least one mitosis while late pro-B cells may undergo 3 sequential mitoses. This appears to be followed by a considerable loss of cells, possibly representing the removal of cells having ineffective or unacceptable Ig gene rearrangements.

The proliferation of precursor B cells in the bone marrow can be modified by systemic factors. Although bone marrow B lymphocytopoiesis appears to be independent of feedback controls from the peripheral B cell pool (Osmond 1986), it is amplified by exposure to a variety of foreign agents (Fulop & Osmond 1983). Thus, a single intraperitoneal administration of sheep red blood cells (SRBC) in mice is followed by a wave of increased proliferation of precursor B cells (Fulop, Pietrangeli & Osmond 1986). Experiments with silica-treated and splenectomized mice indicate that this stimulatory effect is mediated by the activation of macrophages in the spleen, probably via the release of circulating factors (Pietrangeli & Osmond 1985, 1987). Repeated administration of SRBC results in a sustained increase in proliferation and expansion of the precursor B cell pool (Fulop, Pietrangeli & Osmond, 1986).

The foregoing findings have led us to postulate the possibility of an etiological link between conditions of chronic macrophage activation and B cell oncogenesis (Osmond, Park & Jacobsen 1988). Phenotypically distinct B cell neoplasias show characteristic genetic changes. A chromosomal translocation juxtaposing the c-myc oncogene to the Ig H chain locus is universally found in mouse plasmacytoma (PCT) and human Burkitt's lymphoma (BL) (Potter, Pumphrey & Walters, 1972, Ohno et al 1979, Potter & Wax 1983, Klein & Klein 1986, Nordan & Potter 1986, Lenoir & Bornkum 1987, Potter et al 1987). Such genetic errors may occur during early B cell differentiation associated with mistakes in V-D-J joining when functional Ig genes are being formed (Tonegawa 1983, Greaves 1986, Kunkel et al 1986, Bertoli et al 1988). Any factor which increases the Ig recombinatorial activity in precursor B cells in vivo could increase the risk of genetic errors, including myc/Ig juxtaposition. This effect could be produced by increasing the turnover rate and number of cells rearranging Ig genes, as occurs as a result of prolonged macrophage activation.

Two conditions of B cell neoplasia, PCT and BL, are associated with prolonged macrophage activation. Susceptible BALB/c mice given intraperitoneal pristane develop chronic granulomata rich in macrophages (Potter, Pumphrey & Walters 1972, Potter et al 1985). In malaria, gross splenomegaly is associated with marked increase in the macrophage population (Freeman & Parish 1978, Wyler, Oppenheim & Koontz 1979, Wyler 1982, Weiss, Geduldig & Weidanz 1986). The macrophages show extensive phagocytosis of pristane oil droplets and parasitized RBC, respectively, and evidence of intense activation with release of a variety of soluble growth factors. In each case, the phase of chronic macrophage activation occurs during a prolonged lag phase which precedes the establishment of B cell neoplasia. The possible effects on early B cell differentiation are unknown. In the present work, therefore, we have examined mice with pristane-induced granuloma and with malarial splenomegaly to determine whether these conditions might exert a central effect on early B precursor cells in the bone marrow, which could be related to the initiation of B cell neoplasias.

PRECURSOR B CELLS IN BONE MARROW OF PRISTANE-CONDITIONED MICE

The effects of pristane-induced granuloma have been studied in the bone marrow of BALB/c PT mice, susceptible to PCT induction, and DBA/2 mice, resistant to PCT induction. All mice were kindly supplied by Dr. M. Potter, Laboratory of Genetics, National Cancer Institute, NIH, Bethesda, MD., and Dr. J. Wax, Hazleton Laboratories, Rockville, MD. After the intraperitoneal injection of a single dose (0.5 ml) of pristane (2,6,10,14-tetramethylpentadecane) to mice at 9-10 wk of age, bone marrow cell suspensions from groups of mice taken at various time intervals were analysed for each of the stages of precursor B cell differentiation. In each case, the actual number of cells per femur was calculated and expressed as a percentage of the normal population size (Table 2).

Table 2. Population size of B lineage cells in bone marrow of pristane-conditioned BALB/c PT mice [ab]

Cell population		Population size (number of cells/femur as % normal value) at intervals of: 7 days	28 days	82 days
Pro-B cells:	early	181	170	132
	intermediate	199	221	100
	late	56	52	24
Pre-B cells		41	35	33
B lymphocytes		25	26	29

[a] Mice were given 0.5 ml pristane ip.
[b] Bone marrow cells from femurs of groups of 3 mice were pooled at various intervals. Populations of precursor B cells were identified by double immunofluorescence labeling methods, their actual numbers per femur were calculated, as described (Opstelten & Osmond 1985, Park & Osmond 1987, 1989a), and were expressed as a percentage of the respective values derived from an original group of normal mice, not given pristane.

The precursor B cells showed marked changes in the bone marrow of pristane-treated BALB/c PT mice. There was a dichotomy in the effect between the early forms expressing TdT and the later forms. Early and intermediate pro-B cells showed a protracted increase in numbers to approximately twice normal levels at one week and one month after pristane injection, decreasing towards normal values again by 3 months (Table 2). In contrast, the late pro-B cells, pre-B cells and B cells all declined in numbers during the first week to 25-60% normal values, and remained reduced in population sizes for the 3 month period.

The number of cells accumulating in metaphase at a fixed interval (2h 40m) after the injection of vincristine sulfate provides an index of the magnitude of cell production within phenotypically-defined precursor B cell populations (Opstelten & Osmond 1985, Park & Osmond 1987, 1989a). By this technique, the early and intermediate pro-B cells showed more than 2.5 times the normal rate of accumulation of cells in metaphase at 1-4 weeks after pristane administration to BALB/c PT mice, while the values for late pro-B cells were also increased more than two-fold (data not shown). Thus, the observed expansion in the population size of early and intermediate pro-B cells represented a true increase in cell production and in the flow of cells per unit time through these differentiation compartments.

To examine the possible genetic restriction of the pristane-induced effect, a similar analysis has been performed on DBA/2 strain mice. These mice showed an essentially similar pattern of response. Early and intermediate pro-B cells were elevated in population size and in total number of cells entering mitosis per unit time, while later stages tended to be depressed (Table 3).

Table 3. Population of B lineage cells in bone marrow of pristane-conditioned DBA/2 mice [ab]

Cell population		Population size (number of cells/femur as % normal value) at intervals of: 7 days	28 days	82 days
Pro-B cells:	early	141	177	90
	intermediate	267	184	90
	late	59	25	14
Pre-B cells		39	34	56
B lymphocytes		38	34	53

[ab] See Table 2

PRECURSOR B CELLS IN BONE MARROW OF MALARIA-INFECTED MICE

To induce a chronic malarial infection, 5 wk-old C3H/HeJ mice (Jackson Labs, Bar Harbor, ME) were given a single intraperitoneal injection of 10^7 mouse RBC infected with plasmodium yoellii (non-lethal strain). Samples of infected mouse RBC, kindly provided by Dr. H. Shear, Department of Parasitology, New York University, NY., were passaged once in C3H/HeJ mice before use.

Infected mice showed a well-marked but transient parasitemia. Approximately 30% RBC in peripheral blood were infected at 2-3 wk, but no parasites were detectable in the blood by 4 wk onwards. On the other hand, there was a protracted hepatosplenomegaly. Spleen weight and cellularity increased ten-fold in the first wk, after which both these values and the liver weight remained at approximately double normal levels throughout.

In the bone marrow, the early and intermediate pro-B cells had increased in population size by 1 wk and reached 2.3-3.5 times normal values at 1 month (Table 4). The intermediate pro-B cells were still twice as numerous as normal after nearly 6 months (158d). The TdT^+ cells showed an absolute increase in production, the total number of cells entering mitosis per unit time being approximately 2.5 times normal values (data not shown). The later proliferative precursor B cell stages, on the other hand, were all reduced in number at 1 wk and showed no comparable increases thereafter (Table 4).

Table 4. Population size of B lineage cells in bone marrow of malaria-infected C3H/HeJ mice[ab]

Cell population		Population size (number of cells/femur as % normal value) at intervals of:		
		7 days	29 days	158 days
Pro-B cells:	early	136	227	90
	intermediate	127	350	210
	late	60	106	121
Pre-B cells		53	100	69
B lymphocytes		75	183	85

[a] Mice were given 10^7 mouse RBC infected with plasmodium yoellii (non-lethal strain) ip.
[b] See Table 2.

PRISTANE- AND MALARIA-INDUCED CHANGES IN PRECURSOR B CELL PROLIFERATION IN RELATION TO B CELL NEOPLASIAS

The two widely disparate agents, pristane oil and malaria parasites can produce a basically similar end-result on the B cell lineage, viz malignant neoplasia characterised by a c-myc translocation. This raises the possibility that pristane and malaria may share a common oncogenesis-initiating mechanism. Despite their physical and biological differences, the response to these two agents shows one particular feature in common. They both produce massive recruitment of macrophage populations which phagocytose extensively and show evidence of prolonged activation and cytokine production. Our previous work has established that various foreign agents can stimulate primary B cell genesis in the bone marrow of mice as a result of macrophage activation in the spleen. We have proposed the hypothesis, therefore that conditions of chronic macrophage activation may be associated with chronic stimulation of precursor B cells in the bone marrow and that this may predispose to the initiation of the oncogenic process (Osmond Park & Jacobsen 1988).

The present work has revealed that pristane treatment and malaria infection both do exert marked effects on the central genesis of B cells in mouse bone marrow. Despite the disparity in the nature of the two stimulatory agents, their effects on the population dynamics of precursor B cells are remarkably similar. The earliest defined stages of B cell genesis, pro-B cells expressing TdT, show sustained increases in the level of cell production and in the total number of cells flowing through cell cycle per unit time. The subsequent stages of development, late pro-B cells and pre-B cells, on the other hand, show only normal or decreased levels of production. Both the effects on pro-B cells and those on later differentiation stages persist for substantial lengths of time. The increased cell production at the early and intermediate pro-B cell stages would normally be associated with a comparable increase in production at the subsequent stages of development. Such is the case during the regenerative wave of B cell genesis after sublethal γ-irradiation (Park & Osmond 1989b). The observed dichotomy in precursor B cell production levels in both the pristane-treated and malaria-infected mice suggests that the elevated degree of cell production at the early and intermediate pro-B cell stages is followed by an unusually high degree of cell loss at subsequent stages of development.

A substantial loss of cells is a feature of normal B cell genesis in both the mouse (Opstelten & Osmond 1985, Park & Osmond 1989a) and rat (Deenen, Hunt & Opstelten 1986). The cell loss is thought to reflect the elimination of cells that had undergone a defective rearrangement of Ig genes. In pristane-conditioned and malaria-infected mice the major stimulatory effect is evident at the stage of pro-B cells which, as evidenced by their expression of TdT, appear to be in the process of rearranging V_H genes. This process is normally prone to genetic errors. The results raise the possibility that the stimulated levels of pro-B cell proliferation in pristane-conditioned and malaria-infected mice are associated with an increased incidence of aberrant cells, many of which are then eliminated from the bone marrow. We have found that in scid mutant mice the B lineage cells in the bone marrow are deleted completely as they reach the late pro-B cell stage of development, as a result of the faulty V_H gene rearrangements occurring at the earlier pro-B cell stages (Osmond et al 1990). In both pristane-conditioned and malaria-infected mice, however, many B cells are still produced. It may be speculated that a proportion of these cells may also carry genetic errors which, while not being lethal to the cell, may constitute the first of a series of oncogenic events. During the prelymphomatous phase of Eμ-myc transgenic mice there is also a discrepancy between the elevated production of precursors and the apparent output of B cells (Harris et al 1988, D.G. Osmond, C. Sidman, K. Jacobsen & S. Rico-Vargas, unpublished data).

While pristane treatment and malaria infection may share a common mechanism predisposing to the onset of B cell neoplasia in the bone marrow, further oncogenic steps would be responsible for the development of clonal malignancies "fixed" at the particular phenotypic stages characteristic for each condition, ie. PCT associated with intense plasmacytopoiesis in pristane-induced granuloma and B cell lymphoma related to EBV infection in human endemic malaria. The similarity in the bone marrow B lineage responses to pristane observed in 2 mouse strains that were PCT-susceptible (BALB/c PT) and PCT-resistant (DBA/2), suggests that the genetic restriction of PCT induction may operate at the level of the secondary oncogenic events or of tumor cell immunosurveillance.

SUMMARY AND CONCLUSION

Two widely different agents implicated in the etiology of neoplasias of the B cell lineage, pristane and malaria, have both been found to produce a prolonged increase in the level of proliferative activity and cell production by early B lymphocyte precursor cells in mouse bone marrow. This apparently leads to an elevated level of cell loss, suggesting the production of many aberrant early cells. The mechanism and significance of this effect remain to be determined. However, the present findings focus attention on the early stages of B cell genesis in the bone marrow as possible target cells for the initiation of genetic events leading to neoplasia. Together with previous work, the results suggest that pathologically elevated levels of macrophage activation may play a role in predisposing to various B cell neoplasias.

Acknowledgements. Supported by the National Cancer Institute of Canada. We are most grateful to Dr. Michael Potter and Dr. Judith Wax for the supply of BALB/c PT mice and DBA/2 mice and for discussions concerning the pristane/PCT model and to Dr. Hannah Shear for the supply of plasmodium yoellii.

REFERENCES

Bertoli LF, Kubagawa H, Borzillo GV, Burrows PD, Schreeder MT, Carroll AJ, Cooper MD (1988) Bone marrow origin of a B-cell lymphoma. Blood 72:94-101.

Deenen GJ, Hunt SV, Opstelten D (1987) A stathmokinetic study of B lymphocytopoiesis in rat bone marrow: proliferation of cells containing cytoplasmic μ-chain terminal deoxynucleotidyl transferase and carrying HIS24 antigen. J Immunol 139:702-710.

Freeman RR, Parish CR (1978) Spleen cell changes during fatal and self-limiting malarial infections of mice. Immunology 35:479-484

Fulop GM, Osmond DG (1983) Regulation of bone marrow lymphocyte production. III. Increased production of B and non-B lymphocytes after administering systemic antigens. Cell Immunol 75:80-90

Fulop GM, Pietrangeli CE, Osmond DG (1986) Regulation of bone marrow production. IV. Altered kinetic steady state of lymphocyte production after chronic changes in exogenous stimuli. Exp. Hemat 14:27-34

Greaves MF (1986) Differentiation-linked leukemogenesis in lymphocytes. Science 234:697-704

Harris AW, Pinkert CA, Crawford M, Langdon WY, Brinster RL, Adams JM (1988) The Eμ-myc transgenic mouse: A model for high-incidence spontaneous lymphoma and leukemia of early B cells. J Exp Med 167:353-371

Klein G, Klein E (1986) Conditioned tumorigenicity of activated oncogenes. Cancer Res 46:3211-3224

Kunkel TA, Gopinathan KP, Dube KP, Snow ET, Loeb LA (1986) Rearrangements of DNA mediated by terminal transferase. Proc Nat Acad Sci USA 83:1867-1871

Lenoir GM, Bornkum GW (1987) Burkitt's lymphoma, a human cancer model for the study of the multistep model of cancer: proposal for a new scenario. Adv in Viral Oncology 7:173-206

Nordan RP, Potter M (1986) A macrophage-derived factor required by plasmacytomas for survival and proliferation in vitro. Science 233:566-569

Ohno S, Babonits M, Wiener F, Spiro J, Klein G (1979) Non random chromosome changes involving the Ig gene-carrying chromosomes 12 and 6 in pristane-induced mouse plasmacytomas. Cell 18:1001-1007

Opstelten D, Osmond DG (1985) Regulation of pre-B cell proliferation in bone marrow: immunofluorescence stathmokinetic studies of cytoplasmic μ chain-bearing cells in anti-IgM-treated mice, hematologically deficient mutant mice and mice given sheep red blood cells. Eur J Immunol 15:599-605

Osmond DG (1986) Population dynamics of bone marrow B lymphocytes. Immunol Rev 93:104-124

Osmond DG, Park YH (1987) B lymphocyte progenitors in mouse bone marrow. Int Rev Immunol 2:241:261

Osmond DG, Park YH, Jacobsen K (1988) B cell precursors in bone marrow: in vivo proliferation, localization, stimulation by activated macrophages and implications for oncogenesis. In: Potter M, Melchers F (eds) Current Topics in Immunology: mechanisns of B cell neoplasia, vol 141. Heidelberg: Springer, p 2

Osmond DG, Phillips RA, Kim N, Manoukian R, Jacobsen K, Rico-Vargas SA (1990) Dynamics and localization of early B progenitor cells (Pro-B cells) in the bone marrow of mice with severe combined immunodeficiency (SCID) J Cell Biochem Supp 14D, p 208

Park YH, Osmond DG (1987) Phenotype and proliferation of early precursor cells in mouse bone marrow. J Exp Med 165:444-458

Park YH, Osmond DG (1989a) Dynamics of early B lymphocyte precursor cells in mouse bone marrow: proliferation of cells containing terminal deoxynucleotidyl transferase. Eur J Immunol 19:2139-2144

Park YH, Osmond DG (1989b) Post-irradiation regeneration of early B-lymphocyte precursor cells in mouse bone marrow. Immunology 66:343-347

Pietrangeli CE, Osmond DG (1985) Regulation of B-lymphocyte production in the bone marrow: role of macrophages and the spleen in mediating responses to exogenous agents. Cell Immunol 94:147-158

Pietrangeli CE, Osmond DG (1987) Regulation of B lymphocyte production in the bone marrow: mediation of the effects of exogenous stimulants by adoptively transferred spleen cells. Cell Immunol 107:348-357

Potter M, Mushinski JF, Mushinski EB, Brust S, Wax, JS, Wiener F, Babonits M, Rapp UR, Morse HC (1987) Avian v-myc replaces chromosomal translocation in murine plasmacytomagenesis. Science 235:787-789

Potter M, Pumphrey JC, Walters JL (1972) Growth of primary plasmacytomas in the mineral oil-conditioned peritoneal environment. JNCI 49:305-308

Potter M, Wax JS (1983) Peritoneal plasmacytomagenesis in mice: Comparison of different pristane dose regimens. JNCI 71:391-395

Potter M, Wax LS, Anderson AO, Nordan RP (1985) Inhibition of plasmacytoma development in BALB/c mice by indomethacin. J Exp Med 161:996-1012

Tonegawa S (1983) Somatic generation of antibody diversity. Nature 302:575-581

Weiss L, Geduldig U, Weidanz W (1986) Mechanisms of splenic control of murine malaria: Reticular cell activation and the development of a blood-spleen barrier. Am J Anat 176:251-285

Wyler DJ (1982) Malaria: Host-pathogen biology. Rev Infect Diseases 4:785-797

Wyler DJ, Oppenheim JJ, Koontz LC (1979) Influence of malaria infection on the elaboration of soluble mediators by adherent mononuclear cells. Infect. Immunol 24:151-159

The Human CBL Oncogene

W.Y. Langdon and T.J. Blake.

Division of Human Immunology, Institute of Medical and Veterinary Science, Adelaide, South Australia 5000, Australia

INTRODUCTION

The v-cbl oncogene was identified as the transforming gene of the murine Cas NS-1 retrovirus (Langdon et al. 1989a). This novel viral oncogene transforms fibroblasts in vitro and induces pre-B, pro-B and myeloid tumors in vivo. Probes prepared from v-cbl have identified its cellular homolog (c-cbl) in mouse, rat and human DNA, and it is expressed as an 11 kb mRNA in a range of murine and human hemopoietic tumor cell lines. In normal cells highest levels of c-cbl mRNA are found in the testis and thymus (Langdon et al. 1989b). The c-cbl product is a 135 kD protein which is localized in the nucleus, as in the 100 kD gag-cbl fusion protein (Langdon et al. 1989b).

Chromosomal mapping has placed c-cbl on mouse chromosome 9, closely linked to Thy-1 (Regnier et al. 1989). The homologous region in humans is located on the long arm of chromosome 11 in a region which is altered by deletions and translocations in some acute lymphocytic leukemias (ALLs) and acute non-lymphocytic leukemias (ANLLs) (Yunis and Brunning,1986).

This report describes the cloning and characterization of the human c-cbl oncogene, its relationship to v-cbl and its alteration in the human cutaneous T cell lymphoma line, HUT 78.

Cloning and Sequencing of Human c-cbl

cDNA libraries were prepared from HUT 78 (Gazdar et al. 1980) and CCRF-CEM cells (a T cell ALL, Foley et al. 1965) using poly $(A)^+$ mRNA and a combination of random and oligo (dT) primers (Clontech Laboratories). The libraries were initially screened with v-cbl probes and subsequently with isolated fragments from c-cbl cDNA clones. The restriction maps of the largest c-cbl clones, and their relative positions to v-cbl and a murine c-cbl cDNA clone (from a 70Z pre-B cell library) are shown in Fig 1. The entire human c-cbl coding region was determined by sequencing HUT 78 clone 1 and CEM clones 32 and 42. The sequence of human c-cbl showed a short 5′ untranslated region that extends 150 bp upstream from an AUG initiation codon. It is likely that these 150 bases represent the extent of the 5′ untranslated sequence since three independent clones all terminated at the same 5′ position. The first AUG from the 5′ end of c-cbl has the Kozak consensus sequence of a G at position -3 and a G at position +4 making it highly probable that this codon is used for the initiation of c-cbl translation. The open reading frame for c-cbl extends for a further 907 amino acids (aa) until a TAG stop codon. This would result in the translation of a protein with an estimated molecular weight of a 100 kD, which is smaller than the 135 kD protein observed by the Western blotting of nuclear extracts with v-cbl antibodies (Langdon et al 1989b). However,

Current Topics in Microbiology and Immunology, Vol. 166

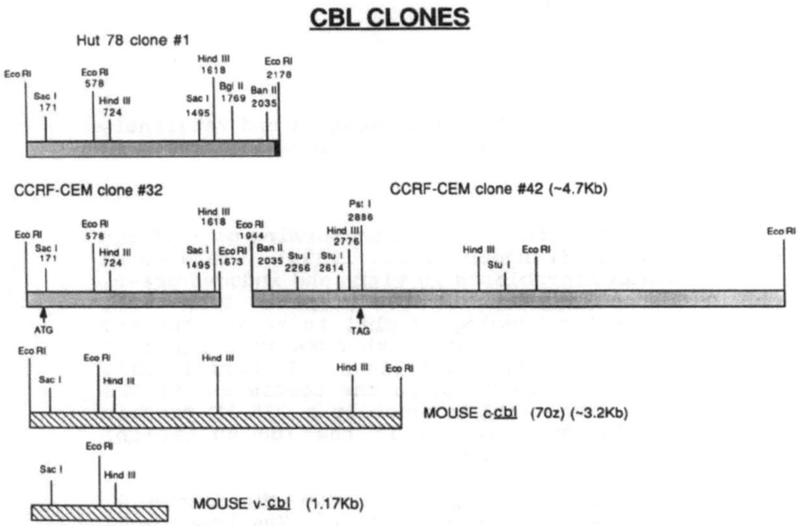

Figure 1 Restriction maps of human c-cbl clones, a murine c-cbl clone and v-cbl.

```
M A G N V K K S S G A G G G[T]G S G G S G[S]G G L I G L M K
M A G N V K K S S G A G G G[-]G S G G S G[A]G G L I G L M K

D A F Q P H H H H H H[H]L S P H P P[G]T V D K K M V E K C W
D A F Q P H H H H H H[-]L S P H P P[C]T V D K K M V E K C W

K L M D K V V R L C Q N P N V A L K N S P P Y I L D L L P D
K L M D K V V R L C Q N P N V A L K N S P P Y I L D L L P D

T Y Q H L R T[I]L S R Y E G K M E T L G E N E Y F R V F M E
T Y Q H L R T[V]L S R Y E G K M E T L G E N E Y F R V F M E

N L M K K T K Q T I S L F K E G K E R M Y E E N S Q P R R N
N L M K K T K Q T I S L F K E G K E R M Y E E N S Q P R R N

L T K L S L I F S H M L A E L K G I F P S G L F Q G D T F R
L T K L S L I F S H M L A E L K G I F P S G L F Q G D T F R

I T K A D A A E F W R K A F G E K T I V P W K S F R Q A L H
I T K A D A A E F W R K A F G E K T I V P W K S F R Q A L H

E V H P I S S G L E A M A L K S T I D L T C N D Y I S V F E
E V H P I S S G L E A M A L K S T I D L T C N D Y I S V F E

F D I F T R L F Q P W S S L L R N W N S L A V T F P G Y M A
F D I F T R L F Q P W S S L L R N W N S L A V T F P G Y M A

F L T Y D E V K A R L Q K F I H K P G S Y I F R L S C T R L
F L T Y D E V K A R L Q K F I H K P G S Y I F R L S C T R L

G Q W A I G Y V T A D G N I L Q T I P H N K P L F Q A L I D
G Q W A I G Y V T A D G N I L Q T I P H N K P L F Q A L I D

G F R E G F Y L F P D G R N Q N P D L T G L C E P T P[Q D H]
G F R E G F Y L F P D G G N Q N P D L T G L C E P T P[H F S]

[I]
[*]
```

Figure 2. Amino acid alignments between human c-cbl and murine v-cbl. Differences are shown in boxes and the human sequence is on the upper line.

the sequence contains 4 potential N-glycosylation sites that could contribute to the synthesis of a larger protein. In addition, the c-cbl sequence is rich in serine and threonine residues that may be subject to certain types of modifications such as O-linked glycosylation or phosphorylation.

Comparing the 360 aa of v-cbl with the N-terminus 360 aa of human c-cbl showed that the only differences are an additional threonine and histidine in the human gene and three substitutions (Fig 2). The three C-terminal amino acids of v-cbl showed a complete divergence from the c-cbl sequence indicating that these amino acids were derived from the Cas NS-1 pol gene (Fig 2).

The region of overlap between v-and c-cbl also contains the only putative nuclear localization signal of Lys, Lys, Thr, Lys. This is consistent with the observation that both v-and c-cbl encoded proteins are located in the nucleus (Langdon et al. 1989b).

The striking feature of the remaining 540 aa of c-cbl is the high number of proline residues. In particular, a core region of 208 aa contains 24% prolines (Fig 3) which is an equivalent proportion to that found in proline-rich transcriptional activation domains of a number of transcription factors (Mermod et al. 1989, Bohman and Tjian, 1989). This region was also found to contain a high proportion of serine and threonine residues (18%) which are also present in proline-rich transcription factors. Two acidic domains were observed either side of the proline-rich region (Fig 3).

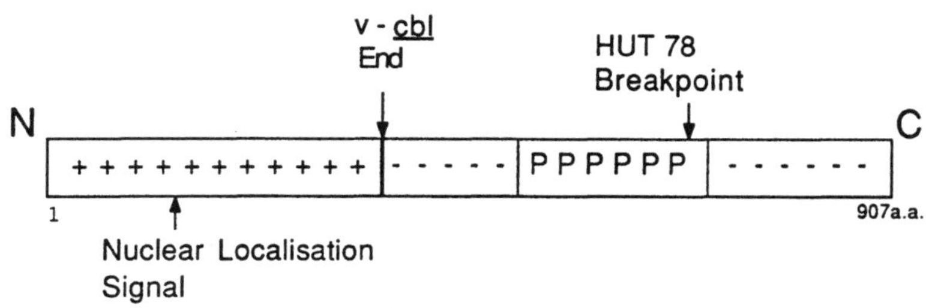

Figure 3. Amino acid domains of human c-cbl. +/-: regions with net positive or negative charges. P: Proline-rich region (24% over 208 aa, also serine and threonine rich - 18% over 208 aa).

Altered CBL in HUT 78

The sequencing of c-cbl clones from the HUT 78 and CEM libraries revealed a divergence of sequence at a position 663 aa from the AUG initiation codon. This sequence is shown in Fig 4 and continues for 11 aa at the 3′ end of HUT 78 clone 1. This sequence was not seen in CEM clones, and on the basis that the CEM clones matched the 70Z mouse c-cbl clone by restriction mapping and hybridization it would appear that HUT 78 has an altered c-cbl gene.

	Cys	Asp	His	Pro	Lys	Ile	Lys	Pro	Ser	Ser	Ser	Ala	Asn	Ala	Ile
CCRF CEM c-cbl	TGT	GAC	CAC	CCC	AAA	ATC	AAA	CCT	TCC	TCA	TCT	GCC	AAT	GCC	ATT
	TGT	GAC	CAC	CCC	TCC	TCT	TCT	TTC	TCT	TCC	TCA	GTG	TCC	ACA	ATA
HUT 78 c-cbl	Cys	Asp	His	Pro	Ser	Ser	Ser	Phe	Ser	Ser	Ser	Val	Ser	Thr	Ile

BREAKPOINT (between Pro and Ser)

Figure 4. Sequence bordering HUT 78 breakpoint.

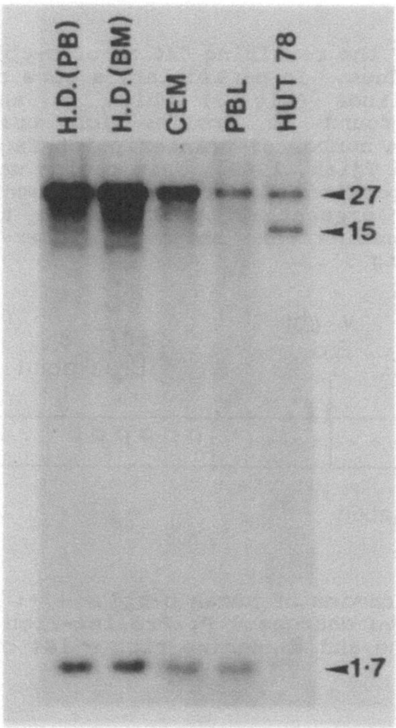

Figure 5. Southern blot of Human DNAs probed with c-cbl probe #42 (2.6 kb EcoRI fragment, Fig. 1). The DNAs were digested with EcoRI. HD is DNA from a patient with a T cell ALL with an 11q23 deletion.

To examine whether this change in the c-cbl sequence is reflected as an alteration in c-cbl's genomic structure, DNA was extracted from HUT 78 and CEM cells, digested with Eco RI and probed with the 2.6 kb Eco RI fragment from CEM clone 42 (see map, Fig 1). This c-cbl fragment spans the position where the sequence change occurs and Southern blotting revealed an additional band of 15 kb that was only observed in HUT 78 DNA (Fig 5). In CEM cells and 20 other human DNA samples the Eco RI fragments were 27 kb and 1.7 kb (Fig 5 and data not shown). The only other DNA that showed an altered pattern was from a T cell ALL patient who had an 11q23 deletion (Fig 5., H.D. - PB and BM). No differences were found between HUT 78 DNA and other human DNA samples when other enzymes such as Pst I, Kpn I, Bgl II, Hind III, Sac I and Xba I were used (data not shown). It is therefore presumed that the large size of the endogenous c-cbl Eco RI fragment made it possible to detect the changes in the HUT 78 and H.D. DNAs. Although it is unlikely that the 15 kb Eco RI fragment seen in HUT 78 is due to an RFLP, the cloning of this genomic fragment and the isolation of additional HUT 78 cDNA clones (which provide more altered sequence) will help eliminate this possibility. Larger HUT 78 cDNA clones with altered c-cbl sequence will also be required to determine if this form encodes a truncated cbl protein.

How is an altered CBL gene transforming?

The sequence data and nuclear localization of the c-cbl product is suggestive that it is functionally involved in transcriptional activation. It would therefore seem probable that the deletions in v-cbl and HUT78 c-cbl would result in a perturbation of this function. A possible transforming mechanism is that the altered cbl product competes with normal cbl and this represses the activation of cbl-responsive genes. For this to result in cellular transformation it is probable that c-cbl-responsive genes are involved in negative growth regulation. A similar mechanism has already been shown for v-erbA which competes with the thyroid-hormone receptor and acts as a repressor of thyroid hormone responsive genes (Damm et al. 1989). It will be of interest to experimentally test this hypothesis for c-cbl.

REFERENCES

Bohmann, D and Tjian R (1989) Biochemical analysis of transcriptional activation by jun. Differential activation of c-and v-jun. Cell 59, 709-717

Damm, K, Thompson CC and Evans RM (1989). Protein encoded by v-erbA functions as a thyroid-hormone receptor antagonist. Nature 339, 593-597

Foley, GE, Lazarus H, Farber S, Uzman, BG, Boone, BA and McCarthy, RE (1965) Continuous culture of human lymphoblasts from peripheral blood of a child with acute leukemia. Cancer 18, 522-529

Gazdar, AF, Carney, DN, Bun, PA, Russell, EK, Jaffe, ES, Schechter, GP, and Guccion, JG (1980) Mitogen requirements for the in vitro propagation of cutaneous T-cell lymphomas Blood 55, 409-417

Langdon, WY, Hartley, JW, Klinken, SP, Ruscetti, SK and Morse, HC III (1989a). v-cbl, an oncogene from a dual-recombinant retrovirus that induces early B-lineage lymphomas. Proc. Natl. Acad. Sci. USA 86, 1168-1172

Langdon, WY, Hyland, CD, Grumont, RJ and Morse, HC III (1989b) The c-cbl proto-oncogene is preferentially expressed in thymus and testis tissue and encodes a nuclear protein J. Virol. 63, 5420-5424

Mermod, N, O'Neill, EA, Kelly, TJ and Tjian, R (1989) The proline-rich transcriptional activator of CTF/NF-1 is distinct from the replication and DNA binding domain. Cell 58, 741-753

Regnier, DC, Kozak, CA, Kingsley DM, Jenkins, NA, Copeland, NG, Langdon, WY and Morse, HC III (1989). Identification of two murine loci homologous to the v-cbl oncogene. J. Virol. 63, 3678-3682

Yunis, JJ and Brunning, RD (1986) Prognostic significance of chromosomal abnormalities in acute leukemias and myelodysplastic syndromes. Clin. Haematol. 15, 597-619.

Lymphoid Tumorigenesis by v-*abl* and BCR-v-*abl* in Transgenic Mice

A.W. Harris, M.L. Bath, H. Rosenbaum, J. McNeall, J.M. Adams and S. Cory

The Walter and Eliza Hall Institute of Medical Research, Melbourne 3050, Australia

INTRODUCTION

Two kinds of fusion gene derivatives of the *abl* oncogene have been associated with lymphoid malignancy (reviewed by Ramakrishnan and Rosenberg 1989). The *gag-abl* (v-*abl*) gene of Abelson mouse leukemia virus enables the virus to rapidly elicit pre-B or T cell lymphomas, depending on the route of virus administration (reviewed by Whitlock and Witte, 1985), and to accelerate the initiation of plasmacytomas in mice of the susceptible BALB/c strain given intraperitoneal injections of pristane (Potter et al 1973). Another variant of *abl*, a *BCR-ABL* fusion gene, is generated by a 9;22 chromosome translocation and is associated with some cases of human acute lymphoblastic leukemia, as well as most cases of chronic myeloid leukemia. To test their oncogenic potential *in vivo*, we have introduced the v-*abl* gene and *bcr-abl* homologues into the germline DNA of mice. To direct expression predominantly to lymphoid cells, the transgene constructs included the transcriptional enhancer from the J_H-Cμ intron, which we term Eμ; the constructs also contained an SV40 or an *Igh*-V promoter. The derived transgenic strains were maintained as (C57BL/6 x SJL)FE2 transgene heterozygotes by serially crossing transgenic animals with normal (C57BL/6 x SJL)F1 hybrids. Here we review our recently published work (Hariharan et al 1989; Rosenbaum et al 1990) and discuss some additional findings on the processes of lymphoid tumorigenesis in these strains.

PLASMACYTOMAGENESIS IN EμSV-v-*abl* MICE

Plasmacytomas in Three Transgenic Strains

Four strains expressing the EμSV-v-*abl* construct were established and found to be highly predisposed to lymphoid malignancy. One developed lymphomas and will be described in a subsequent section of this paper. The other three, designated v-*abl* 2, 8 and 40 developed exclusively plasmacytomas (Rosenbaum et al 1990). From about 3 months of age, these tumors arose sporadically in the three strains, the cumulative incidence reaching 60% (v-*abl* 8 and 40) or 30% (v-*abl* 2) at 12 months (Fig. 1). Similar kinetics and incidence occur in BALB/c mice given intraperitoneal injections of pristane (Potter et al 1984).

In the v-*abl* transgenic strains there was considerable variation in the gross presentation of pathologic tissues in autopsies of terminally ill animals. The major tumor masses were most often located in the mesentery and the mesenteric and pancreatic lymph nodes; the axillary and brachial lymph nodes (unilateral); under the roof of the skull; or in the epidural space of the vertebral column. In addition, most sick animals showed extensive multicentric invasion of bone marrow and some invasion of the splenic red pulp on histologic examination.

Current Topics in Microbiology and Immunology, Vol. 166

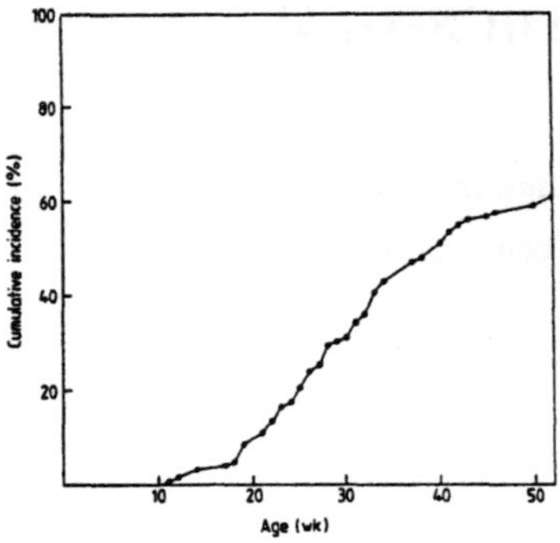

Fig. 1. Kinetics of tumorigenesis in v-*abl* 40 mice. The plot records the age at which animals from an initial group of 125 became ill with malignant disease verified by autopsy and in representative cases by histology and serum electrophoresis.

Tumor cell morphology (Fig. 2) ranged from immunoblastic lymphoma (e.g. Fig. 2E,F,G) to classic plasmacytoma (e.g. Fig. 2A,H), and mitotic frequencies ranged from 3 to 20 mitoses per mm^2 of tumor section amongst 44 cases examined.

The malignant character of v-*abl* tumors was manifest by their ability to grow when transferred into normal histocompatible mice. The prominent homogeneous immunoglobulin species present in the serum of primary (Fig. 3) and transplanted hosts provided the key evidence that the tumors were plasmacytomas. Clonal paraproteins of IgA and IgG (mostly IgG_{2b}) type were about equally frequent. Of 49 proteins analyzed, none was IgM, IgD or IgE. The plasmacytomas induced by pristane or by pristane and Abelson virus are more heavily biased toward the IgA class (Potter et al 1973; 1984). Some of them cause the host kidney to develop large intratubular eosinophilic casts and, likewise, about 20% of the v-*abl* transgenic tumors also caused "myeloma kidney" (Fig. 2I). Amyloidosis, reported not to occur with oil-induced BALB/c tumors (Potter 1972), has been observed in the spleen (Fig. 2J) in about one-fifth of the v-*abl* animals with plasmacytoma.

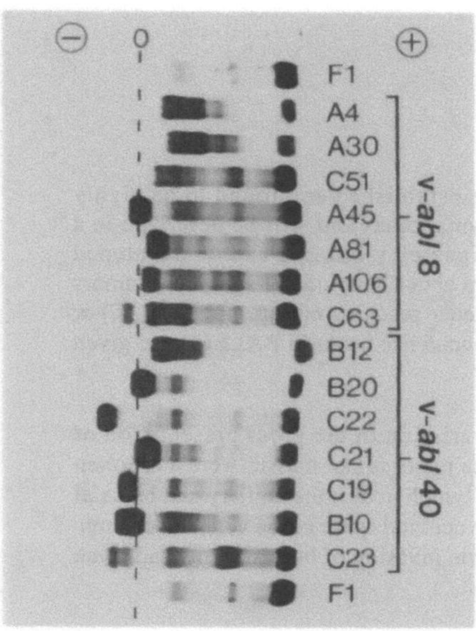

Fig. 3. Clonal immunoglobulins produced by plasmacytoma-bearing EμSV-v-*abl* mice. Serum proteins are shown after electrophoresis in thin-layer agarose gel by Coomassie staining. v-*abl*8-C63, v-*abl*40-B10 and C-23 each contain two clonal immunoglobulins which were typed as IgA and IgG (Rosenbaum et al 1990). F1 indicates serum from normal BJF1 hybrid mice.

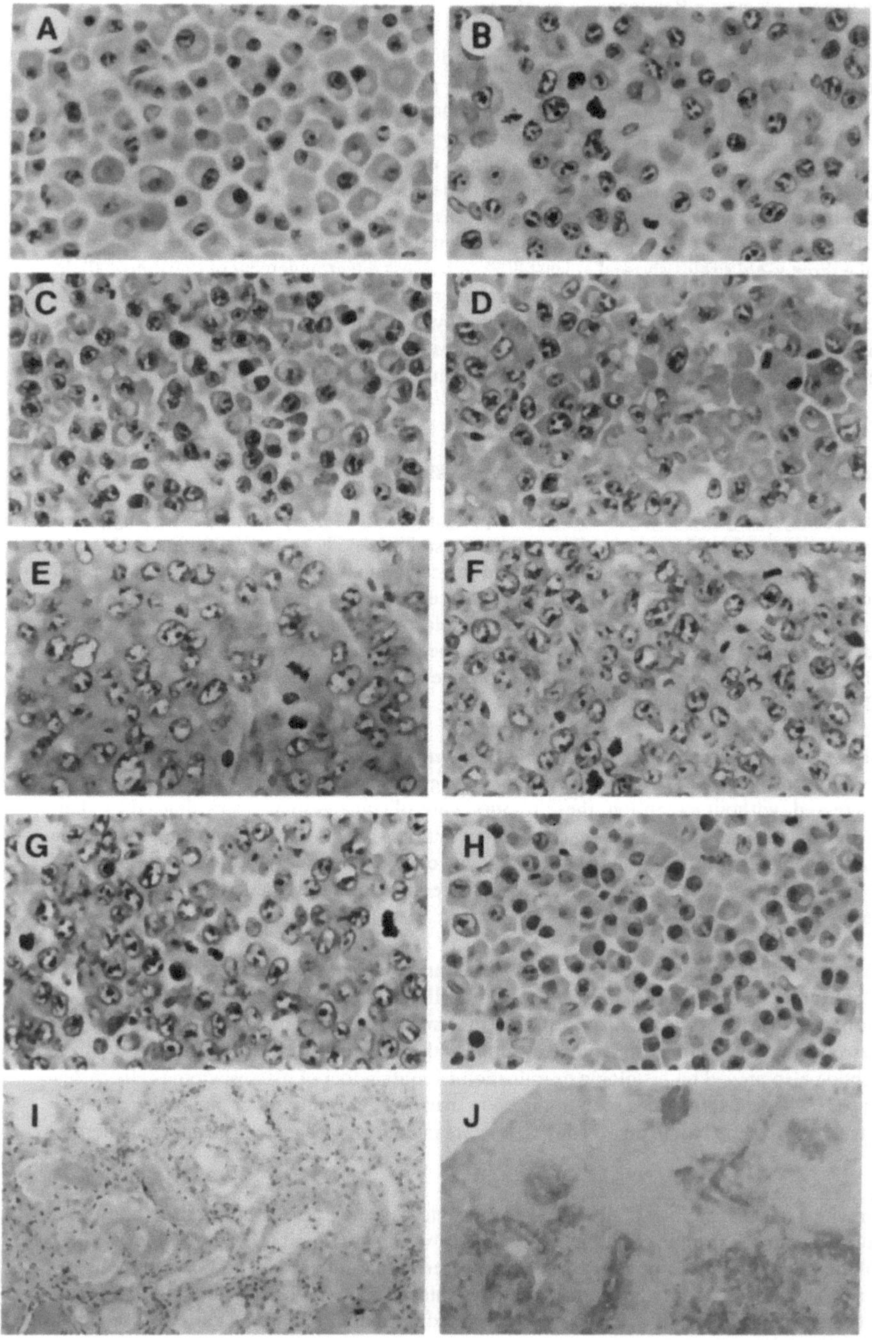
A
B
C
D
E
F
G
H
I
J

The Plasmacytomas Contain a Rearranged c-*myc* Gene

The v-*abl* transgene was expressed in the lymphoid tissues of healthy young mice but no transplantable cells could be detected. Nor was any perturbation discernible in the various lymphocyte subsets. Thus, by itself, EμSV-v-*abl* is not tumorigenic and does not promote lymphoid hyperplasia. Since the plasmacytomas arose sporadically and were of clonal origin, they must have arisen from single Eμ-v-*abl* expressing cells that had undergone some low frequency genetic alteration(s). A clear candidate would be a chromosome translocation involving the c-*myc* gene, because such translocations are found regularly in the plasmacytomas induced by Abelson virus and oil in BALB/c mice (Ohno et al 1984). Indeed, Southern blot analysis of restricted DNA from 32 tumors of the three v-*abl* strains showed at least 80% to possess a rearrangement of one c-*myc* allele (Rosenbaum et al, 1990). At least one half of these had recombined c-*myc* to the *Igh-2*(α) gene, a common target of *myc* translocation in BALB/c plasmacytomas (reviewed by Cory 1986). Thus, a spontaneous chromosomal translocation that deregulates c-*myc* expression in a v-*abl*-expressing, immunoglobulin-secreting cell appears to be the rate-limiting step of plasmacytomagenesis in the transgenic model.

Plasmacytomagenesis is Accelerated by an Eμ-*myc* Transgene

The high frequency of *myc* rearrangements in the spontaneous Eμ-v-*abl* plasmacytomas suggested that the two oncogenes acted in concert. We sought to test this hypothesis by mating Eμ-v-*abl* mice with our Eμ-*myc* strain. Synergy would be expected to accelerate tumorigenesis. But what type of tumor would eventuate? Given the ability of Eμ-*myc* mice to develop plasmacytomas when injected with pristane (Harris et al 1988b), it seemed likely that plasmacytomas would be produced. On the other hand, in view of the lymphomagenic potency of Abelson virus and the invariant occurrence of pre-B and B lymphomas in Eμ-*myc* mice in the absence of pristane, it also seemed possible that pre-B or B lymphomas would instead be obtained.

Three types of cross between Eμ-*myc* and v-*abl* 40 or v-*abl* 8 were performed and the results were qualitatively uniform (Fig. 4). All of the doubly transgenic progeny became ill with tumors by 11 wk of age and, without exception, these tumors were plasmacytomas, as evidenced by their histologic appearance and by the appearance of (often multiple) homogeneous immunoglobulins in terminal serum samples (Fig. 5). Thus, intriguingly, the oncogenic synergy between v-*abl* and *myc* appeared to be stage specific in the B cell lineage, effective in plasma cells but not in pre-B or B cells (see below).

Special features of the tumor pathology in the doubly transgenic animals offered some clues to the origin of the tumors. In virtually all cases terminal illness was abrupt, precipitated by bowel obstruction resulting from intussusception of the small intestine over multiple masses of plasmacytoma in the lamina propria. Regions of the bowel not directly involved in the intussusception also proved on histologic examination to contain foci of plasmacytoma cells. A few such cells (<10%) were sometimes seen in the bone marrow. The only other major site of tumor growth was the mesenteric lymph node. Thus the tumors in doubly transgenic mice appear to derive

Fig. 2. (previous page) Histology of disease in plasmacytomagenic EμSV-v-*abl* strains. Shown are sections of tumor tissue from v-*abl* 8 (A,B,C,E and H), v-*abl* 40 (D and G) and v-*abl* 2 (F) x450; kidney tubules swollen by eosinophilic casts, from a terminal tumor-bearing v-*abl* 40 mouse (I) x112; extensive deposits of amyloid surrounding the splenic white pulp zones in a v-*abl* 8 mouse with advanced plasmacytoma (J) x22. Tissues were fixed in Bouin's solution, embedded, sectioned at 3-4 μm and stained with hematoxylin and eosin.

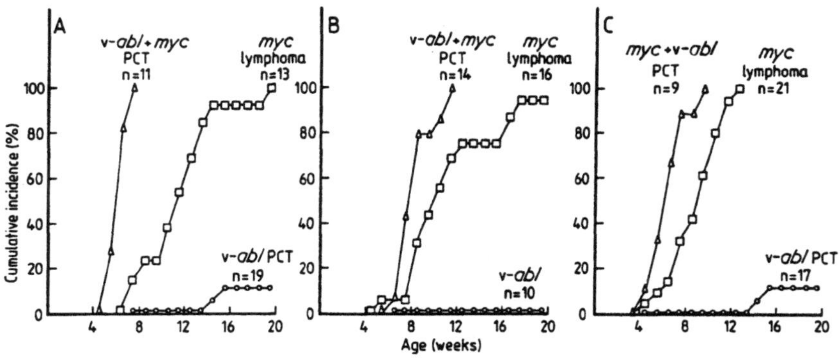

Fig. 4. Accelerated plasmacytomagenesis in EμSV-v-*abl*/Eμ-*myc* doubly transgenic mice. Results are shown for crosses v-*abl*40F x Eμ-*myc*M (A), v-*abl*8 x Eμ-*myc* (B) and Eμ-*myc* x v-*abl*40 (C). Animals were recorded as contributing to tumor incidence when they showed general signs of illness. Tumor types were diagnosed by autopsy, histology and serum electrophoresis. Progeny that inherited neither transgene remained healthy.

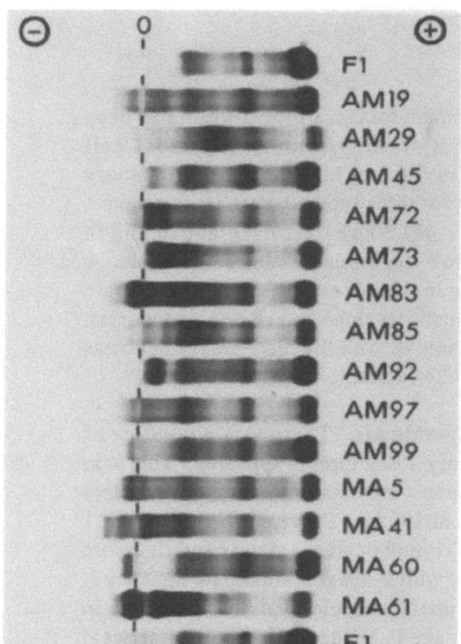

Fig. 5. Immunoglobulins produced by tumors of v-*abl*/*myc* mice. Shown are Coomassie-stained proteins from serum after electrophoresis in thin-layer agarose gel. Serum samples were from representative tumor-bearing doubly transgenic mice generated by the three crosses shown in Fig. 4. F1 indicates serum from normal BJF1 hybrid mice showing a faint smear of polyclonal immunoglobulins in the left-hand one-third of the gel track.

selectively from gut-associated plasma cells. Moreover, their multifocal distribution along the small intestine and the appearance of multiple abnormal immunoglobulins in the serum imply an oligo- or polyclonal initiation of tumors, perhaps triggered by immune responses against antigens associated with food and intestinal microorganisms. Such a scenario suggests that deliberate immunization of very young v-*abl/myc* mice might precipitate tumors whose immunoglobulin products would have specificity for the immunizing antigen.

Plasmacytomas Suppress Overgrowth of pre-B Cells in v-*abl/myc* Mice

The exclusive development of plasmacytomas in the v-*abl/myc* mice was particularly intriguing because some of these tumors arose at times at which lymphoblastic lymphomas would have been expected in the *myc* animals (Fig. 4B and C). We examined lymphoid tissues of the plasmacytoma-bearing animals for accumulations of lymphoblasts, but found none. The bone marrow of two young v-*abl/myc* mice, taken before the onset of plasmacytoma, proved to contain the excess of large $B220^+$,Ig^- pre-B cells (50-60% of nucleated cells) characteristic of the pre-lymphomatous Eμ-*myc* mouse (Langdon et al 1986). Thus, the v-*abl* transgene was not antagonistic to Eμ-*myc*-driven overproliferation. On the other hand, the pre-B cells had dropped to below 10% of bone marrow cells when v-*abl/myc* animals developed their terminal gut-associated plasmacytomas. It appears, then, that the plasma cell tumors somehow suppress the proliferation of pre-B cells, thereby precluding the lymphomagenic process observed in littermates that had inherited only the *myc* transgene. The suppression might represent an end-product feedback inhibition of B lineage development, mediated by a factor produced by immunoglobulin-secreting cells.

LYMPHOMAGENESIS IN EμSV-v-*abl* AND EμV_H-*BCR*-v-*abl* MICE

Pre-B and T Cell Lymphomas in the v-*abl* 1 Strain

While three strains of v-*abl* mice produced exclusively plasmacytomas, a fourth strain has been developed which carries the same transgene construct but develops only pre-B and T cell lymphomas. This strain, designated v-*abl* 1, was segregated from the progeny of a primary transgenic mouse that had integrated the EμSV-v-*abl* construct at three separate sites in its genome. One of the other segregants was the plasmacytomagenic v-*abl* 2 line described above and in Rosenbaum et al (1990). Our initial survey of tumorigenesis in v-*abl* 1 mice is almost complete. A group of 73 observed for almost a year has shown a tumor incidence of 70%, but not a single plasmacytoma has emerged (Table 1). Instead, nine of the animals developed pre-B cell lymphomas, immunophenotyped as $Thy1^-$, $B220^+$, Ig^-, and another 42 became ill with thymic T cell lymphomas with the cell surface phenotype $B220^-$, Ig^-, $Thy1^+$, $CD8^+$ and CD4-variable.

Curiously, the pre-B cell tumors developed only in young animals. There was a window of susceptibility from birth to about 10 wk of age, after which only T cell tumors appeared. This is in sharp contrast to pre-B cell tumorigenesis in the Eμ-*myc* mouse which continues at a more or less constant rate until all the animals have succumbed (Harris et al 1988a). Perhaps when the v-*abl* 1 mouse reaches adulthood the proliferation of pre-B (or pro-B) lymphocytes is slower and therefore the probability that a genetic alteration will occur to cooperate with v-*abl* and initiate malignancy is reduced. If a similar slow-down occurs in conventional animals, it could explain the decline in susceptibility to Abelson viral tumorigenesis that occurs with age in B6 and SJL mice (Risser et al 1978). The BALB/c strain does not show such a decline and it would be interesting to test whether the v-*abl* 1 transgene would evoke many more pre-B tumors when crossed into that strain.

There is some evidence that the most efficient target cell for Abelson viral transformation is a more primitive cell than a pre-B lymphocyte (Tidmarsh et al 1989; Rosenbaum et al 1990). Perhaps susceptibility to lymphoma initiation by v-*abl* 1 is closely coupled to the size and proliferation rate of that particular cell compartment. The difference between v-*abl* 1 and the other three v-*abl* strains might be that, due to particular genomic integration sites, only the v-*abl* 1 strain expresses its transgene in those primitive cells. This would still leave unanswered, however, the question of why some of the v-*abl* 1 mice that escape early lymphomas do not then develop plasmacytomas.

Table 1. Spontaneous tumors of v-*abl* 1 and *BCR*-v-*abl* 32 mice[a]

Strain	Number of mice	Tumor incidence (%) Pre-B lymphomas	T lymphomas
v-*abl* 1	73	12	58
BCR-v-*abl* 32(M)[b]	123	11	55
BCR-v-*abl* 32 (F)[c]	49	70	37

[a]monitored for tumor development to 10-12 months of age
[b]progeny of transgenic father and normal BJF1 mother
[c]progeny of transgenic mother and normal BJF1 father. Some mice developed pre-B and T lymphomas simultaneously.

Pre-B and T Cell Lymphomas in the *BCR*-v-*abl* 32 Strain

In order to test the biological activity of *BCR-ABL* fusion genes, we took advantage of a construct having an 827 amino acid-encoding human *BCR* cDNA segment coupled to the *abl* portion of *gag*-v-*abl* derived from the p120 strain of Abelson virus (Hariharan et al 1988), a gene designated *BCR*-v-*abl* to distinguish it from the *BCR-ABL* genes found in human leukemias. The EμSV-*BCR*-v-*abl* transgene elicited some T cell and pre-B cell lymphomas, but their incidence was low; most primary transgenic mice and most of their transgenic progeny remained healthy and did not detectably express the transgene (Hariharan et al 1989). We have now succeeded, however, in deriving a highly tumorigenic strain. Among the descendants of one primary mouse, *BCR*-v-*abl* 32, all three transgenic members of one litter developed lymphomas at an early age (Hariharan et al 1989), in contrast to large numbers of their cousins. The data implied the existence within the founder of two independent inserts of the transgene, an inactive one transmitted at a high frequency and an active one transmitted at low frequency. By extensive test-breeding over subsequent generations, we isolated a line bearing only the active insert. Curiously, the high-incidence tumorigenesis is modulated by a gender-dependent reversible imprinting effect akin to that described for several other transgenes (reviewed by Solter 1988). In this *BCR*-v-*abl* 32 strain, the progeny of transgenic males developed lymphomas to an incidence of 66% by about 20 wk of age, while all animals bearing a maternally derived transgene developed lymphomas within the first 10 wk of life. There was an accompanying difference in the balance between pre-B and T lymphomas (Table 1). Tumorigenesis in the animals carrying the male-transmitted transgene exhibits strikingly similar kinetics and tumor phenotypes to that found with the v-*abl* 1 strain, implying a common mechanism in the initiation of lymphoid malignancy.

These results show that *BCR*-v-*abl* can be a very potent contributor to lymphocyte transformation. In fact, the maternally transmitted *BCR*-v-*abl* 32 transgene is the most rapidly lymphomagenic transgene reported to date, producing tumors at a rate otherwise seen only as a result of the action of two cooperating oncogenes, such as *myc* and *ras* (Langdon et al 1989).

CONCLUDING REMARKS

By using fusion variants of the *abl* oncogene we have derived strains of transgenic mice with two distinctive and exclusive patterns of lymphoid tumorigenesis. One group gives rise to plasmacytomas in concert with a deregulated c-*myc* gene while the other develops pre-B and T cell lymphomas. The reason for this difference is presently obscure, but it presumably must relate to the level or the cell type specificity of transgene expression. Future experiments will seek to define these parameters and clarify the mechanisms by which the modified *abl* genes contribute to lymphoid cell malignancy.

ACKNOWLEDGEMENTS

We thank Kim Patane for animal husbandry, Steven Mihajlovic for histological preparation and Iras Collins for preparing this manuscript. This work was supported by the National Health and Medical Research Council of Australia and the US National Cancer Institute (CA43540).

REFERENCES

Cory S (1986) Activation of cellular oncogenes in hemopoietic cells by chromosome translocation. Adv Cancer Res 47:189-234

Hariharan JK, Adams JM, Cory S (1988) *bcr-abl* oncogene renders myeloid cell line factor independent: potential autocrine mechanism in chronic myeloid leukemia. Oncogene Res 3:1-13

Hariharan IK, Harris AW, Crawford M, Abud H, Webb E, Cory S, Adams JM (1989) A *bcr*-v-*abl* oncogene induces lymphomas in transgenic mice. Mol Cell Biol 9:2798-2805

Harris AW, Pinkert CA, Crawford M, Langdon WY, Brinster RL, Adams JM (1988a) The Eμ-*myc* transgenic mouse: a model for high-incidence spontaneous lymphoma and leukemia of early B cells. J Exp Med 167:353-371

Harris AW, Langdon WY, Alexander WS, Hariharan IK, Rosenbaum H, Vaux D, Webb E, Bernard O, Crawford M, Abud H, Adams JM, Cory S (1988b) Transgenic mouse models for hematopoietic tumorigenesis. Curr Topics Microbiol Immunol 141:83-93

Langdon WY, Harris AW, Cory S, Adams JM (1986) The c-*myc* oncogene perturbs B lymphocyte development in Eμ-*myc* transgenic mice. Cell 47:11-18

Langdon WY, Harris AW, Cory S (1989) Acceleration of B-lymphoid tumorigenesis in Eμ-*myc* transgenic mice by v-H-*ras* and v-*raf* but not v-*abl*. Oncogene Res 4:253-258

Ohno S, Migita S, Wiener F, Babonits M, Klein G, Mushinski JF, Potter M (1984) Chromosomal translocations activating *myc* sequences and transduction of v-*abl* are critical events in the rapid induction of plasmacytomas by pristane and Abelson virus. J Exp Med 159:1762-1777

Potter M (1972) Immunoglobulin-producing tumors and myeloma proteins of mice. Physiol Rev 52:631-719

Potter M, Sklar MD, Rowe WP (1973) Rapid viral induction of plasmacytomas in pristane-primed BALB/c mice. Science 182:592-594

Potter M, Wiener F, Mushinski JF (1984) Recent developments in plasmacytomagenesis in mice. Adv Viral Oncol 4:139-162

Ramakrishnan L, Rosenberg N (1989) *abl* genes. Biochim Biophys Acta 989:209-224

Risser R, Potter M, Rowe WP (1978) Abelson virus-induced lymphomagenesis in mice. J Exp Med 148:714-726

Rosenbaum H, Harris AW, Bath ML, McNeall J, Webb E, Adams JM, Cory S (1990) An Eμ-v-*abl* transgene elicits plasmacytomas in concert with an activated *myc* gene. EMBO J 9:897-905
Solter D (1988) Differential imprinting and expression of maternal and paternal genomes. Ann Rev Genet 22:127-146
Tidmarsh GF, Heimfeld S, Whitlock CA, Weissman IL, Müller-Sieburg CE (1989) Identification of a novel bone marrow-derived B-cell progenitor population that coexpresses B220 and Thy-1 and is highly enriched for Abelson leukemia virus targets. Mol Cell Biol 9:2665-2671
Whitlock CA, Witte ON (1985) The complexity of virus-cell interactions in Abelson virus infection of lymphoid and other hematopoietic cells. Adv Immunol 37:73-98

Abnormalities of the Immune System Induced by Dysregulated *bcl*-2 Expression in Transgenic Mice

A. Strasser, A.W. Harris, D.L. Vaux, E. Webb, M.L. Bath, J.M. Adams and S. Cory
The Walter and Eliza Hall Institute of Medical Research, Melbourne 3050, Australia.

INTRODUCTION

Dysregulated expression of the putative cellular oncogene *bcl*-2 by chromosomal translocation has been strongly implicated in follicular center B cell lymphoma, one of the most common hematologic malignancies in humans. These tumors usually contain a 14;18 chromosomal translocation (Fukuhara et al 1979) that juxtaposes *bcl*-2 with the immunoglobulin heavy chain (*Igh*) locus (Tsujimoto et al 1984; Cleary and Sklar 1985; Bakshi et al 1985). The recombination presumably subjects *bcl*-2 to the control of *Igh* regulatory sequences which enforce its constitutive expression in B lymphoid cells. The *bcl*-2 gene encodes a 24 kD non-glycosylated protein located on the cytoplasmic face of the plasma membrane (Tsujimoto et al 1987; Chen-Levy et al 1989). It is normally expressed in pre-B cells, quiescent in resting B cells, expressed again in activated B cells and then downregulated upon their terminal differentiation (Reed et al 1987; Gurfinkel et al 1987; Chen-Levy et al 1989). The first indication of the function of this putative oncogene was revealed by our previous studies on interleukin-3-dependent cell lines infected with a *bcl*-2 retrovirus. The cells survived withdrawal of growth factor, but entered a Go state, suggesting that constitutive *bcl*-2 expression promotes cell survival rather than proliferation (Vaux et al 1988).

Transgenic mice offer the possibility of generating a mouse model for follicular B cell lymphoma. Korsmeyer and his colleagues have produced transgenic mice carrying a *bcl*-2 mini-gene fused to a γ_1 Ig heavy chain gene, a construct mimicking the 14;18 translocation (McDonnell et al 1989). These mice developed splenomegaly and abnormal splenic architecture and their B cells exhibited a marked survival advantage *in vitro*. We have used a human *bcl*-2 cDNA under the control of the 5′ *Igh* enhancer to generate multiple strains of transgenic mice. These mice exhibit a variety of phenotypes, with effects in both the B and T lineages. This paper briefly summarises their salient characteristics.

Eμ-*bcl*-2 TRANSGENIC MICE

We used two vectors to generate *bcl*-2 transgenic mice, both of which contain the mouse 5′ *Igh* enhancer, which we term Eμ (Fig.1). The first vector, pEμSV, has been utilised extensively in our previous work (Rosenbaum et al 1989; Hariharan et al 1989), and includes an SV40 promoter, splice and polyadenylation signal. The second vector, pEμIg, was constructed in an attempt to achieve a higher proportion of active transgenic inserts. It was derived from a functional Ig μ gene shown previously to be expressed efficiently in B and T cells of transgenic mice (Grosschedl et al 1984), by insertion of a polylinker between the V_H promoter and coding sequence. Thus the remaining portion of the Cμ gene, including the large J_H-Cμ1 intron, lies downstream from and out of frame with the inserted *bcl*-2 cDNA.

We generated 28 primary *bcl*-2 transgenic mice on a (C57BL/6 x SJL)F2 background, 11 with the EμSV construct and 17 with the EμIg construct (Table 1). Since we did not detect any consistent differences in expression and phenotype between mice containing those two constructs, we conclude

Current Topics in Microbiology and Immunology, Vol. 166

Table 1. Properties of Eμ-*bcl*-2 Transgenic Mice

Primary mouse	Sex	Transgene transmission	Increased WBC[a]	Increased serum Ig[b]	Lymphoid disease[c]	Increased b.m. B cells[d]	Increased survival *in vitro*[e] B cells	T cells	Classification[f]	Strain maintained
EμSV-*bcl*-2										
2	F	5/67			--	--	--	--	0	--
10	M	55/107	--		--	--	--	--	0	--
22	M	27/66[g]	+	+	+	+	+	--	B	+
23	F	4/35			--	--			0	--
25	M	34/107	--	--	+[h]	±	--	+	T	+
34	M	31/63	--		--[i]	±	+	--	B	--
35	M	9/63	±		--[i]	±	+	--	B	--
36	F	24/38[g]	+	+	+[h]	+	+	+	B+T	+
40	F				?					--
43	M		±		?					--
44	M		±		--					--
EμIg-*bcl*-2										
15	M	46/96	+	+	+[h]	+	+	+	B+T	+
45	M	57/111	--		--	±	--	--	0	--
54	F	1/25	--		--					--
58	F	17/47	--	+	+[i]	±	+	--	B	+
59	M	51/105	+	+	--[i]	+	+	--	B	--
69	F	25/45	--		--	--	--	--	B	--
73	M	27/99	±	--	--	--	--	--	0	+
77	M	19/66	+	+	--[i]	+	+	--	B	--
82	M			+	--				B	--
92	F			+	+				B	--
114	M		--	+	+				B	--
127	F	0/2	--	+	+				B	--
128	F		--	--	--				0	--
136	F		--	+	+				B	--
145	F		±	+	+				B	--
149	M		-	--	--				0	--
151	M	21/51	+	+	+	+	+	--	B	--

[a] -- = 5-10, ± = 11-13, + = >20 x 10^6/ml; [b] judged from Coomassie-stained electropherogram; [c] lymphadenopathy and kidney disease; [d] -- = 5-10%, ± = 11-15%, + = >15% $B220^+Ig^+$ cells in bone marrow; [e] + = >10x normal after 7d; [f] see text; [g] two transgene inserts present, more active one perpetuated; [h] also occasional T lymphoma; [i] primary mouse healthy to 1 yr, progeny not kept to assess possible disease.

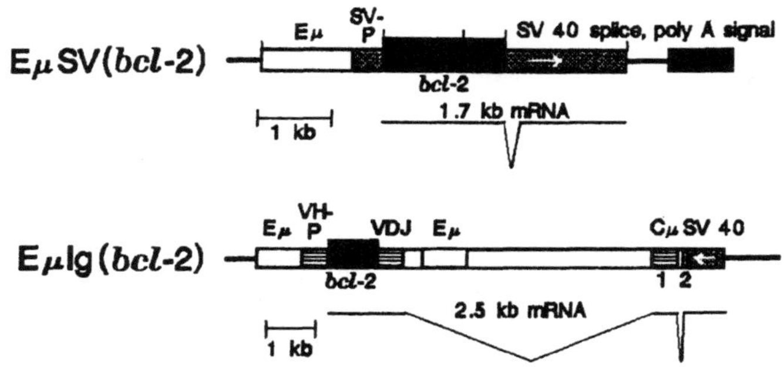

Fig. 1. Eμ-*bcl*-2 transgene constructs. pEμSV has been described elsewhere (Rosenbaum et al 1989). pEμIg was derived from pEμV_H (Hariharan et al 1989) by insertion into the polylinker of the 6.5 kb NdeI/BamH1 fragment from pμ (Fig. 1 of Grosschedl et al 1984). An additional modification was the inversion of the region containing the SV40 splice (but not that containing the polyadenylation signal); thus all the splices expected are those from the immunoglobulin gene. The *bcl*-2 sequence comprises residues 1-939 of the clone isolated by Cleary, Smith and Sklar (1986), which was inserted into the SmaI site of the polylinker as a blunted EcoRI-TaqI fragment.

that both vectors function with similar efficiency *in vivo* and both types of lines will be referred to as Eμ-*bcl*-2 mice in the rest of this paper. Eighteen of the 28 primary transgenics were bred with normal (C57BL/6 x SJL)F1 hybrid mice. Seven produced significantly less than 50% positive progeny and hence were probably mosaic; eight produced transgenic offspring with ~50% frequency, as expected from integration of the transgene at a single site; two others each yielded two sublines with different transgene inserts. The primary mice and seventeen independent breeding lines were surveyed for abnormalities of the blood and hematopoietic organs using several criteria, including histology, flow cytometry and *in vitro* culture. Table 1 summarises these results and shows that the mice could be categorised into four types. Fifteen 1° mice and nine lines exhibited B lymphoid abnormalities. One 1° mouse and the line derived from it exhibited T lymphoid abnormalities. Another two exhibited abnormalities of both the B and T compartments. Finally, 7 primary mice and five lines appeared to be normal, despite evidence of transcription of the transgene in the one line analysed. Two of the "B" lines (EμSV-*bcl*-2-22 and EμIg-*bcl*-2-58) have been perpetuated for more detailed scrutiny, as have the two "B/T" lines (EμSV-*bcl*-2-36 and EμSV-*bcl*-2-15) and the sole "T" line (EμSV-*bcl*-2-25).

B LYMPHOID ABNORMALITIES INDUCED BY CONSTITUTIVE *bcl*-2 EXPRESSION

The most readily diagnosed abnormality of the "B" strains was the increased number of B lymphocytes (2- to 5-fold, depending on the strain) in their spleen, lymph nodes and bone marrow (Table 1). Some lines also displayed elevated levels of white blood cells and FACS analysis established that this was due to excess B cells. Most of the Eμ-*bcl*-2 B cells were slightly smaller than normal B cells and mitoses were infrequent in tissue sections. They expressed normal levels of B220, IgM, IgD and class II MHC molecules, as shown in Fig. 2 for splenocytes from our best studied strain, Eμ-*bcl*-2-22. The B cell increase seemed to be polyclonal, since no individual rearranged J_H fragments were present at high enough concentration in spleen or lymph node DNA to be detected by Southern blot analysis (data not shown).

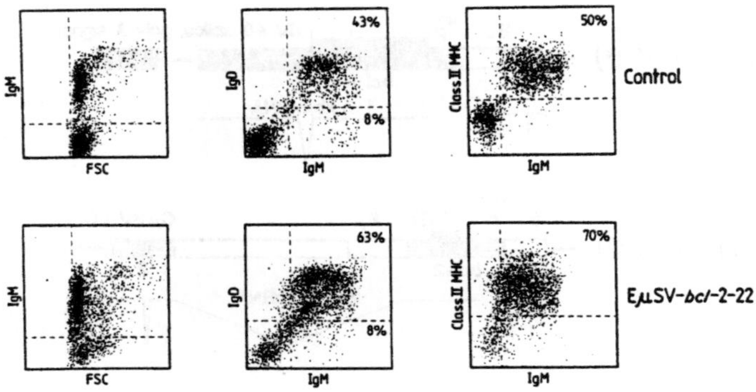

Fig. 2. Excess B cells in "B" type *bcl*-2 transgenic strains express normal cell surface markers. Spleen cells from EμSV-*bcl*-2-22 mice and negative littermates were stained with fluorochrome labeled monoclonal antibodies specific for IgM, IgD and class II MHC and analyzed by flow cytometry.

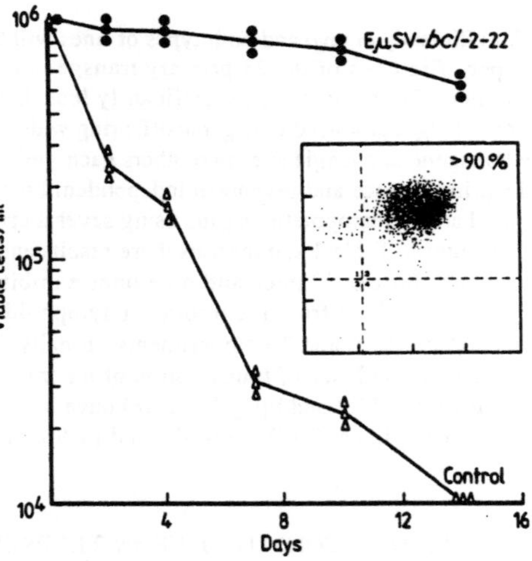

Fig. 3. Prolonged *in vitro* survival of B cells from "B" type *bcl*-2 transgenic strains. Spleen cells from three EμSV-*bcl*-2-22 mice and three negative littermates were cultured at a density of 1 x 10^6 cells/ml in normal tissue culture medium without addition of cytokines or mitogens. The number of viable cells was determined by microscopy using a hemocytometer and trypan blue exclusion. After 10 days in culture, EμSV-*bcl*-2-22 spleen cells were stained with fluorochrome-labeled monoclonal antibodies specific for B220 (vertical axis) and IgM (horizontal axis) and analyzed by flow cytometry.

Just as anticipated from our earlier studies (Vaux et al 1988), Eμ-*bcl*-2 expressing cells showed enhanced survival capacity. This property became apparent when cell suspensions prepared from the spleen, lymph nodes, bone marrow or peripheral blood were cultured *in vitro* in the absence of added growth factors, as shown in Fig. 3 for splenocytes from Eμ-*bcl*-2-22 mice. After 10 days, when B cell viability had dropped to 1% or less in control cultures, between 20% and 80% of B cells from "B"strain transgenics, depending on the strain, were still viable, but they did not proliferate. Similar results have been reported by McDonnell et al (1989) for their *bcl*-2 transgenic mice. Constitutive *bcl*-2 expression appears to promote extended survival of cells from all stages of B lymphocyte development, since pre-B cells were readily apparent in long term cultures of Eμ-*bcl*-2 bone marrow cells, as were hemolytic plaque forming cells in cultures of spleen cells from immunized mice (Strasser et al in preparation).

In view of these observations, we believe that the B cell excess in Eμ-*bcl*-2 mice is due to prolonged survival rather than to lymphoproliferation. If so, the mice might be expected to exhibit abnormally prolonged immune responses *in vitro* and *in vivo*. This is indeed the case, as will be described elsewhere (Strasser et al in preparation).

In contrast to our expectations and to the results of McDonnell et al (1989), our Eμ-*bcl*-2 mice did not show follicular hyperplasia or develop B cell lymphoma. However, more than 40 mice of the Eμ-*bcl*-2-22 strain and several primary mice have developed a systemic autoimmune-like disease (Strasser et al in preparation). The basis for the difference between our mice and those of McDonnell et al (1989) is not yet clear, but might be due to differences in the genetic background or in the transgene constructs.

T LYMPHOID ABNORMALITIES INDUCED BY *bcl*-2

One strain, Eμ-*bcl*-2-25, displayed no perturbation of the B lymphoid compartment but did exhibit a two-fold excess of T cells in peripheral lymphoid organs. Moreover, T cells from these mice, derived from either the thymus or the periphery, exhibited prolonged survival when cultured in the absence of exogenously added growth factors. None of the mice have developed autoimmune-like disease. However, a few have succumbed to clonal T lymphomas (Strasser et al in preparation).

Mice from the two "B/T" type strains (Eμ-*bcl*-2-36 and Eμ-*bcl*-2-15) demonstrated both kinds of abnormality, i.e. excess B and T lymphocytes, increased serum immunoglobulin and prolonged survival of B as well as T cells in culture. Some of these mice have developed the systemic autoimmune-like disease and a few have succumbed to clonal T cell tumors (Strasser et al, in preparation).

COLLUSION OF *bcl*-2 WITH OTHER ONCOGENES IN TUMORIGENESIS

The availability in our laboratory of transgenic mice carrying a range of oncogenes permits us to perform crosses to test the capacity of *bcl*-2 to synergize with other oncogenes in transforming hematopioetic cells. To date, crosses have been made using strain 22 of the "B" type Eμ-*bcl*-2 mice and either Eμ-N-*ras* transgenic mice, which develop thymic T cell lymphomas or histiocytic sarcomas (Harris et al 1988a), or with Eμ-*myc* transgenic mice, which develop pre-B and B cell lymphomas (Adams et al 1985; Harris et al 1988b). Our preliminary results have revealed no synergy between *bcl*-2 and N-*ras*, but a striking synergy was revealed with c-*myc*, since mice bearing both oncogenes developed tumors much more rapidly than with Eμ-*myc* alone. Intriguingly, the tumors were all of a novel immature phenotype, as defined by cell surface marker expression and lack of *Igh* and *Tcr* rearrangements. They may therefore represent a neoplastic counterpart of a lymphoid progenitor or stem cell (Strasser et al in preparation). These data clearly show that *bcl*-2 can contribute to tumorigenesis.

In summary, our data indicate that the *bcl*-2 gene has the novel capacity of promoting the survival but not the proliferation of B and T lymphoid cells. It is therefore not surprising to find that constitutive *bcl*-2 expression perturbs lymphoid homeostasis and the immune response, particularly within the B lymphoid compartment. T lymphoid tumors have arisen at low frequency in certain strains but, to date, no B lymphoid tumors. These results imply that, by itself, constitutive *bcl*-2 expression is insufficient to cause lymphoid tumors, although it can be lethal by promoting a systemic autoimmune disease. Coupled with constitutive *myc* expression, however, a deregulated *bcl*-2 gene becomes profoundly tumorigenic for primitive lymphoid cells. This gene combination also seems likely to be a fundamental cause of those aggressive B lymphomas which carry both a 14;18 and an 8;14 translocation (Gauwerky et al 1988).

ACKNOWLEDGEMENTS

We thank Dr M Cleary for the generous gift of the *bcl*-2 cDNA, Kim Patane for animal husbandry and Iras Collins for preparing this manuscript. A.S. is the recipient of a fellowship from the Swiss National Foundation. This work was supported by the National Health and Medical Research Council of Australia and the US National Cancer Institute (CA43540).

REFERENCES

Adams JM, Harris AW, Pinkert CA, Corcoran LM, Alexander WS, Cory S, Palmiter RD, Brinster RL (1985) The c-*myc* oncogene driven by immunoglobulin enhancers induces lymphoid malignancy in transgenic mice. Nature 318:533-538

Bakshi A, Jensen JP, Goldman P, Wright JJ, McBride OW, Epstein AL, Korsmeyer SJ. (1985) Cloning of the chromosomal breakpoint of t(14;18) human lymphomas: clustering around J_H on chromosome 14 and near a transcriptional unit on 18. Cell 41:889-906

Chen-Levy Z, Nourse J, Cleary M (1989) The *bcl*-2 candidate proto-oncogene produces a 24 kilodalton integral membrane protein highly expressed in lymphoid cell lines and lymphomas carrying the t(14;18) translocation. Mol Cell Biol 9:701-710

Cleary ML, Sklar J (1985) Nucleotide sequence of t(14;18) chromosomal breakpoint in follicular lymphoma and demonstration of a breakpoint cluster region near a transcriptionally active locus on chromosome 18. Proc Natl Acad Sci USA 82:7439-7443

Cleary ML, Smith SD, Sklar J (1986) Cloning and structural analysis of cDNAs for *bcl*-2 and a hybrid *bcl*-2/immunoglobulin transcript resulting from the t(14;18) translocation. Cell 47:19-28

Fukuhara S, Rowley JD, Variakojis D, Golomb HM (1979) Chromosome abnormalities in poorly differentiated lymphocytic lymphoma. Cancer Res 39:3119-3128

Gauwerky CE, Haluska FG, Tsujimoto Y, Nowell PC, Croce CM (1988) Evolution of B cell malignancy: pre-B cell leukemia resulting from *myc* activation in a B cell neoplasm with a rearranged *BCL*-2 gene. Proc Natl Acad Sci USA 85:8548-8552

Grosschedl RD, Weaver D, Baltimore D, Costantini F (1984) Introduction of a μ immunoglobulin gene into the mouse germline: specific expression in lymphoid cells and synthesis of functional antibody. Cell 38:647-658

Gurfinkel N, Unger T, Givol D, Mushinski JF (1987) Expression of the *bcl*-2 gene in mouse B lymphocyte cell lines is differentiation stage specific. Eur J Immunol. 17:567-570

Hariharan IK, Harris AW, Crawford M, Abud H, Webb E, Cory S, Adams JM (1989) A *bcr*-v-*abl* oncogene induces lymphomas in transgenic mice. Mol Cell Biol 9:2798-2805

Harris AW, Langdon WY, Alexander WS, Hariharan IK, Rosenbaum H, Vaux D, Webb E, Bernard O, Crawford M, Abud H, Adams JM, Cory S (1988a) Transgenic mouse models for hematopoietic tumorigenesis. Curr Topics Microbiol Immunol 141:83-93

Harris AW, Pinkert CA, Crawford M, Langdon WY, Brinster RL, Adams JM (1988b) The Eμ-*myc* transgenic mouse: a model for high-incidence spontaneous lymphoma and leukemia of early B cells. J Exp Med 167:353-371

McDonnell TJ, Deane N, Platt FM, Nunez G, Jaeger U, McKearn JP, Korsmeyer SJ (1989) *bcl*-2 immunoglobulin transgenic mice demonstrate extended B cell survival and follicular lymphoproliferation. Cell 57:79-88

Reed JC, Tsujimoto Y, Alpers JD, Croce CM, Nowell PC (1987) Regulation of *bcl*-2 proto-oncogene expression during normal human lymphocyte proliferation. Science 236:1295-1299

Rosenbaum H, Webb E, Adams JM, Cory S, Harris AW (1989) N-*myc* transgene promotes B lymphoid proliferation, elicits lymphomas and reveals cross-regulation with c-*myc*. EMBO J 8:749-755

Tsujimoto Y, Finger LR, Yunis J, Nowell PC, Croce CM (1984) Cloning of the chromosome breakpoint of neoplastic B cells with the t(14;18) chromosome translocation. Science 226:1097-1099

Tsujimoto Y, Ikegaki N, Croce CM (1987) Characterization of the protein product of *bcl*-2, the gene involved in human follicular lymphoma. Oncogene 2:3-7

Vaux DL, Cory S, Adams JM (1988) *Bcl*-2 gene promotes haemopoietic cell survival and cooperates with c-*myc* to immortalize pre-B cells. Nature 335:440-442

Growth Regulation to B-Cells: Immortalization

Immortalization of Primary Murine B Lymphocytes with Oncogene-Containing Retroviral Vectors

R.W. Overell, K.E. Weisser, B. Hess, R. Goodwin, L. Clark and K.H. Grabstein

Immunex Corporation, 51 University Street, Seattle, WA98101

INTRODUCTION

Various oncogenes have been implicated in B cell neoplasia, and the use of transgenic mouse models and in vivo experiments with oncogene-containing retroviral vectors has started to assign cooperative functions to these oncogenes in B cell transformation (Adams et al.,1985; Potter et al.,1987). In addition, much has been learned through the in vitro transformation of pre-B cells by retrovirally transferred oncogenes (Schwartz et al.,1986). However, the ability of oncogenes to immortalize primary mature (sIg^+) B cells has not been systematically explored. Indeed, it has been proposed that mature lymphocytes are refractory to infection with retroviruses. In these studies we have shown that mature B cells purified from murine spleen are readily infectable with ecotropic retroviruses, and have assayed a panel of oncogene-containing constructs for their ability to immortalize these primary cells in vitro.

RESULTS AND DISCUSSION

To demonstrate that mature B cells were infectable with retroviral vectors packaged by the psi-2 cell line (Mann et al.,1983), we initially carried out infections with the deltaH retroviral vector (see Overell et al.,1989). Using the higher-titer H2S retroviral vector, which is similar to the N2 vector described by Keller et al.(1985) but contains the *hph* drug resistance marker, it was possible to demonstrate gene transfer frequencies approaching 20% in primary B lymphocytes, using a reverse plaque assay to detect hygromycin B (HMB)-resistant B cells (Table 1).

Table 1. Efficient infection of mature B cells by the H2S retroviral vector

Infecting vector[a]	[hygromycin B] (mg/ml)	PFC/culture	Infection frequency(%)
psi-2	0	48,000	
	1	36	0.08
	2	16	0.03
H2S	0	42,000	
	1	7,720	18.4
	2	3420	8.1

[a]Primary splenic B cells were cocultivated with psi-2 cells for 20 hr as previously described (Overell et al.,1989). Psi-2, untransfected psi-2 cells. The titer of the H2S vector produced from the clone of psi-2 cells used was $5x10^6$ HMB^r CFU/ml, as measured on NIH/3T3 cells.

Current Topics in Microbiology and Immunology, Vol. 166

In these experiments, hygromycin B was added immediately after the infection period and the cultures assayed for plaque forming cells (PFC) after 48hr of incubation by means of a reverse plaque assay. At the higher concentration of hygromycin B (2 mg/ml) a lower infection frequency was observed (Table 1), perhaps indicating that the level of expression of the drug resistance marker was too low in many cells for them to be scored as drug resistant in this assay.

Having established that primary B cells could be efficiently infected with these vectors, we tested a large number of retroviral vectors which expressed oncogenes alone or in combination for their ability to immortalize these cells. The discussion here will focus primarily on the results obtained with two vectors, one co-expressing v-Ha-*ras* and v-*myc* (deltaRM, Overell et al., 1988) and the other expressing the v-*abl* oncogene. These vectors had similar titers, as measured by focus-formation on NIH/3T3 cells. We have previously shown that infection of primary B cells with the deltaRM retroviral vector led to a transient increase in proliferation, and that these primary cells expressed the appropriate vector-specific transcripts to high levels (Overell et al., 1989). That these transiently proliferating cells were indeed B cells was shown by flow cytometry (Fig. 1), using a probe specific for surface immunoglobulin (sIg). Despite these transient effects on B cell proliferation, no cell lines of a B cell phenotype were obtained from these experiments. Rare cell lines which were obtained were found to be of a myeloid phenotype.

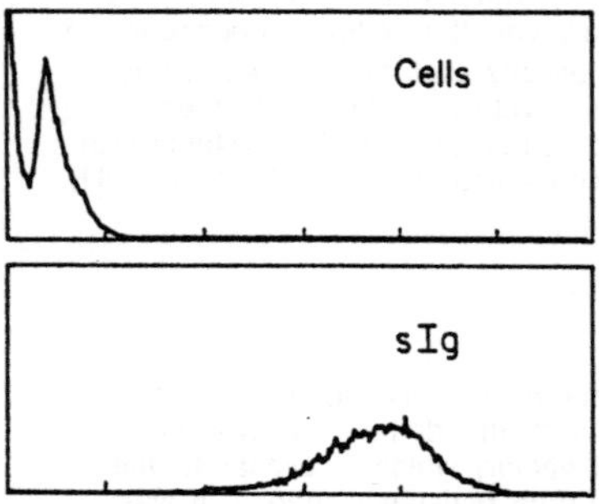

Fig. 1. Cell surface phenotype of primary B cell cultures 5 days after infection with the deltaRM retrovirus. Upper panel, cells alone; lower panel, cells treated with directly FITC-conjugated $F(ab')_2$ rabbit anti-mouse IgM.

We next tested retroviral vectors which expressed the v-*abl* oncogene. These vectors expressed either the wild-type P160 protein or a carboxy-terminal truncation of this molecule, which is approximately 100kd in size and which we call v-trabl. The trabl vector was made because of problems which we initially encountered in making high-titer P160 abl retroviruses in psi-2 cells. We reasoned that these problems might be circumvented by using the trabl vector, since the toxicity of the abl protein has been ascribed to the carboxyl portion of the v-abl protein (Ziegler et al.,1981). Both of these *abl* oncogenes were expressed in the deltaH retroviral vector (Overell et al.,1988) under the transcriptional control of the retroviral LTR. The abl and trabl vectors ultimately had similar titers for focus formation on NIH/3T3 cells (~10^6 FFU/ml), and since their behaviour in the mature B cell system was similar, the results obtained with the two vectors will be discussed together. Following cocultivation of primary B cells with psi-2 cells producing the abl viruses, cell lines emerged after a lag of 2-3 weeks . The frequency of emergence of these cell lines was maximally 1 in $2X10^6$ exposed primary B cells, as judged from experiments in which immortal cell lines emerged from wells seeded at dilutions limiting for the tranformation event. Following their emergence from primary culture, these lines could be readily grown and have been in continuous culture for 3 months with no sign of a reduced rate of growth. We therefore consider these cells to be immortal. These abl/trabl cell lines were found to be of a B cell phenotype as judged by flow cytometry (Fig.2).

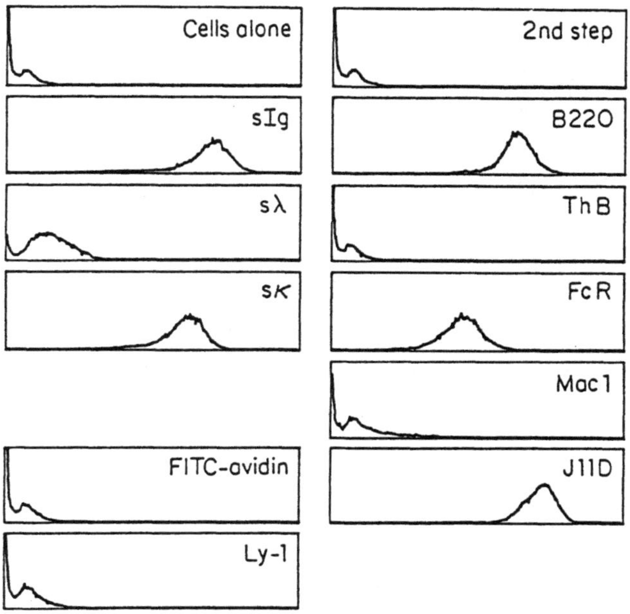

Fig. 2. Cell surface phenotype of a representative cell line derived from a primary B cell culture infected with the deltaH*abl* retrovirus. Flow cytometry was carried out with the indicated reagents as previously described (Overell et al.,1989), except the Ly-1 reagent which was obtained from Beckton-Dickinson.

These cells thus had the surface phenotype of mature B cells. They were not of the CD5(Ly-1)$^+$ B cell lineage, as judged by staining with the Ly-1 antibody (Fig. 1). The abl/trabl B cell lines were derived in the presence of LPS and were absolutely dependent on the presence of LPS in the culture medium for their continued survival and proliferation. The cell lines, which contained multiple proviral insertions but expressed low or non-detectable levels of transcripts encoding the abl oncogenes, were non-tumorigenic in nude mice, even after prolonged growth in culture. These cell lines therefore differ from the A-MuLV-induced B lines generated and characterized by Serunian and Rosenberg (1986), which contained integrated copies of the A-MuLV provirus and were independent of LPS for growth.

These results raise some interesting issues concerning the transformation of B lineage cells by oncogenes. Firstly, both deltaRM and A-MuLV transform pre-B cells efficiently, yet neither is sufficient to immortize the mature B cells. This represents an interesting parallel in the transforming potential of v-abl and (ras+myc). Secondly, it is of importance to determine why these vectors are unable to transform primary B cells. It is quite possible that the vectors that we have used are efficient gene transfer vehicles, but are not efficient expression vectors in the B cells. This explanation does not appear to be the case with the deltaRM vector, since high-level expression at the RNA level was observed in the target cells (Overell et al.,1989). Moreover, using a vector which coexpressed *ras* and *myc* under the control of an immunoglobulin enhancer, Clynes et al.(1988) were able to induce the rapid onset of plasmacytomas in mice, but the kinetics and clonality of tumor formation suggested that secondary events could be needed for the transformation of mature B cells in vivo.The mature B cell lines obtained with the abl retroviruses in our studies emerged at a frequency which was approximately 10^5-fold lower than the infection frequency (Table 1), indicating either that the target cell for the transformation event was very rare, or that additional secondary events were necessary for the immortalization of the primary cells. It seems likely that other cooperating genetic events are necessary for the immortalization of mature B cells, and it is at present unclear whether these are dominant or recessive in nature. Thirdly, regarding this latter point, it will be of interest to co-infect the primary cells with RM and other vectors, in particular with the SV40 T-antigen expressing retrovirus. The latter vector, if expressed to sufficiently high levels in the B cells, might be expected to neutralize the p53 and RB proteins, both of which have been shown to act in a dominant fashion to restrain cell growth in other mammalian systems. Finally, it will be important to determine whether the immortal B cell lines induced by the abl viruses can be further transformed to LPS independence and/or tumorigenicity by other oncogenes. These immortal B cell lines are analogous to the NIH/3T3 cell line, in that they are immortal but non-tumorigenic. However, unlike NIH/3T3 cells, it would appear that the abl/trabl B cell lines are not transformable by *ras* oncogenes, since our initial experiments have indicated that activated *ras* oncogenes are unable to confer LPS independence on these cells. These B cell lines could therefore provide the basis for a useful assay system for the detection of novel oncogenes which might be involved in the control of B cell growth and the generation of B cell malignancies.

REFERENCES

Adams JM, Harris AW, Pinkert CA, Corcoran LM, Alexander WS, Cory S, Palmiter RD, Brinster RL (1985) The c-myc oncogene driven by immunoglobulin enhancers induces lymphoid malignancy in transgenic mice. Nature 318: 533-538

Clynes R, Wax J, Stanton LW, Smith-Gill S, Potter M, Marcu KB (1988) Rapid induction of IgM-secreting murine plasmacytomas by pristane and an immunoglobulin heavy-chain promoter/enhancer-driven c-myc/v-Ha-ras retrovirus. Proc Natl Acad Sci USA 85:6067-6071

Keller G, Paige C, Gilboa E, Wagner EF (1985) Expression of a foreign gene in myeloid and lymphoid cells derived from multipotent haematopoietic precursors. Nature 318: 149-154

Overell RW, Weisser KE, Cosman D (1988) Stably transmitted triple-promoter retroviral vectors and their use in transformation of primary mammalian cells. Mol Cell Biol 8:1803-1808

Overell RW, Weisser KE, Hess B, Namen AE, Grabstein KH (1989) Stage-specific transformation of murine B lineage cells by ras and myc. Oncogene 4:1425-1432

Potter M, Mushinski JF, Mushinski EB, Brust S, Wax JS, Wiener F, Babonits M, Rapp UR, Morse HC (1987) Avian v-myc replaces chromosomal translocation in murine plasmacytomagenesis. Science 235:787-789

Serunian LA and Rosenberg N (1986) Abelson virus potentiates long-term growth of mature B lymphocytes. Mol Cell Biol 6:183-194

Ziegler SF, Whitlock CA, Goff SP, Gifford A, Witte ON (1981) Lethal effect of the Abelson murine leukemia virus transforming gene product. Cell 27:477-486

Functionality of Clonal Lymphoid Progenitor Cells Expressing the P210 BCR/ABL Oncogene

P.A. Scherle* and O.N. Witte

Department of Microbiology, Molecular Biology Institute and Howard Hughes Medical Institute, University of California Los Angeles, 405 Hilgard Avenue, Los Angeles, CA, 90024

We have developed an in vitro system in which clonal progenitor cell lines which express the P210 BCR/ABL oncogene and are committed to the B cell lineage can be derived from murine bone marrow and maintained in culture. These cells retain immunoglobulin heavy and light chain genes in a predominantly germline configuration. They express high levels of TdT and sterile u transcripts and undergo diverse DJ rearrangements at the Ig heavy chain locus in vitro. Although growth stimulated by BCR/ABL, the progenitor cells retain full differentiative capacity in vivo. Transfer of these cells to immunodeficient SCID mice leads to reconstitution of functional B cells which respond in a specific fashion to both T-independent and T-dependent antigens and which display diverse V_H family utilization. Tumors have not been observed in the reconstituted mice for at least 60 days. This system demonstrates that P210 can growth stimulate an early progenitor cell without preventing differentiation and should be useful for studying both the B cell developmental process and leukemogenesis.

The P210 BCR/ABL gene product is found in > 90% of patients with chronic myeloid leukemia (CML) and results from a translocation which juxtaposes portions of the BCR gene on chromosome 22 near the second exon of c-ABL on chromosome 9 (Rosenberg and Witte 1988). Several lines of evidence suggest that transformation by P210 may provide a means to growth stimulate early hematopoietic progenitor cells without arresting differentiation. The BCR/ABL translocation appears to occur in a pluripotent stem cell (Champlain and Golde 1985). In the chronic phase of the disease, predominantly myeloid proliferation occurs although the lymphoid lineage can also be affected (Champlain and Golde 1985). Importantly, differentiation is not completely blocked. In vitro, P210-expressing hematopoietic cells and fibroblasts have altered growth characteristics but are not fully transformed (McLaughlin et al 1987; Lugo and Witte 1989).

* Present address: Metabolism Branch, National Cancer Institute, National Institutes of Health, 9000 Rockville Pike, Bethesda, MD 20814.

Current Topics in Microbiology and Immunology, Vol. 166

By modifying conditions established by Whitlock and Witte (1982) for long term bone marrow culture and by using the BCR/ABL oncogene as a transforming agent, we have derived clonal lymphoid progenitor cell lines (LPC) which retain their differentiative capacity (Scherle et al 1990). Critical to the establishment of the LPC is the use of the stromal cell line, S17 (Collins and Dorshkind 1987). Infection of freshly isolated bone marrow with a P210-containing retroviral construct (McLaughlin et al 1987) and culture on a confluent S17 stromal layer results in the outgrowth of cells with a more immature phenotype than cells cultured on the heterogeneous adherent cell layer which normally forms in long term lymphoid bone marrow cultures (LBMC) (Table 1). Cells in both cultures are clonal or pauciclonal based on the number of retroviral integration sites (McLaughlin et al 1987; Scherle et al 1990), are > 80% B220$^+$, < 5% MAC-1$^+$ and < 5% cytoplasmic u$^+$, and do not contain precursors of the myeloid or erythroid lineages. LPC cells, in contrast to LBMC cells, retain both heavy and light chain genes in a germline configuration. P210-expressing LBMC cells have undergone rearrangement at both alleles of the heavy chain locus. Both LPC and LBMC cells are growth stimulated by P210, maintaining a cell density 5-10 times higher than uninfected control cultures. The LPC cells, however, remain dependent on the S17 stroma for growth whereas the P210-expressing LBMC cells become adherent cell independent after several weeks in culture.

Although the LPC cells have not undergone rearrangement at the Ig heavy chain locus, they appear to be committed to the B cell lineage. They express high levels of terminal deoxynucleotidyl transferase (TdT), sterile u transcripts and the B lymphoid specific gene, B29, (Hermanson et al 1988) a member of the Ig supergene family (Fig. 1). Further, they spontaneously undergo rearrangements at the heavy chain locus in vitro (Scherle et al 1990). Kappa light chain alleles remain germline, characteristic of early pre-B cells.

Table 1. Comparison of P210-expressing cultures established on S17 (LPC) versus heterogeneous stroma (LBMC)

	LPC (P210)	LBMC (P210)
Ig - heavy	G(R)*	R
light	G	G
B220	+	+
MAC-1	-	-
cytoplasmic u	-	+
adherent layer dependent	+	+
B cell reconstitution in vivo	+	-

* G = germline; R = rearranged.

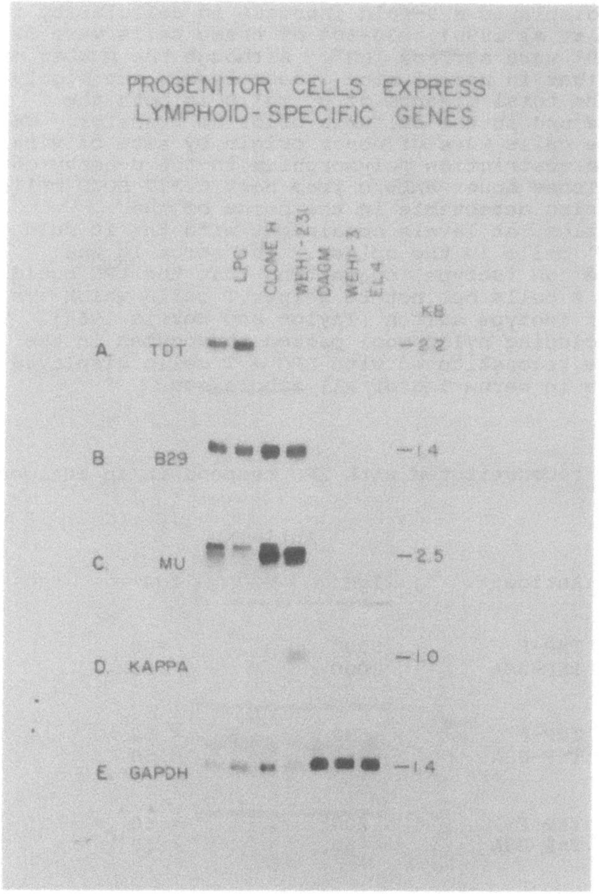

Figure 1. Gene expression was determined by Northern analysis using specific probes. LPC = lymphoid progenitor culture cells; clone H = pre-B cell; WEHI-231 = mature B cell; WEHI-3, DAGM = myeloid cell lines; EL-4 = T cell.

The P210-expressing LPC undergo further differentiation upon transfer to an in vivo environment. For these studies, the severe combined immunodeficient (SCID) mouse (Bosma et al 1983) was used as a host since it has no mature T or B cells and no serum Ig. 30-60 days post transfer of 1 x 10^6 LPC cells, the mice were examined for B cell reconstitution of lymphoid organs and serum Ig. LPC reconstituted mice displayed a 3-fold increase in cellularity in the spleen (Scherle et al 1990). 30-50% of these cells were B220$^+$ and approximately 10% were surface IgM$^+$. Although the number of u$^+$ cells is 10-20% of that in normal mice, these levels are highly significant given the total absence of u$^+$ cells in both the unreconstituted mice and in the cultures prior to transfer. We confirmed that these cells were of donor origin by site of viral integration and by a restriction polymorphism in the u-heavy chain gene which distinguishes donor BALB/c from host C.B17 SCID cells. Immunoglobulin was also detectable in the serum of the reconstituted SCID mice, at levels consistent with the 10-fold lower number of IgM$^+$ cells in the spleen. The serum Ig was predominantly of the IgM isotype, suggesting that the LPC could give rise to mature B cells but not to helper T cells which are in general required for isotype switch (Taylor and Wortis 1968). This was confirmed by including nylon-wool passed thymocytes in the inoculum. SCID mice reconstituted with LPC + T cells displayed a substantial increase in serum IgG of all subclasses.

Table 2. SCID mice reconstituted with LPC respond in an antigen-specific fashion

		Anti-TNP	
	Antigen	IgM	IgG
BALB/c	TNP-F	5000	800
	TNP-BSA	2000	100
SCID	TNP-F	< 50	< 50
	TNP-BSA	< 50	< 50
SCID + LPC	TNP-F	200	< 50
	TNP-BSA	< 50	< 50
SCID + LPC + T	TNP-F	ND	ND
	TNP-BSA	250	50
SCID + T	TNP-F	ND	ND
	TNP-BSA	< 50	< 50

Anti-TNP titers are expressed as the reciprocal of the serum dilution to give 50% maximum response in an ELISA. Data represent the average of 3-6 mice 5-7 days after secondary immunization. TNP-F = TNP-Ficoll; ND = not determined.

The mature B cells to which the P210-expressing LPC give rise in vivo are functional in an antigen-specific response (Table 2). SCID mice, 30-60 days after reconstitution with LPC or LPC + T, were immunized with the T-independent antigen, TNP-Ficoll, or the

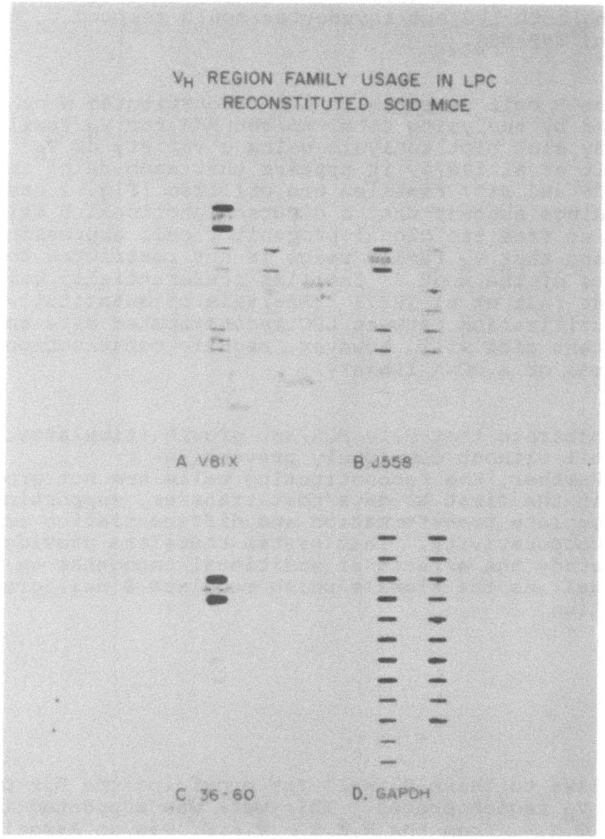

Figure 2. V_H family utilization was determined by slot blot analysis of total RNA isolated from spleen of BALB/c mice (spleen) unreconstituted SCID mice (SCID) or SCID mice reconstituted with LPC (SCID + LPC). Probes were specific for each V_H family (Alt et al 1987). RNA isolated from hybridomas of known V_H family usage served as controls and are in the upper left corner of each blot.

T-dependent antigen, TNP-BSA (Sharon et al 1975). Consistent with their inefficient class switching and thus, probable absence of T cells, SCID mice reconstituted with LPC alone produced TNP-specific antibodies upon immunization with TNP-Ficoll but not with TNP-BSA.

This was not due to an inherent defect in the LPC derived B cells since mice receiving both LPC and thymocytes could respond to the T-dependent antigen, TNP-BSA.

The diversity of the B cell response in the reconstituted mice was further investigated by analyzing total spleen RNA for V_H family gene expression. By slot blot analysis using a variety of V_H specific probes (Alt et al 1987), it appears that members of the 81X, Q52, J558, J606 and S107 families are utilized (Fig. 2 and not shown). These findings suggest that a diverse functional B cell compartment can arise from the clonal progenitor cell expressing the P210 oncogene and that V_H family usage is not restricted to one family or to members of the most 3' families preferentially used early in development (Alt et al 1987). Analysis of quantitative differences in V_H utilization between LPC reconstituted mice and normal immunocompetent mice will, however, require construction of a panel of hybridomas or a cDNA library.

These findings demonstrate that P210 BCR/ABL growth stimulates an early progenitor cell without completely preventing differentiation. Further, the reconstituting cells are not grossly leukemic at least in the first 60 days post transfer, supporting the concept that complete transformation and differentiation arrest requires oncogene cooperativity. This system therefore provides a model in which to study the effects of additional oncogenes on leukemogenesis as well as the signals which regulate B cell growth versus differentiation.

ACKNOWLEDGEMENTS

The authors would like to thank R. Wall for providing the B29 probe and F. Alt for the V_H region probes. This work was supported in part by a grant to O.N.W. from the N.I.H. P.A.S. was an Associate of the Howard Hughes Medical Institute. O.N.W. is an Investigator of the Howard Hughes Medical Institute.

REFERENCES

Alt FW, Blackwell TK, Yancopoulos, GD (1987) Development of the primary antibody repertoire. Nature (Lond) 238:1079-1087
Bosma GC, Custer RP, Bosma MJ (1983) A severe combined immunodeficiency mutation in the mouse. Nature (Lond) 301:527-530
Champlain RE, Golde DW (1985) Chronic myelogenous leukemia (CML): recent advances. Blood 65:1039-1047
Collins LS, Dorshkind K (1987) A stromal cell line from myeloid long-term bone marrow cultures can support myelopoiesis and B lymphopoiesis. J. Immunol. 138:1082-1087
Hermanson GC, Eisenberg D, Kincade PW, Wall R (1988) B29: a member of the immunoglobulin supergene family exclusively expressed on B lineage cells. Proc. Natl. Acad. Sci. USA 85:6890-6894

Lugo TG, Witte ON (1989) The BCR-ABL oncogene transforms Rat-1 cells and cooperates with v-myc. Mol. Cell. Biol. 9:1263-1270

McLaughlin J, Chianese E, Witte, ON (1987) In vitro transformation of immature hematopoietic cells by the P210 BCR/ABL oncogene product of the Philadelphia chromosome. Proc. Natl. Acad. Sci. USA 84:6558-6562

Rosenberg N, Witte O (1988) The viral and cellular forms of the Abelson (abl) oncogene. Adv. Vir. Res. 35:39-81

Scherle PA, Dorshkind K, Witte, ON (1990) Clonal lymphoid progenitor cell lines expressing the BCR/ABL oncogene retain full differentiative capacity. Proc. Natl. Acad. Sci. USA 87:1908-1912

Sharon R, McMaster PR, Kask AM, Owens JD, Paul WE (1975) DNP-Lys-Ficoll: a T-independent antigen which elicits both IgM and IgG anti-DNP antibody-secreting cells. J. Immunol. 114:1585-1589

Taylor RB, Wortis HH (1968) Thymus dependence of antibody: variation with dose of antigen and class of antibody. Nature (Lond) 220:927-929

Whitlock CA, Witte ON (1982) Long-term culture of B lymphocytes and their precursors from murine bone marrow. Proc. Natl. Acad. Sci. USA 79:3608-3612

Constitutive and Cell Cycle Regulated Expression of c-*myc* mRNA is Related to the State of Differentiation in Murine B-Lymphoid Tumors

K.M. Catron[1], C.R. Toth[1], J. Purkerson[2], P. Isakson[2] and T.P. Bender[1]

Departments of Microbiology[1] and Pharmacology[2], University of Virginia, Charlottesville, VA. 22908

INTRODUCTION

The c-*myb* proto-oncogene encodes a nuclear, DNA binding protein (Klempnauer 1984; Luscher 1990) which has recently been shown to have the ability to act as a transcriptional transactivator (Westin 1989). The expression of c-*myb* mRNA has been predominantly associated with normal tissue and tumor cell lines of immature hematopoietic origin (Westin 1982; Gonda 1982) and the chemically induced differentiation of several hematopoietic tumor cell lines *in vitro* results in the down-regulation of c-*myb* mRNA levels (Gonda 1984; Kirsch 1986). In addition, c-*myb* expression has recently been reported to be down-regulated in neuroblastomas which were induced to differentiate with retinoic acid (Thiele 1988). We have previously shown in murine B-lymphoid tumors that pre-B cell lymphomas contain 10 to greater than 100 fold more c-*myb* mRNA than B cell lymphomas and plasmacytomas (Bender 1987a) and that this difference is maintained primarily by a premature block to transcription elongation (attenuation) which occurs in the first intron of the c-*myb* gene (Bender 1987b). Attenuation has also been shown to be involved in the regulation of c-*myb* mRNA levels during the chemically induced differentiation of a murine erythroleukemia cell line (Watson 1988). We have recently examined the basis for high versus low level expression of c-*myb* mRNA in murine B-lymphoid tumors and report that pre-B cell lymphomas constitutively express c-*myb* mRNA while B cell lymphomas regulate c-*myb* mRNA levels during the cell cycle. In addition, we describe a protein/DNA interaction, detected near the putative site of attenuation, that correlates with active attenuation in these tumor cell lines.

RELATIONSHIP BETWEEN c-*myb* mRNA EXPRESSION AND THE CELL CYCLE IN MURINE B-LYMPHOID TUMORS

The down-regulation of c-*myb* mRNA expression has been associated with differentiation in the erythroid (Kirsch 1986), monocyte (Gonda 1984) and B and T-lymphoid (Bender 1987; Thompson 1986) lineages. In the T cell lineage, c-*myb* mRNA has been found to be constitutively expressed in avian thymocytes (Thompson 1986) but to be regulated during the cell cycle in an avian T cell lymphoma (Thompson 1986), human peripheral blood T cells (Reed 1986) and cloned murine T cell lines (Reed 1987). To examine the relationship between c-*myb* mRNA expression and the cell cycle in murine B cell lymphomas we have developed a protocol, using the BCL_1 lymphoma, to select for cells that appear to be in the G_1 stage of the cell cycle. This cell line is a tumor representative of the mature B cell stage of development and contains approximately 50-fold less steady state c-*myb* mRNA than the pre-B cell lymphoma 70Z/3.12. BCL_1 grows as an adherent cell line *in vitro* in Dulbecco's modified Eagle's medium (DMEM) supplemented with 5% fetal calf serum (FCS) and non-essential amino acids (NEAA), which are required for growth. We have found that if NEAA are removed from the growth medium, BCL_1 cells cease to proliferate and dividing cells round up and detach from the plastic. After gentle agitation and replacement of the medium lacking NEAA with complete growth medium the remaining cells are viable and adherent. Greater than 97% of this adherent population is 2N in DNA content as measured by fluorescence activated cell sorting (FACS) after staining with Hoechst stain 33258 (Fig. 1, 0hr). As shown in Fig. 1, between eight and twelve hours after the replacement of NEAA approximately 20% of these cells begin to synthesize DNA and enter S-phase (a detailed description of this protocol will be described elsewhere; K. Catron, manuscript in preparation). We feel that this protocol selects for cells in the G_1 phase rather than G_O as we do not detect expression of c-*fos* mRNA between five minutes and one hour after addition of NEAA and expression of c-*myc* mRNA does not change at any time during

Current Topics in Microbiology and Immunology, Vol. 166

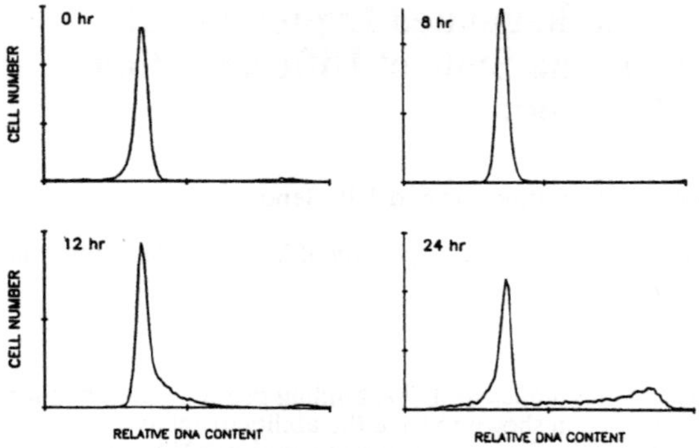

Figure 1. Relative DNA content of NEAA starved BCL_1 cells. BCL_1 cells were grown to approximately three-quarters confluency on 75 cm^2 plastic tissue culture flasks in DMEM supplemented with 5% FCS and NEAA (100mM). To select for G_1 BCL_1 cells the complete growth medium was removed. The cells were washed once with phosphate buffered saline and replaced with growth medium lacking NEAA. After 12 hours without NEAA, the cells were gently agitated, the medium was removed by suction and replaced with complete growth medium. At the indicated times cells were harvested and an aliquote ($\sim 10^6$ cells) was stained with Hoechst stain 33258. Stained cells were analyzed by FACS for relative DNA content (Shapiro 1989).

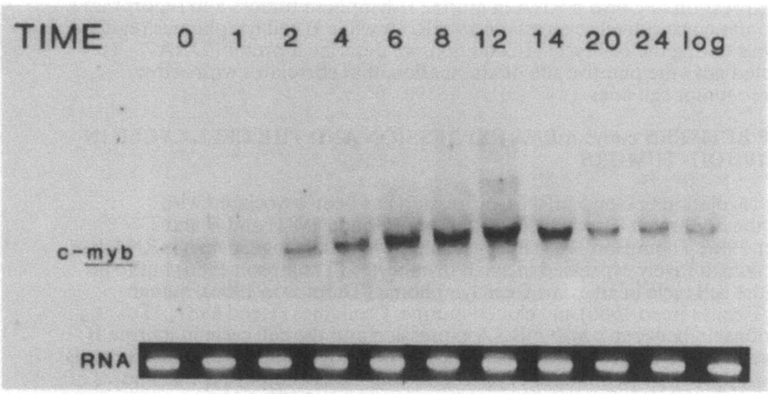

Figure 2. Expression of c-*myb* mRNA in BCL_1 cells released from NEAA starvation. NEAA starved BCL_1 cells were selected as described in the legend to Fig. 1. At the time points shown, after replenishing starved cells with NEAA, cells were harvested and an aliquote from each time point was stained with Hoechst stain 33258 and analyzed by FACS for relative DNA content. The rest of the cells were used to prepare total cellular RNA (Chirgwin 1979). Ten micrograms of total cellular RNA from each time point was fractionated on a 1% agarose gel containing 0.22M formaldehyde, transferred to nitrocellulose and hybridized to a nick translated 2.4 kb murine c-*myb* cDNA probe as previously described (Bender 1986; Bender 1987a). One microgram of total cellular RNA was fractionated on a 1% non-denaturing agarose gel to confirm that an equivalent amount

the experiment (Greenberg 1984; Kelly 1983). In contrast histone 2b mRNA, which is predominantly expressed during S-phase, begins to increase at ten hours, becomes maximal at approximately 16 hours and by 24 hours is equivalent to the expression detected in a logarithmically growing population of BCL_1 cells. As shown in Fig. 2, c-*myb* mRNA expression is very low or undetectable at T_0. However, c-*myb* mRNA expression begins to increase at two hours after addition of NEAA and is maximal at 12-14 hours after which it decreases to that of a logarithmically growing population of BCL_1 cells as the cells enter G_2/M and lose synchrony. The detection of c-*myb* mRNA expression during this time slightly precedes the increase in histone 2b mRNA expression indicating that expression of c-*myb* mRNA may begin to increase late in G_1 and continue into S-phase. If NEAA are added back to G_1 selected BCL_1 cells in the presence of aphidicolin (5μg/ml), which inhibits DNA synthesis, the cells do not enter S-phase as measured by FACS analysis and tritiated thymidine uptake. In addition, expression of histone 2b mRNA does not increase. However, aphidicolin does not inhibit the expression of c-*myb* mRNA during this time course indicating that c-*myb* mRNA expression is independent of DNA synthesis and probably begins during late G_1. These results are in basic agreement with findings relating to c-*myb* mRNA expression during the cell cycle in T cells (Thompson 1986; Reed 1986, 1987) which indicate that c-*myb* mRNA is expressed during late G_1/early S-phase. However, it has been reported during the activation of resting human T cells that expression of c-*myb* protein is blocked by treatment of the cells with aphidicolin (Lipsick 1987). This is in contrast to our findings using the BCL_1 cell line where we do not find the inhibition of c-*myb* mRNA expression by aphidicolin. The reason for this difference is not clear at present. Although it is possible that this reflects a fundamental difference between B and T cells it brings up the interesting possibility that substantial levels of c-*myb* mRNA appear in the cells before it is translated.

In our attempts to examine the regulation of c-*myb* mRNA during the cell cycle in pre-B cell lymphomas and to compare findings in these cells with B cell lymphomas we have been unsuccessful in using metabolic blocks to obtain a synchronous population of cells other than with BCL_1. As an alternative, we have used counter flow centrifugation (elutriation) to separate logarithmically growing cells based on size. Elutriation has the advantages of not requiring metabolic manipulation of the cells and allowing one to examine gene expression at various stages of the cell cycle in actively cycling cells (Thompson 1985). We have used this technique to compare c-*myb* mRNA expression in the pre-B cell lymphoma 70Z/3.12 and the mature B cell lymphoma A20.1.11. The BCL_1 cell line was found to be too fragile to separate by elutriation. During elutriation experiments an aliquote of cells was taken from each fraction and analyzed for DNA content by FACS analysis after staining with Hoechst stain 33258 to measure enrichment for cells at a given point in the cell cycle. Each cell line was examined at different rotor speeds to maximize separation. As we found in our studies with NEAA starved BCL_1 cells c-*myb* mRNA levels vary during the cell cycle in the B cell lymphoma with maximal expression occurring in late G_1/early S-phase. In contrast, we found no variation in c-*myb* mRNA levels during the cell cycle in 70Z/3.12. We note in elutriation fractions obtained with B cell lymphoma cells that the maximal levels of c-*myb* mRNA expression are still not equivalent to the level detected in the pre-B cell lymphoma even when one takes into account the amount of enrichment for given stages of the cell cycle in each fraction. Analogous results were obtained in elutriation experiments with a murine erythroleukemia cell line (C19) indicating that constitutive expression of c-*myb* mRNA may be a general property of immature hematopoietic cell lines where c-*myb* mRNA is expressed at a high level. These results indicate that c-*myb* mRNA is constitutively expressed in pre-B cell lymphomas, as opposed to B cell lymphomas, and suggests that a switch may occur during normal B cell development from constitutive to a cell cycle regulated mode of expression.

We have examined the expression of c-*myb* mRNA in Balb/c 3T3 cells after release from serum starvation to compare our results in B-lymphoid tumors to a non-hematopoietic cell type. We do not detect expression of c-*myb* mRNA at any point up to 36 hours after release from serum starvation or in log phase Balb/c 3T3 cells even after extensive autoradiography. By contrast, c-*fos* and c-*myc* mRNA are readily detected at appropriate times after release from serum starvation (Greenberg 1984). Thus, steadystate expression of c-*myb* mRNA during the cell cycle does not occur in all cell types. This result is in contrast to previous data where it was suggested that c-*myb* mRNA was expressed in these cells after release from serum starvation (Greenberg 1984). However, these authors used a v-*myb* target, containing viral sequences, in a nuclear run-on assay which may explain this discrepancy. Alternatively, it is possible that read through transcription occurs during the cell cycle in Balb/c 3T3 cells but that steady state levels of c-*myb* mRNA do not accumulate.

A PROTEIN/DNA INTERACTION CORRELATES WITH ATTENUATION

We have previously shown that the differentially regulated expression of c-*myb* mRNA in murine B-lymphoid tumors appears to be primarily maintained by attenuation which occurs in the first intron of the murine c-*myb* locus (Bender 1987b). As shown diagrammatically in Fig. 3, we have approximately mapped the site of attenuation to a 300bp BstEII/XbaI fragment using a nuclear run-on assay. Since this fragment is the 3' most target to which nuclear run-on probes derived from B cell lymphomas (A20.2J and BCL_1) and plasmacytomas (S194 and MPC-11) hybridize it remains possible that attenuation occurs slightly 3' of the XbaI site and that hybridization is poorly detected. Interestingly, we have reported a DNase I hypersensitive site (marked with an arrow in Fig. 3) which is more prominent in B cell lymphomas than pre-B cell lymphomas (Bender 1987b) that maps to approximately the same position as attenuation. Since DNase I hypersensitive sites are often associated with sites of protein/DNA interactions we have used a gel retardation assay (Singh 1986) and a modified lambda exnuclease assay (Takimoto 1989) to search for protein/DNA interactions in this region. As shown in Fig. 3 we have concentrated our efforts on an approximately 1kb region which includes the putative site of attenuation. The nucleic acid sequence of this region is quite highly conserved in the mouse and human systems (>60%). This high degree of sequence homology is unusual in introns and has previously been associated with functional regulatory domains (Emorine 1983). We have detected a number of protein/DNA interactions in this region which have not consistently correlated with either high or low levels of c-*myb* mRNA expression. However, using fragment 4 (Fig.3) which is a 155bp BglII/Bam H1 fragment a very strong gel shift was detected in nuclear extracts from a B cell lymphoma (WEHI 231) and a plasmacytoma (S194) but was only weakly detected in a pre-B cell lymphoma (70Z/3.12). This result was confirmed using the exonuclease assay and has been extended to three additional pre-B cell lymphomas ($HAFTL_1$, 1881 and LS8), a murine erythroleukemia cell line (C19), a B cell lymphoma (A20.1.11), an additional plasmacytoma (MOPC21) and a fibroblast cell line (Balb/c 3T3). In all cases, this interaction was strongly detected using the exonuclease assay in cell lines which express low or undetectable levels of c-*myb* mRNA but was either not detected or very weakly detected in the pre-B cell lymphomas and the murine erythroleukemia cell line. Thus, this protein/DNA interaction correlates with active attenuation in murine B-lymphoid tumors. We have recently obtained a DNase I footprint of this interaction and confirmed the gel retardation results with an oligonucleotide which includes the DNase I protected sequence. At present we have no information relating to a potential functional relationship between the protein/DNA interaction detected using fragment 4 and transcriptional attenuation. We are investigating this possibility.

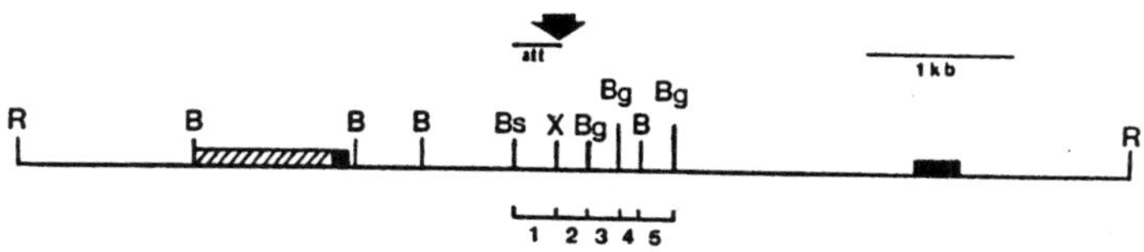

Figure 3. Diagrammatic representation of the 7.6 kb EcoR1 genomic DNA fragment containing the murine
c-*myb* attenuator region. The 300bp Bst EII/Xba I fragment used to approximately map the position of attenuation is underlined and the differentially detected DNase I hypersensitive site IV (Bender 1987a) is marked with an arrow. Fragment 4, is the 155kb Bgl II/Bam HI fragment used to detect a protein/DNA interaction in B cell lymphomas and plasmacytomas. Restriction sites are R (EcoRI), B (Bam HI), Bs (BstE II), X (Xba I) and Bg (Bgl II).

ACKNOWLEDGEMENTS
The authors would like to thank Ms. Marie Burley for help in preparation of this manuscript. This work was supported by grant 5P01CA40042-05 from the National Institutes of Health and grant CRT2663 from the Council For Tobacco Research, USA to T.P.B.

REFERENCES

Bender TP and Kuehl WM (1986) Murine *myb* Proto-oncogene messenger RNA: cDNA sequence and evidence for 5' heterogeneity. PNAS 83:3204-3208
Bender TP and Kuehl WM (1987a) Differential expression of the c-*myb*proto-oncogene marks in the pre-B cell/B cell junction in murine B lymphoid tumors. J Immunol 139:3822-3827
Bender TP, Thompson CB and Kuehl WM (1987b) Differential expression of c-*myb* mRNA in murine B lymphomas by a block to transcription elongation. Science 237:1473-1476
Chirgwin J, Aeybyle A, McDonald R and Rutter W (1979) Isolation of biologically active ribonucleic acid from sources enriched in ribonuclease. Biochemistry 18:5294-5299
Emorine L, Kuehl WM, Weir L, Max E and Leder P (1983) A conserved sequence in the immunoglobulin J_i-C_i intron: possible enhancer element. Nalture 304:447
Gonda TJ and Metcalf D (1984) Expression of *myb*, *myc* and *fos* proto-oncogenes during the differentiation of a murine myeloid leukemia. Nature 310:249-251
Gonda TJ, Sheiness DK and Bishop JM (1982) Transcripts from the cellular homologues of retroviral oncogenes: Distribution among chicken tissues. Mol Cell Biol 2:617-624
Greenberg ME and Ziff EB (1984) Stimulation of 3T3 cells induces transcription of the c-*fos* proto-oncogene. Nature 311:433
Kelly K, Cochran BH, Stiles CD and Leder P (1983) Cell-specific regulation of the c-*myc* gene by lymphocyte mitogens and platlet - derived growth factor. Cell 35:603-610
Kirsch IR, Bertness V, Silver J and Hollis GF (1986) Regulated expression of the c-*myb* and c-*myc* oncogenes during erythroid differentiation. J Cellular Bioch 32:11-21
Klempnauer K, Seymonds J, Evan GI and Bishop JM (1984) Subcellular localization of proteins encoded by oncogenes of avian myeloblastosis virus and avian E26 and by the chicken c-*myb* gene. Cell 37:537-547

Lipsick JS and Boyle WJ (1987) c-*myb* protein expression is a late event during T cell activation. Mol Cell Biol 7:3358-3360

Luscher B, Christenson E, Litchfield DW, Krebs EG and Eisenman RN (1990) Myb DNA binding inhibited by phosporylaiton at a site deleted during oncogenic activation. Nature 344:517-522

Reed JC, Alpers JD, Nowell PC and Hoover RG (1986) Sequential expression of proto-oncogenes during lectin-stimulated mitogenesis of normal human lymphocytes. PNAS 83:3982-3987

Reed JC, Alpers JD, Scherle PA, Hoover RG, Nowell PC and Prystowsky MB (1987) Proto-oncogene expression in cloned T lymphocytes: mitogenes and growth factors induce different patterns of expression. Oncogne 1:223-228

Shapiro HM (1988) Practerical Flow cytometry. 2nd edition Alan R Liss, Inc New York, p 141

Singh H, Sen R Baltimore D and Sharp PA (1986) A nuclear factor that binds to a conserved sequence motif in transcriptional control elements of immunoglobulin genes. Nature 319:154-158

Takimoto M, Quinn JP, Farina AR, Staudt LM and Levens D (1989) *fos/jun* and octamer-binding protein interact with a common site in a negative element of the human c-*myc* gene. J Biol Chem 264:8992-8999

Thiele C, Cohen PS and Israel MA (1988) Regulation of c-*myb* expression in human neuroblastoma cells during retinoic acid induced differentiation. Mol Cell Biol 8:1677-1683

Thompson CB, Challoner PB Neiman PE and Groudine M (1985) Levels of c-*myc* oncogene mRNA are invariant throughout the cell cycle. Nature 314:363-366

Thompson CB, Challoner PB, Neiman PE and Groudine M (1986) Expression of the c-*myb* proto-oncogene during cellular proliferation. Nature 319:374-379

Watson RJ (1989) Expression of the c-*myb* and c-*myc* genes is regulated independently in differentiating mouse erythroleukemia cells by common processes of premature transcription arrest and increased mRNA turnover. Molec Cell Biol 8:3938-3942

Westin EH, Gallo RC, Arya SK, Eva A, Souza LM, Baluda MA, Aaronson SA and Wong-Staal F (1982) Differential expression of the *amv* gene in human hematopoietic cells. Proc Natl Acad Sci 79:2194-2198

Weston K and Bishop JM (1989) Transcriptional transaction by the v-*myb* oncogene and its cellular progenitor, c-*myb*. Cell 58:85-93

Bcl-2: B Cell Life, Death and Neoplasia

Stanley J. Korsmeyer, Timothy J. McDonnell, Gabriel Nunez, David Hockenbery and Robert Young

Howard Hughes Medical Institute, Department of Medicine and Molecular Microbiology, Washington University School of Medicine, 660 South Euclid, Box 8045, St. Louis, MO 63110

The t(14;18)(q32;q21) is the most common translocation associated with human lymphoid malignancies. Approximately 85% of follicular and 20% of diffuse B cell lymphomas possess this translocation (Fukuhara 1979, Yunis 1987). Follicular lymphoma is initially a low grade disease consisting of small resting B cells. Over time, a histologic conversion to diffuse large cell architecture and aggressive high grade lymphoma occurs frequently in patients (Horning 1989). Molecular cloning of the t(14;18) breakpoint revealed a putative proto-oncogene, Bcl-2 at 18q21 (Bakhshi 1985, Cleary 1985, Tsujimoto 1985). The translocation occurs early in pre B cell development during Ig gene rearrangement and juxtaposes Bcl-2 with the Ig heavy chain locus at 14q32 (Bakhshi 1987). This creates a Bcl-2-Ig fusion gene that is markedly deregulated resulting in inappropriately elevated levels of Bcl-2-Ig RNA and the 25 kd Bcl-2 protein (Graninger 1987, Seto 1988). Bcl-2 encodes a 25 kd integral membrane protein of an intracellular organelle (Hockenbery, submitted). We have taken several approaches to examining the functional role of Bcl-2 including overexpressing Bcl-2 in lymphoblastoid lines, growth factor dependent lines and transgenic mice.

We used the N2 series of retroviral vectors from Dr. Eli Gilboa which possess a Neo^R gene for selection and inserted either the human or murine Bcl-2 cDNA under the control of the SV40 promoter. In an initial set of experiments done in collaboration with Dr. Riccardo Dalla-Favera's lab we placed deregulated human Bcl-2 (N2-H-Bcl-2) into lymphoblastoid cell lines (LCL) immortalized with EBV. The presence of N2-H-Bcl-2 produced a consistent 3-fold improvement in clonogenicity as assessed in soft agar (Nunez 1989). Deregulated Bcl-2 as a single agent was insufficient to confer tumorigenicity to LCL, however Bcl-2 was able to complement myc in this system. These findings suggested a role for Bcl-2 in B cell growth and neoplasia. This prompted us to examine if Bcl-2 could be directly implicated in a growth factor pathway. To address this, the murine N2-M-Bcl-2 vector was placed into a series of interleukin dependent cell lines to determine if overproduced Bcl-2 would spare the need for a specific ligand receptor interaction. No long-term growth factor independent lines resulted from deregulated Bcl-2 in IL-3, IL-2 or IL-6 requiring cell lines (Nunez 1990). However, Bcl-2 consistently spared the death and extended the survival of certain interleukin deprived cells. This was particularly true for IL-3 dependent pro B lymphocytes (FL5.12), promyelocytes (FDC P1) and mast cells (32D). Although viable, IL-3 deprived cells

Current Topics in Microbiology and Immunology, Vol. 166

bearing deregulated Bcl-2 from either N2-M-Bcl-2 or a transfected SFFV LTR based expression vector were in G_0. Bcl-2 did not influence cell cycle progression nor did it alter the dose response range to limiting concentrations of IL-3. This result was not restricted to the IL-3/IL-3 receptor signal transduction pathway in that promyeloid cells maintained in GM-CSF or IL-4 displayed a similar response. Yet, Bcl-2 enhanced cell survival was not universal as neither IL-2 dependent T cell lines nor an IL-6 dependent myeloma line have demonstrated a consistent effect upon IL withdrawal. Therefore, Bcl-2 interferes with cell death but in a cell type or factor-restricted fashion.

These observations have prompted a detailed study of the mechanism of death of IL-3 deprived cells. Serial examination of FL5.12 following IL-3 deprivation revealed a death by apoptosis in which plasma membrane blebbing and volume loss was followed by nuclear condensation and an endonucleolytic cleavage of DNA into oligo-nucleosomal length fragments. The presence of deregulated Bcl-2 blocks this programmed cell death. Cells return to G_0 but did not die and can be rescued by addition of IL-3 30 days following deprivation (Hockenbery, submitted).

Bcl-2's role as an antidote to programmed cell death prompted us to examine Bcl-2's distribution in normal lymphoid tissue. Immunohistochemical examination of well defined secondary germinal centers revealed that the follicular mantle zone, comprised of long-lived recirculating IgM/IgD B cells, was the most intensely stained for Bcl-2 (Fig. 1). Of note, Bcl-2 protein was essentially absent from the centroblast populated dark zone and the basal light zone where centrocytes are dying rapidly by apoptosis. Provocatively, Bcl-2 returns in the apical light zone where it has been proposed that a subset of B cells survive due to the presence of residual antigen presented by follicular dendritic cells. This remarkable topographic distribution indicates a prominent role for Bcl-2 in a B cell survival pathway.

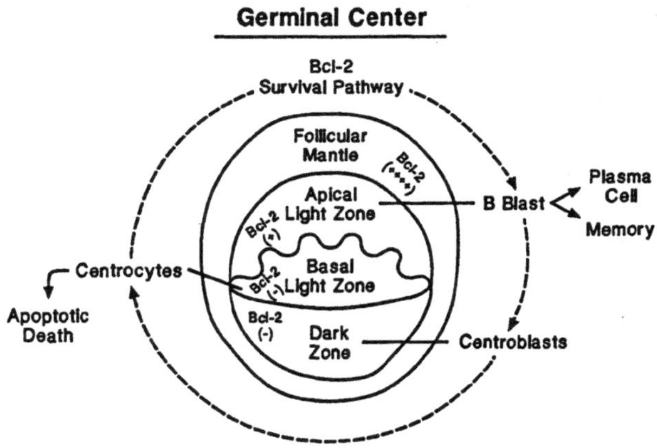

Fig. 1 Distribution of Bcl-2 protein within germinal centers as determined by immunoperoxidase staining using a monospecific monoclonal antibody.

To study the role of Bcl-2 in an intact immune system we generated transgenic mice bearing a Bcl-2-Ig minigene that recapitulates the pathologic consequence of the t(14;18) (McDonnell 1989). The deregulated Bcl-2-Ig transgene was overexpressed relative to endogenous mouse Bcl-2 in spleen and thymus (McDonnell 1990). The human 25 kd Bcl-2 protein, which arose from the transgene, was overproduced in spleen greater than thymus and also within the expanded B cell subset. Despite the generalized deregulation, two color flow cytometry and density gradient centrifugation indicated the expanded lymphocytes were predominantly small, resting B cells co-expressing B220, IgM, IgD, Ia and K but not CD5. Cell cycle analysis confirmed that about 97% of the expressed B cells reside in G_0/G_1. Transgenics possess a 3-4 fold excess of small resting B cells in spleen, lymph node and bone marrow. These recirculating IgM/IgD B cells accumulate due to extended survival rather than increased proliferation. While resting these cells are not terminal, but proliferate in response to lipopolysaccharide and anti-IgM and demonstrated normal B cell colony formation in soft agar (McDonnell 1990). Moreover, these mice proved able to manifest primary and secondary responses to the T-dependent antigen FITC-KLH. The initial primary and classic secondary response was relatively normal. However, Bcl-2-Ig transgenics demonstrate an extended secondary immune response from the effect of Bcl-2 upon enhancing cell survival.

Bcl-2-Ig transgenic mice are initially healthy with expanded follicular center B cell zones of the spleen and fusion of adjacent regions of white pulp. Since the deregulated Bcl-2-Ig transgene is present in every B cell this initial expansion is polyclonal. Autopsies of asymptomatic mice revealed evidence for oligoclonality of B cells by one year of age, but none of these tissues showed evidence of lymphoma. However Bcl-2-Ig transgenic mice progress to malignant lymphoma after a long latency period that now averages 14 mos, but may increase with time (McDonnell submitted). The most common lymphoma was a high grade diffuse large cell that was immunoblastic (DHL-I). DHL-I presented as a primary mesenteric lymphoma with frequent dissemination to splenic red pulp, periportal liver, and kidney. These high grade B cell tumors are all monoclonal and have a high percentage of cells in cycle. This progression from polyclonal hyperplasia to monoclonal high grade lymphoma with a long latency period implicates the acquisition of secondary genetic changes. Half of the DHL-I tumors have demonstrated a rearranged c-myc gene. This indicates that transgenic mice prospectively select a gene which promotes proliferation to complement their inherent survival advantage. Overall 15 high grade DHLs and 3 low grade poorly differentiated lymphocytic lymphomas (PDLL) have developed in Bcl-2-Ig transgenic mice. In contrast a comparable number of non-transgenic littermates have demonstrated 2 lymphomas that were of a different histologic type and confined to the spleen. The low grade PDLLs effaced white pulp architecture and invaded red pulp. However they were not monoclonal possessing only a minority population of clonal B cells. This suggests that deregulated Bcl-2 may be sufficient for the generation of these low grade neoplasms (McDonnell, submitted).

These studies indicate a unique role for Bcl-2 as a proto-oncogene that extends cell survival by blocking programmed cell death independent of promoting proliferation. Bcl-2-Ig transgenic mice establish that prolonged B cell life is tumorigenic.

Bakhshi A, Jensen JP, Goldman P, Wright JJ, McBride OW, Epstein AL, Korsmeyer SJ (1985) Cloning the chromosomal breakpoint of the t(14;18) human lymphomas: clustering around J_H on chromosome 14 and near a transcriptional unit on 18. Cell 41:889-906.

Bakhshi A, Wright JJ, Graninger W, Seto M, Owens J, Cossman J, Jenson JP, Goldman P, Korsmeyer SJ (1987) Mechanism of the t(14;18) chromosomal translocation: structural analysis of both derivative 14 and 18 reciprocal partners. Proc Natl Acad Sci USA 84:2396-2400.

Cleary ML, Sklar J (1985) Nucleotide sequence of a t(14;18) chromosomal breakpoint in follicular lymphoma and demonstration of a breakpoint cluster region near a transcriptionally active locus on chromosome 18. Proc Natl Acad Sci USA 82:7439-7443.

Fukuhara S, Rowley JD, Varrakojis D, Golomb HM (1979) Chromosome abnormalities in poorly differentiated lymphocytic lymphoma. Cancer Res 39:3119-3128.

Graninger WB, Seto M, Boutain B, Goldman P, Korsmeyer SJ (1987) Expression of Bcl-2 and Bcl-2-Ig fusion transcripts in normal and neoplastic cells J Clin Invest 80:1512-1515.

Hockenbery D, Nunez G, Milliman C, Nahm M, Schreiber R, Korsmeyer SJ Bcl-2 protein blocks programmed cell death. Submitted 1990.

Horning SJ, Rosenberg SA (1984) The natural history of initially untreated low-grade non-Hodgkin's lymphomas. N Engl J Med 311:1471-1475.

McDonnell TJ, Deane N, Platt FM, Nunez G, Jaeger U, McKearn JP, Korsmeyer SJ (1989) Bcl-2-immunoglobulin transgenic mice demonstrate extended B cell survival and follicular lymphoproliferation. Cell 57:79-88.
McDonnell TJ, Nunez G, Platt FM, Hockenbery D, London L, McKearn JP, Korsmeyer SJ (1990) Deregulated Bcl-2-immunoglobulin transgene expands a resting but responsive immunoblogulin M and D-expressing B-cell population. Mol Cell Biol 10:1901-1907.
McDonnell TJ, Korsmeyer SJ Bcl-2-Ig transgenic mice progress from benign lymphoid hyperplasia to high grade malignant lymphoma. Submitted 1990.
Nunez G, Seto M, Seremetis D, Ferrero F, Grignani F, Korsmeyer SJ, Dalla-Favera R (1989) Growth and tumor-promoting effects of deregulated Bcl-2 in human B lymphoblastoid cells. Proc Natl Acad Sci USA 86:4589-4593.
Nunez G, London L, Hockenbery D, Alexander M, McKearn JP, Korsmeyer SJ (1990) Deregulated Bcl-2 gene expression selectively prolongs survival of growth factor-deprived hemopoietic cell lines. J Immunol 144:3602-3610.
Seto M, Jaeger U, Hockett RD, Graninger W, Bennet S, Goldman P, Korsmeyer SJ (1988) Alternative promoters and exons, somatic mutation and transcriptional deregulation of the Bcl-2-Ig fusion gene in lymphoma. EMBO J 7:123-131.
Tsujimoto Y, Gorham J, Cossman J, Jaffe E, Croce CM (1985). The t(14;18) chromosome translocations involved in B-cell neoplasms result from mistakes in VDJ joining. Science 229: 1390-1393.
Yunis JJ, Frizzera G, Oken MM, McKenna J, Theologides A, Arnesen M (1987) Multiple recurrent genomic defects in follicular lymphoma: a possible model for cancer. N Engl J Med 316:79-84.

C-*myc* Genes in B-Cell Neoplasia

Binding of NF-KB-like Factors to Regulatory Sequences of the c-*myc* Gene

M.P. Duyao*, D.J. Kessler*, D.B. Spicer*, and G.E. Sonenshein

Department of Biochemistry, Boston University School of Medicine, Boston, MA 02118 USA

INTRODUCTION

Changes in expression of the c-*myc* proto-oncogene during the response of cells to growth or differentiation are often mediated at the transcriptional level, including control of the rates of initiation and elongation. Regulation of the level of initiation and elongation of c-*myc* gene transcription is likely to be mediated by the interaction of specific proteins with regulatory elements near or within the gene. Both positive and negative regulatory elements have been identified for the murine c-*myc* gene (Remmers et.al. 1986; Bentley and Groudine, 1988). To elucidate the mechanisms involved in negative control of c-*myc* transcription, we have employed the WEHI 231 early B lymphoma cell line as a model system, since a major component of regulation of c-*myc* gene expression in these murine cells occurs at the level of transcription. Proliferation of WEHI 231 cells can be arrested within 24-48 hours following incubation with an antiserum against the expressed surface Ig, such as a goat anti-mouse Ig preparation (GaMIg) (McCormack et. al. 1984) or by treatment with phorbol ester (Levine et.al. 1986). Previously we demonstrated that a selective 5- to 10- fold decrease in c-*myc* mRNA expression occurs within 24 hours of treatment; this drop correlated with a decrease in the transcription of the c-*myc* gene (Levine et. al. 1986). We reasoned that control of this decreased rate of transcription could relate either to increased factor binding to a negative element or decreased binding to a positive element. In either case, a change in binding might be apparent. Here we report on the mapping of changes in the complex interaction of proteins at two sites within the c-*myc* locus during the down modulation of c-*myc* gene transcription in growth arrested WEHI 231 cells. The sequences mediating this binding are interacting with proteins within the NF-KB family of factors, implicating these factors in the regulation of c-*myc* gene transcription.

RESULTS

Changes in Binding Map to Upstream Regulatory and Internal Sequences.

To characterize protein-DNA interactions, crude nuclear extracts were prepared from WEHI 231 cells in exponential growth and following 24 hours of treatment with goat anti-mouse immunoglobulin antiserum (GaMIg) or phorbol ester (TPA) at which time the transcription of c-*myc* is diminished. DNA fragments including upstream regulatory elements and exon 1 sequences gave multiple band shift patterns upon incubation with extracts from exponentially proliferating cells (E), (data not shown). When nuclear extracts from cells treated with GaMIg or TPA for 24 hours (24) were compared to those of exponentially proliferating cells, striking changes in binding profile were observed with two fragments, A and B, see Fig. 1a.

a) Upstream Regulatory Element: To map the specific site(s) of interaction within these regions, subfragments of A were employed independently in electrophoretic mobility-shift assays

* These authors made equal contributions to this manuscript.

Current Topics in Microbiology and Immunology, Vol. 166

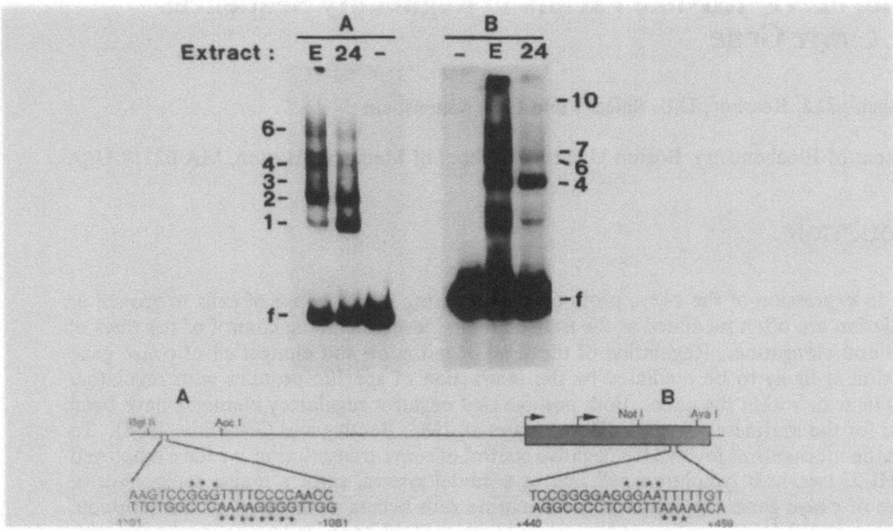

Fig. 1. A) Electrophoretic mobility-shift analysis on P^{32}-labeled fragments A and B. Binding assay with fragment A was performed using nuclear extracts from either exponentially proliferating (E) or GaMIg-treated (24) WEHI 231 cells. EMSA with fragment B was performed using nuclear extracts from WEHI cells in exponential growth (E) or WEHI 231 cells treated with TPA (24). B) Schematic diagram showing the location of fragments A and B and the protected sequences, URE and IRE. Asteriks show contact points as indicated by methylation interference analysis.

(EMSA) with nuclear extracts from exponentially proliferating WEHI 231 cells (data not shown). The results with fragment A indicated there are at least two distinct binding domains. The majority of the complexes (bands 1, 2, 3, 4 and 6) mapped to one site while band 5 mapped to a second site. DNAase I protection of bands 1 to 6 and copper O-phenanthroline chemical cleavage analysis of either band 1 or band 2 yielded protection of the sequence, 5'-AAGTCCGGGTTTTCCCCAACC-3', which spans from 1101 to 1081 b.p. upstream of the c-*myc* P1 start site. To define the nucleotides directly involved in this protein-DNA interaction, methylation interference experiments were performed using nuclear extracts from exponentially growing WEHI 231 cells and DNA labelled on either strand. All the complexes gave a similar methylation interference pattern (see Fig. 1) (data not shown). The complexes observed require contact with eleven bases within the sequence GGGTTTTCCCC and its complementary strand. This protected domain will be referred to as the upstream regulatory element (URE).

b) <u>Internal Sequence within Exon 1</u>: The 180 b.p. Not I-Ava I fragment within exon 1, fragment B, was similarly subjected to EMSA using nuclear extracts from WEHI 231 cells in exponential growth. Ten bands could be distinguished and these were given numbers 1 through 10 (data not shown). The fragment B was divided into three domains (sites 1, 2, and 3). The sequences mediating formation of bands 4, 6, 7, and 10 mapped to site 3 (data not shown). When nuclear extracts from cells treated for 24 hours with TPA were employed, we observed a reduction in formation of bands 6, 7, and 10, whereas the intensity of band 4 appeared essentially unchanged (Fig. 1). (In some experiments the extent of decline of band 6 was less than observed in Fig.

1b, and may be a reflection of the slight variability in extent of inhibition induced by the phorbol ester treatment.) The 1,10 phenanthroline/copper cleavage footprint of band 4 yielded a 20 b.p. protected region: 5′-TCCGGGGAGGGAATTTTTGT-3′. Additional sequences extending 5′ of this site also displayed some protection but to a much lesser extent (data not shown). Methylation interference analysis performed with band 4 DNA (data not shown), demonstrated protein interaction with two guanine residues as well as with two adenine residues on the coding strand, and three adenine residues on the non-coding strand (as indicated in Fig. 1). This protected region within fragment B has been termed the internal regulatory element (IRE).

The Protected Sequences Contain an NF-KB-like Binding Motif.

We noted that the protected regions within the upstream and internal regulatory elements of c-*myc* have homology with each other and with sequences mediating binding of NF-KB-like factors. To test the possibility that these two regions interact with similar proteins, a competition experiment was performed using labelled fragment B DNA (Fig. 2). A 100-fold molar excess of an oligonucleotide containing the URE sequence, competed for bands 4, 6, 7 and 10. (To a lesser extent, formation of band 8, which maps to site 2, was also reduced.) As expected, an oligonucleotide containing the protected internal site within exon 1 (IRE) effectively competed for formation of bands 4, 6, 7 and 10. In contrast, addition of a non-related oligonucleotide (CK-1), containing the sequences of the rat myosin light chain enhancer domain, failed to compete for binding. Similarly the IRE oligonucleotide can effectively compete for binding to the c-*myc* upstream regulatory region (data not shown). Thus these two sequences may effectively interact with the same or similar proteins.

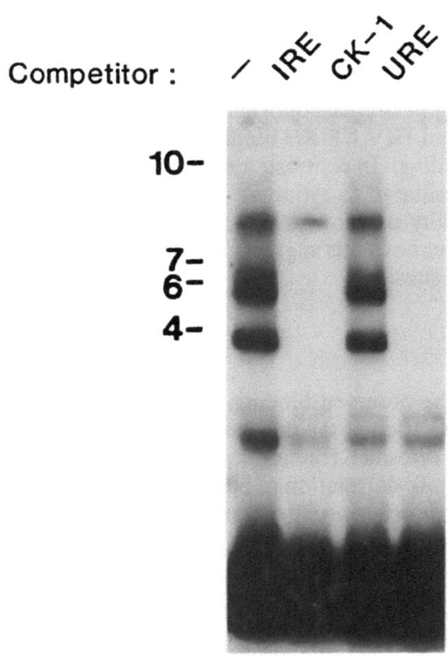

Fig. 2. Competition of IRE protein complexes by the URE.

Table 1. Synthetic Oligonucleotide Sequences Containing NF-KB Binding Motifs

Gene	Oligonucleotide Sequence*
SV40	GTTAGGGTGTGGAAAGTCCCCAGGCTCCCCA
HIV LTR	GATCCAGGGACTTTCC
human IL-2R α	GATCCGGCAGGGGAATCTCCCT
human IL-2	GATCCACAAAGAGGGATTTCACCTACATCC

*sequences involved in NF-KB binding are underlined

Comparison with the binding sites of known nuclear factors revealed homology of the two c-*myc* binding sequences to sequences mediating binding of NF-KB (GGGGACTTTCC) and members of the NF-KB family. Furthermore, the pattern of methylation interference is strikingly similar to that seen with NF-KB, and a related family member H2TF1(See review Lenardo and Baltimore, 1989).

As an initial test of whether an NF-KB-like factor is binding, competition analysis was performed using several double-stranded synthetic oligonucleotides representing various NF-KB binding sites (listed in Table 1). The results for fragments A and B are seen in Fig. 3a and 3b, respectively. Competition with a 50-fold molar excess of either the IL-2 or IL-2 receptor (IL-2R) NF-KB-like binding motifs effectively prevented the formation of complexes with fragment A represented by diminuition of bands 1, 2, 3, 4, and 6, but not band 5, which maps to another site. NF-KB binding sequences in SV40 and HIV (which are identical to the site in the kappa light chain gene enhancer) also competed; however, the efficiency of competition varied for the different complexes, with bands 1 and 3 displaying the most sensitivity to competition. An oligonucleotide containing an AP2 binding site derived from the metallothionein (MT) gene, which does not contain an NF-KB binding site, failed to compete for any binding. As seen in Fig. 3b, a 50-fold molar excess of the NF-KB-like binding sites within the IL-2 and IL-2R genes competes for binding to the IRE within exon 1, resulting in decreased formation of bands 4, 6, 7, and 10. A slightly higher molar excess of the SV40 and HIV NF-KB binding sites is needed for these oligonucleotides to effectively compete for binding. In contrast, a mutated version of this sequence, with the two internal guanine residues mutated to cytidines, failed to compete for binding even at a 200-fold molar excess. The ability of known NF-KB binding sites to compete for interaction of factors from WEHI 231 nuclear extracts suggests involvement of an NF-KB-like protein in complex formation at these two sites.

The Two c-*myc* Sequences Bind NF-KB Protein.

To assess the potential interaction of NF-KB itself in the observed binding, five criteria were employed: a) induction of binding activity upon differentiation of the murine pre-B 70Z/3 line to B cells; b) detergent release of binding in the cytosolic fraction of non-B cells; c) enhancement of binding activity in the presence of GTP; d) correlation of band formation with nuclear extracts from cells in which NF-KB is constitutively present or absent; e) lack of binding with mutation of the site.

a) Analysis of Nuclear Extracts from 70Z/3 pre-B Cells: Sen and Baltimore (1986b) have shown that NF-KB activity is induced upon differentiation of 70Z/3 pre-B cells to B cells following treatment with LPS. This induction of binding activity by LPS treatment is considered diagnostic of NF-KB (Lenardo and Baltimore, 1989). To evaluate whether NF-KB itself can bind to the upstream regulatory element, we tested the binding to fragment A of nuclear

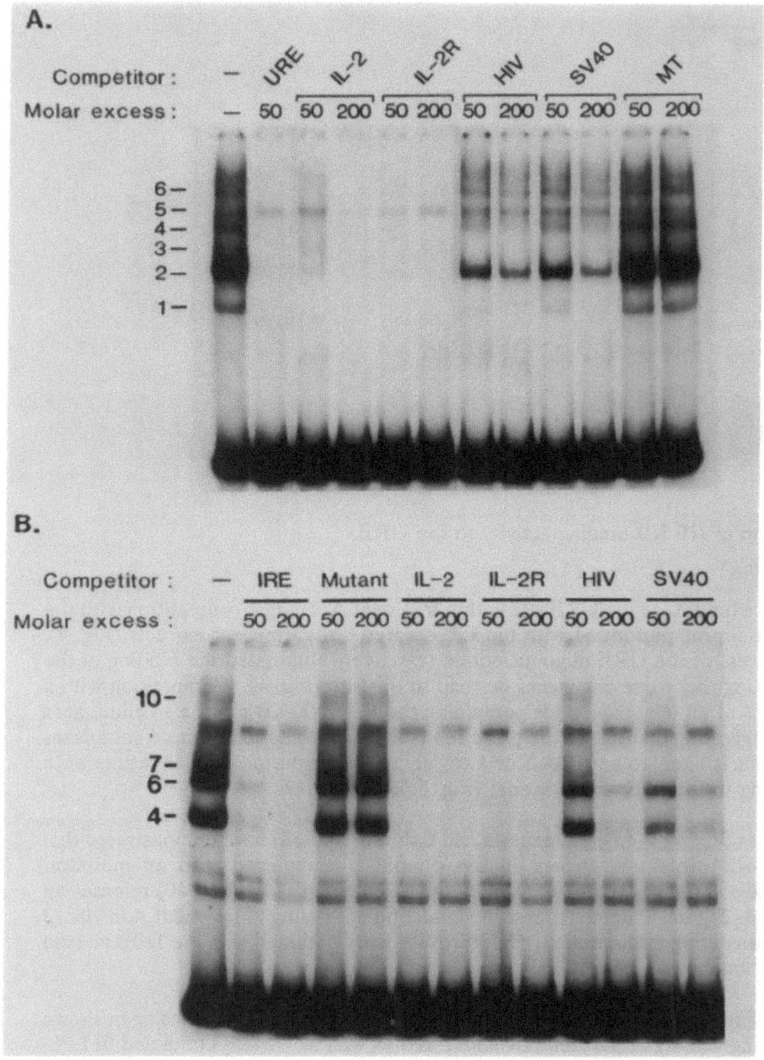

Fig. 3. Competition for binding of fragment A (A) and fragment B (B) by NK-KB Motifs from different genes.

extracts from 70Z/3 cells before or after 2 hours of LPS treatment (Fig. 4). In uninduced cells, the major binding detected appeared to co-migrate with band 5, although it is not known whether this complex is the same as that observed with WEHI 231 extracts. After treatment with LPS, the presence of three new complexes was observed. The major complex co-migrated with the complex represented by band 3 in WEHI 231 extract and the two minor complexes co-migrated with bands 1 and 2 (positions indicated in the figure). To demonstrate that these

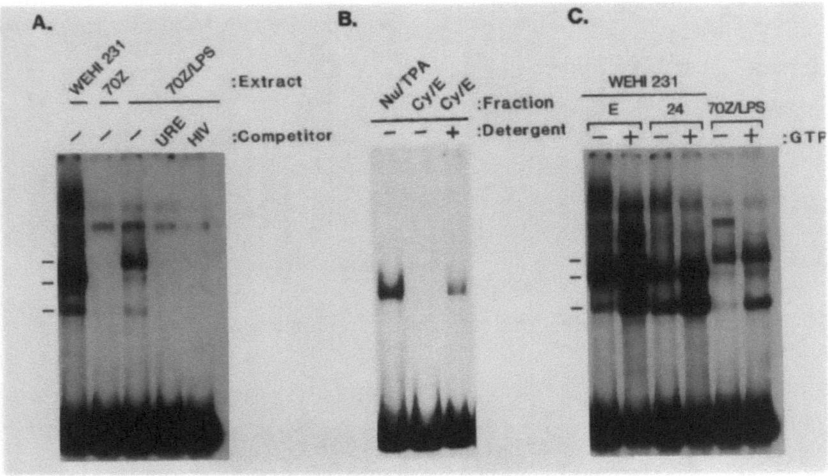

Fig. 4. Demonstration of NF-KB binding activity to the URE.

protein complexes are binding to the URE site within fragment A, binding competition with the specific oligonucleotide was performed with the LPS-treated 70Z/3 nuclear extracts (Fig. 4). A 100-fold molar excess of the URE oligonucleotide selectively eliminated the binding of the three bands, confirming that these complexes do map to this site (Fig. 4). Competition with a 100-fold molar excess of an oligonucleotide containing the HIV NF-KB motif also eliminated these three bands, suggesting that one of the proteins involved in formation of these complexes in induced 70Z/3 cells is likely to be NF-KB or a closely related protein. Similar results were obtained with binding to the IRE within exon 1 (Fig. 5c).

b) Detergent Releases Binding Activity: Baeuerle and Baltimore (1988) have demonstrated that in non-B cells, NF-KB protein is present in the cytoplasm complexed with an inhibitor. Treatment with 0.2% sodium deoxycholate (DOC) and 1.2% Nonidet P40 (NP40) releases an active NF-KB protein. Figure 4b demonstrates the induction of binding to fragment A in 70Z/3 cytosolic fraction upon detergent treatment; similar results were obtained with the 180 b.p. exon 1 fragment B (data not shown).

c) Presence of GTP Enhances Binding: The addition of GTP in binding assays has been shown to enhance the binding activity of purified bovine NF-KB as well as NF-KB contained in LPS-treated 70Z/3 nuclear extracts (Lenardo et. al. 1988). The intensities of two of the complexes (bands 1 and 3) induced by LPS treatment of 70Z/3 cells, which bind to the URE sequences, were markedly enhanced with GTP. Addition of GTP to binding reactions with nuclear extracts from exponentially growing WEHI 231 cells greatly enhanced the formation of all of the fragment A URE site complexes, bands 1, 2, 3, 4, and 6 (Fig. 4). Similar enhancement of binding was seen with the exon 1 region (Fig. 5). With extracts from GaMIg-treated WEHI 231 cells, formation of bands 1 and 2 was clearly enhanced; formation of the upper complexes was not as greatly affected.

d) Correlation of Co-migrating Complexes with the Presence of NF-KB Activity: Sen and Baltimore (1986a) have demonstrated that NF-KB binding activity correlates with expression

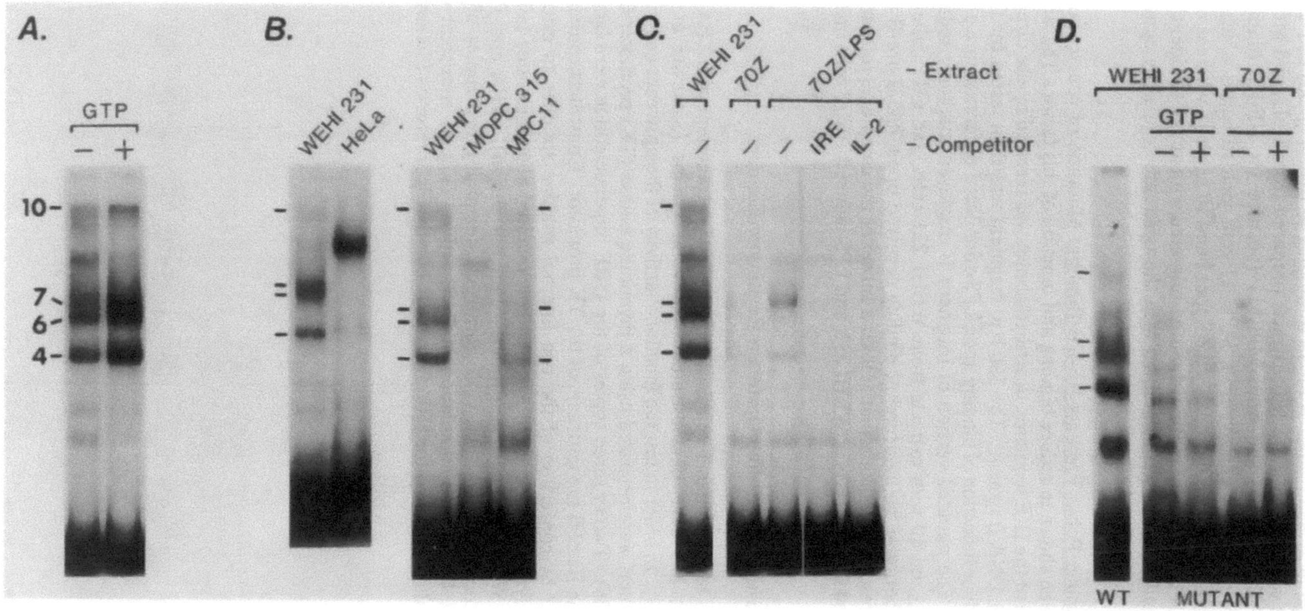

Fig. 5. Demonstration of NF-KB binding activity to the IRE.

of the immunoglobulin kappa light chain gene. Analysis of binding to IRE (Fig. 5b) and URE (data not shown) sequences was made with nuclear extracts prepared from HeLa and MOPC 315 cells (negative for NF-KB binding) and from MPC 11 (positive for NF-KB binding). Interaction of fragment B with nuclear extracts from MPC 11 cells yielded bands which co-migrate with WEHI 231 bands 4 and 7 (and probably the upper band 10 as well), whereas none of the major bands with MOPC 315 or HeLa cell extracts co-migrated with any of the identified IRE WEHI 231 bands. Similar results were obtained with the upstream regulatory element (data not shown).

e) Mutation of Internal Contact Points Prevents NF-KB Binding: Mutation of contact points within an NF-KB site has been shown to reduce binding and function of this factor. Therefore, the mutated oligonucleotide of the IRE sequence, described above, was inserted by site directed mutagenesis into the exon 1 DNA fragment. The 180 b.p. fragment was then tested for GTP enhancement of binding and for induction of binding upon LPS-stimulated differentiation of the 70Z/3 cell line (Fig. 5d). The mutated version of fragment B displayed markedly reduced formation of bands 4, 6, 7, and 10 with extracts from WEHI 231 cells, and GTP failed to enhance this binding. Furthermore, no induction of binding occurred upon 70Z/3 pre-B to B cell differentiation. These results all indicate that at least a subset of the complexes interacting with c-*myc* sequences derived from extracts of WEHI 231, and LPS-induced 70Z/3 cells include NF-KB.

Functional Activity of the c-*myc* NF-KB-like Upstream Regulatory Element.

Induction of NF-KB activity in the human T cell Jurkat line following treatment with TPA and phytohemagglutinin (PHA) has been demonstrated by several groups (Siekevitz et. al., 1987;Leung and Nabel, 1988). In order to test the functional activity of the upstream c-*myc* NF-KB-like binding site, constructs were prepared using a thymidine kinase (TK) promoter-CAT construct (Mason et. al. 1986). Two or three copies of the URE oligonucleotide were ligated, in either orientation, into the BamHI site upstream of the TK promoter. Constructs containing an oligonucleotide mutated by conversion of the two internal guanine residues to cytosine residues were similarly analyzed (see legend to Fig. 6). This mutated oligonucleotide fails to form any complexes with WEHI 231 nuclear extracts (data not shown). TPA and PHA treatment of Jurkat cells transfected with TK-CAT constructs containing two copies of the normal URE resulted in a 4- to 9-fold stimulation in CAT activity (Fig. 6) consistent with induced CAT conversion values obtained for HIV and IL-2R by other laboratories (Siekevitz,

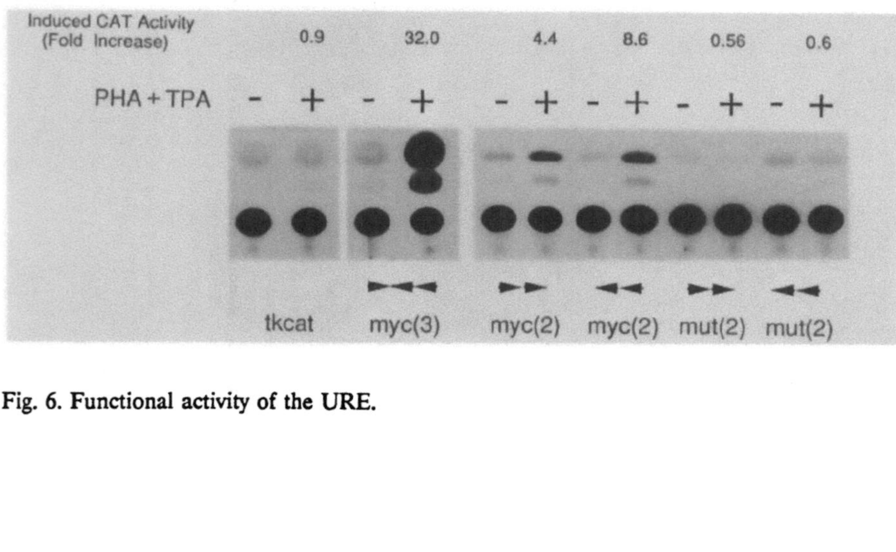

Fig. 6. Functional activity of the URE.

et. al. 1987;Bohnlein et. al. 1988). Three copies of the site resulted in over a 30-fold activation of CAT activity. The constructs containing no URE site or mutated site oligonucleotides did not display stimulation. Thus the URE site, located upstream of the c-*myc* promoter, can bind NF-KB-like factors and function to modulate transcriptional activity of a heterologous promoter.

DISCUSSION

We have identified two sites within the murine c-*myc* gene which interacts with NF-KB-like factors in the formation of multiple complexes. One site is located 1101 to 1081 b.p. upstream of the P1 initiation site. The second site extends from +441 to +460 downstream of the P1 start site and is within the attenuation control domain identified by Bentley and Groudine (1988). Decreased formation of large complexes with these sites is seen with the decrease in c-*myc* gene transcription during the anti-Ig or TPA induced growth arrest of WEHI 231 cells. Involvement of NF-KB itself has been demonstrated based on the observed induction of binding to these sites upon differentiation of 70Z/3 pre-B to B cells, the detergent release of binding activity in cytoplasmic extracts, the GTP enhancement of binding, and the mutant study. However, the multiplicity of complexes formed in mobility shift analyses is greater than that observed typically with members of the NF-KB family, including binding to the kappa enhancer B site (Sen and Baltimore, 1986b). This result implies either that i) multiple members of the NF-KB family interact with these sites, ii) that modified forms of the factors can bind, and/or iii) that additional proteins are present in the complexes. Proteins analyses are required to distinguish between these possibilities.

While NF-KB was originally implicated in control of kappa light chain expression in B cells, this factor and other related family members have been shown to be involved in the activation of a growing number of genes (Lenardo and Baltimore, 1989). In non-B cells, NF-KB is in an inactive cytoplasmic form complexed with an inhibitor protein (Bauerle and Baltimore, 1988). Activation of NF-KB binding in these cells can be mediated by such diverse agents as phorbol ester, mitogens, interleukins, viral infection, and tumor necrosis factor (Sen and Baltimore, 1986; Bohnlein et. al., 1988; Lenardo et. al., 1989; Osborn et. al., 1989). These agents are also often known to induce c-*myc* expression in diverse cell types, suggesting that in some cases induction of c-*myc* expression is related directly to NF-KB activiation. The transfection experiments in Jurkat cells presented here suggest a possible role of this family of factors during induction of c-*myc* expression. On the other hand, there is clearly not a complete concordance between c-*myc* expression and NF-KB activity. For example, NF-KB activity is constitutively present in nuclei of B cells regardless of their proliferative status and the level of c-*myc* expression. Experiments are in progress to test the functional significance of this binding site in the regulation of c-*myc* gene transcription.

Growth arrest of WEHI 231 cells correlates with a drop in the rate of c-*myc* gene transcription that is accompanied by several changes in the formation of complexes with these sites involving NF-KB-like factors. A loss of ability to form band 3 as well as the larger complexes (bands 4-6) and the concomitant increase in ability to form the highest mobility complex (band 1) was observed with the URE. A decrease in ability to form larger complexes was also seen with the IRE within exon 1. Interestingly, the IRE sequence encodes the U-rich stretch and upstream region found at the 3′ ends of the prematurely terminated transcripts following microinjection of c-*myc* DNA into oocytes (Bentley and Groudine, 1988). If altered binding to these sites has a functional role in the transcriptional down-modulation of c-*myc*, the results with WEHI 231 cells are unusual in that alteration in binding of the NF-KB family of factors is involved in a negative regulatory event.

ACKNOWLEDGEMENTS

This research was supported by Public Health Service Grant CA 35355.

REFERENCES

Baeuerle PA and Baltimore D (1988) IKB: A specific inhibitor of the NF-KB transcription factor. Science 242:540-546.

Baldwin AS and Sharp PA (1988) Two transcription factors, NF-KB and H2TF1, interact with a single regulatory sequence in the class I major histocompatibility complex promoter. Proc. Natl. Acad. Sci. USA 85:723-727.

Bentley DL and Groudine M (1988) Sequence requirements for premature termination of transcription in the human c-*myc* gene. Cell 53:245-256.

Bohnlein E, Lowenthal JW, Siekevitz M, Ballard DW, Franza BR and Greene WC (1988) The same inducible nuclear proteins regulates mitogen activation of both the interleukin-2 receptor-alpha gene and type 1 HIV. Cell 53:827-836.

Lenardo MJ and Baltimore D (1989) NF-KB: A pleiotropic mediator of inducible and tissue-specific gene control. Cell 58:227-229.

Lenardo MJ, Kuang A, Gifford A and Baltimore D (1988) NF-KB protein purification from bovine spleen: nucleotide stimulation and binding site specificity. Proc. Natl. Acad. Sci. USA 85:8825-8829.

Leung K and Nabel GJ (1988) HTLV-1 transactivator induces interleukin-2 receptor expression through an NF-KB-like factor. Nature 333:776-778.

Levine RA, McCormack JE, Buckler A and Sonenshein GE (1986) Transcriptional and post transcriptional control of c-*myc* gene expression in WEHI 231 cells. Mol. Cell. Biol. 6:4112-4116.

Mason P, Elkington J, Lloyd M, Jones M and Williams J (1986) Mutations downstream of the polyadenylation site of a xenopus β-globin mRNA affect the position but not the efficiency of 3′ processing. Cell 46:263-270.

McCormack JE, Pepe VH, Kent RB, Dean M, Marshak-Rothstein A and Sonenshein GE (1984) Specific regulation of c-myc oncogene expression in a murine B-cell lymphoma. Proc. Natl. Acad. Sci. 81:5546-5550.

Osborn L, Kunkel S and Nabel GJ (1988) Tumor necrosis factor α and interleukin 1 stimulate the human immunodeficiency virus enhancer by activation of the nuclear factor KB. Proc. Natl. Acad. Sci. USA 86:2336-2340.

Remmers EF, Yang, JQ and Marcus KB (1986) A negative transcriptional control element located upstream of the murine c-*myc* gene. EMBO J. 5:899-904.

Sen R and Baltimore D (1986a) Multiple nuclear factors interact with the immunoglobulin enhancer sequences. Cell 46:705-716.

Sen R and Baltimore D (1986b) Inducibility of K immunoglobulin enhancer-binding protein NF-KB by a posttranslational mechanism. Cell 47:921-928.

Siekevitz M, Josephs SF, Dukovich M, Peffer N, Wong-Staal F, Greene WC (1987) Activation of the HTLV-1 LTR by T cell mitogens and the trans-activator protein of HTLV-I. Nature 238: 1575-1578.

Tumorigenesis in Transgenic Mice Expressing the c-*myc* Oncogene with Various Lymphoid Enhancer Elements

Richard H. Scheuermann and Steven R. Bauer. Basel Institute for Immunology, Basel, Switzerland

INTRODUCTION

The c-*myc* oncogene initiates tumorigenesis of many, but not all tissues when introduced as a transgene and expressed under the influence of a strong transcriptional control element (for review see Cleveland et al, 1988). Under control of enhancers for the heavy chain and the κ light chain Ig genes the tumors were of preB and B cell origin reflecting the lineage specificity and presumably the developmental stage at which these enhancers are active (Adams et al, 1985, Schmidt et al, 1988). Recently we have identified a nuclear protein, NF-μNR that binds to multiple sites flanking the IgH enhancer and which may function as a suppressive element silencing expression of the IgH enhancer in inappropriate cells and at inappropriate stages of early B-cell development (Scheuermann & Chen, 1989). Given the ability of the c-*myc* oncogene to transform lymphoid cells we decided to examine the effects of NF-μNR deletion mutations on expression of c-*myc* transgenic constructs assessed at the level of transgene expression and tumor phenotypes. Due to the influence of the NF-μNR deletions we hoped to observe tumors arising at earlier stages and originating from a wider spectrum of lymphoid development than was observed in previous studies. In keeping with these goals we also have included a c-*myc* transgenic construct expressed under the influence of the T cell receptor (TCR) β enhancer (Krimpenfort, et al., 1989). The tumors which develop in these transgenic mice have allowed us to define specific cell types and what their relationships are during lymphoid differentiation.

RESULTS

The DNA constructs that were used to generate transgenic mice are described in Figure 1. The transgene is expressed from its own promoters, P1 and P2, and contains a piece of phage lambda inserted into the 3' noncoding section of the gene as a specific tag. We have designed several constructs in which the mouse c-*myc* gene is driven by a number of different transcriptional enhancers in order to extend its expression to earlier B lineage and non-B lineage cells. PB mice contain the putative enhancer for the VpreB & lambda 5 genes. IGH mice contain the wild type 1.0kb heavy chain enhancer. DB and DBX mice contain mutant versions of the heavy chain enhancer in which negative regulatory elements have been removed to allow the enhancer to function in a wider spectrum of cell types. Finally, BET mice have c-*myc* driven by the TCR β chain enhancer.

Northern analysis of preneoplastic mice has revealed that the transgene is expressed exclusively in lymphoid tissues. PB mice show very little expression of the transgene, with small amounts of specific RNA seen in the bone marrow of some founder lines but not others. Indeed after over 8 months none of these mice have come down with tumors. Northern analysis of preneoplastic tissues from transgenic mice containing the other constructs has demonstrated that the transgenic c-*myc* is expressed at high levels in all

Current Topics in Microbiology and Immunology, Vol. 166

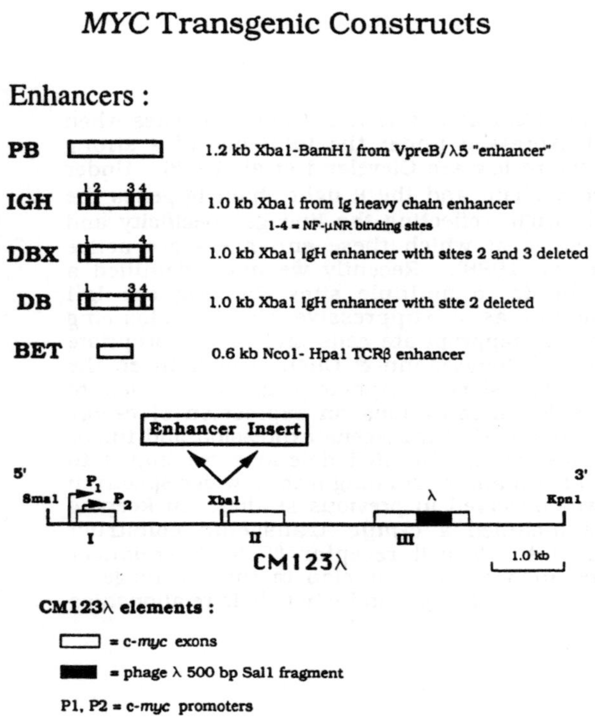

Figure 1. *MYC* transgenic constructs. Various enhancer containing fragments were cloned into a 7.2 kb SmaI-KpnI genomic c-*myc* clone (gift of Dr. Marc Piechaczyk, Laboratory of Molecular Biology, INSERM, Montpelier, France) directly or by linker addition into the XbaI site just upstream of *myc* exon 2. Names of the constructs and their derivative transgenic strains are shown in bold letters to the left of each enhancer diagram. The enhancer in construct **PB** is described in Kudo et al, 1989. The **IGH**, **DBX** and **DB** enhancer inserts are described in Scheuermann & Chen, 1989. The **BET** enhancer insert is described in Krimpenfort et al, 1988. The phage λ insert was a 500 bp SalI fragment inserted into the XhoI site just downstream of the termination codon in *myc* exon 3.

lymphoid tissues including lymph node, spleen, thymus, and bone marrow (unpublished results); expression in BET mice is predominantly found in the thymus, in the lymph nodes of IGH mice, and in all lymphoid tissues of DB mice. Northern analysis of bone marrow samples (Fig. 2) illustrates several points. The transgene, as analyzed with a lambda or c-*myc* probe, is not expressed in this PB mouse. Some expression is observed in the IGH mouse, and considerably higher expression in the two DB mice; this observation supports the idea that we have removed negative elements from the enhancer in these constructs. However, it is not clear whether this represents expression in a wider range of cell types or just higher expression in the same cell type. Expression of the endogenous c-*myc* gene is seen as a smaller RNA species in the PB sample. In support of its auto-regulatory function (Alexander

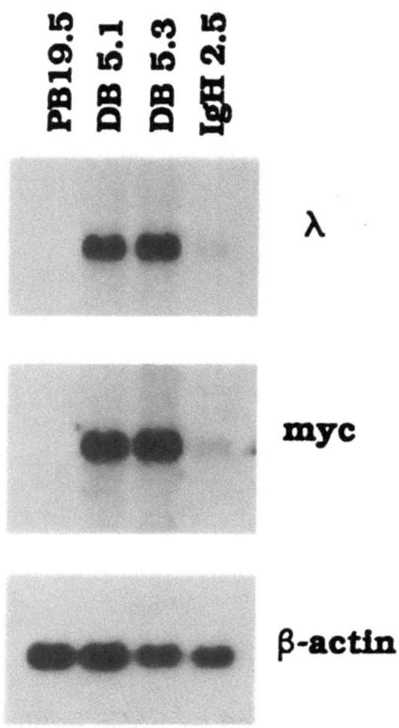

Figure 2. Preneoplastic *myc* transgene expression. A Northern blot of total bone marrow RNA from individual preneoplastic transgenic mice was successively probed with the transgene-specific 500 bp λSal1 fragment (Fig.1), with a 1.4kb Sac1-Hind3 fragment of myc cDNA from pMc-myc 54 (Stanton et al., 1983) spanning exons 2 and 3, and with β-actin (Alonso et al, 1986) as a control for equal RNA loading. The letters in the sample names designate the construct while the numbers designate individual mice.

et al., 1987) expression of the transgene represses expression of the endogenous gene in the IGH and DB mice. Based on these analyses and those described below the combination of DB, DBX and BET mice is extending expression of the transgene to new cell types as compared with mice containing c-*myc* driven by the wild type IgH enhancer.

A summary of the tumors induced in these mice is shown in Figure 3. Most of the mice come down with tumors located in any and/or all lymphoid organs. Except for the BET mice most of the tumors represent cells of the B cell lineage, but many mice have other types of tumors. The percent of tumors outside the B lineage (defined as not staining with antibodies recognizing the B220 antigen) is presented as a measure of relaxed specificity. 10% of the IGH mice had tumors with non-B cells. With the DB mice non-B tumors represent 35% of the tumors; this also supports the idea that we have relieved transcriptional suppression with the enhancer deletion. Interestingly, 80% (4/5) DB founder mice had non-B tumors suggesting that backcrossing against C57/B6 may favor tumorigenesis of B lineage cells. Other reports describing transgenic mice with mouse or human IgH enhancer-*myc* constructs also indicate that the genetic background of the recipient strain can influence the onset and lineage of tumors (Harris, et al., 1988; Yukawa, et al., 1989). All of the BET mice had non-B tumors. Many of the mice had tumors of multiple cell types as judged by non-homogeneous staining of the cells during FACS analysis. 10% of the IGH mice showed this phenomenon. On the other hand 35% and 40% of the DBX and DB mice contained multiple tumors, maybe as a result of decreased enhancer suppression. 88% of the BET mice contained multiple tumors. The implications that these multiple tumors have for tumorigenesis and differentiation is discussed below. The average and median ages of the mice at the time of sacrifice was between 9-12 weeks for all the different constructs and were not statistically different for the number of mice we have examined. However, the BET mice have generally come down with

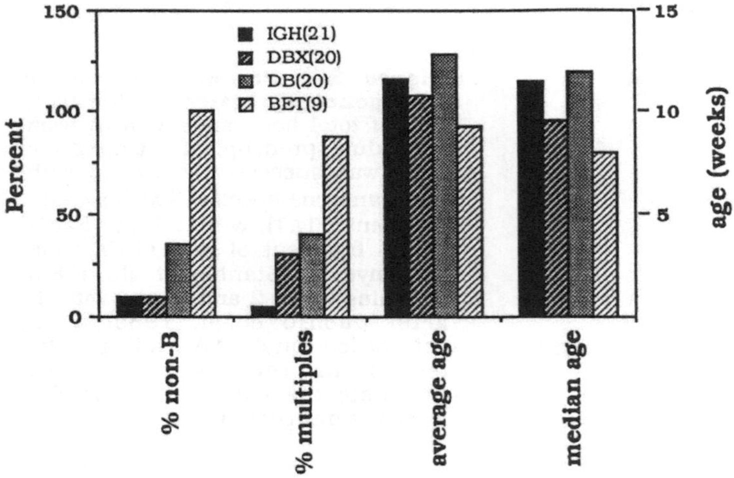

Figure 3. Tumor summary. Tumor phenotypes determined by FACS analysis are represented on the left as % of total tumors analyzed for a given construct versus occurrence of non-B (B220 negative) lineage tumors. Next to this is shown the % of tumors for a given construct which were of multiple types where distinct populations could be discerned arising in different lymphoid organs or within the same organ but bearing a distinct set of cell surface markers. Median and average ages of tumor onset are also shown on the right. The inset key gives the number of tumors analyzed for each construct.

tumors at an earlier age and since they initially affect the thymus they are usually only diagnosed at a later stage in tumor progression. The fact that these mice succumb to these tumors earlier probably reflects an earlier activation of the ß enhancer as compared to the IgH enhancer or that the early precursor T cells expand earlier during ontogeny.

Some interesting results have come from the analysis of the cell surface staining patterns of the tumors with regards to differentiation in normal and tumorigenic animals. Figure 4 shows the staining pattern of five different tumors with a selected set of antibody reagents. These tumors are defined as being within the B cell lineage based on their staining with B220. The first tumor (DBX11.8.2) stained positive with B220 but not with F4/80, IA, or Ig reagents. The next tumor (IGH2.5) has acquired low level staining with F4/80 and exhibits a multiple phenotype - 30% of the cells have cell surface expression of IA although all cells are $B220^{hi}$ and $F4/80^{lo}$. Thus, it is likely that these two types of cells, IA^+ and IA^-, represent adjacent cells in a differentiative pathway in which the expression of MHC classII is induced. The next tumor (DBX11.8.3) has characteristics of this more mature cell which has now become IA^+ in addition to B220 and F4/80. Next is another

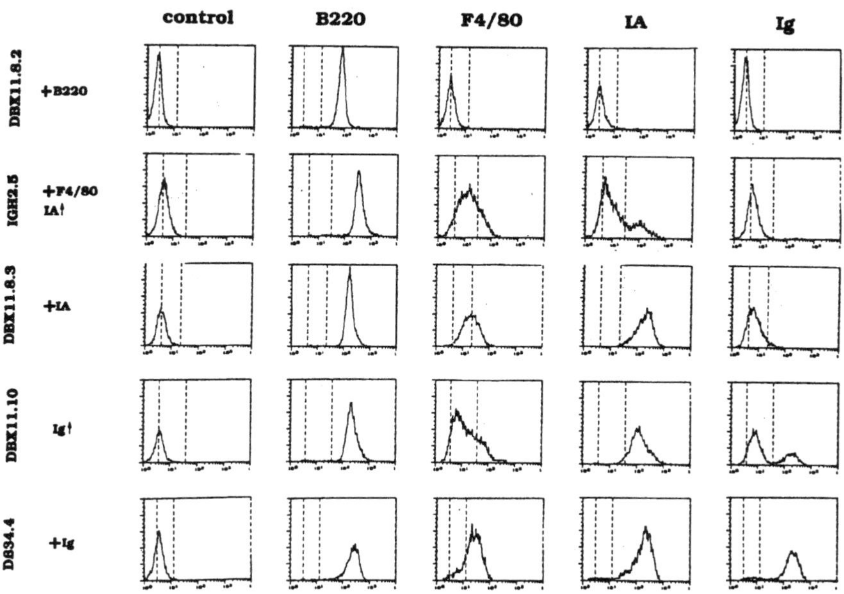

Figure 4. Cell surface markers in B-lineage tumors. FACS analysis of the tumors indicated at the left of the figure is shown as cell counts (vertical axis) versus log fluorescence intensity. The characteristics appearing in each more differentiated tumor type moving from top to bottom are indicated to the left of each FACS profile. The following reagents were used in this and following figures: B220 was detected with antibody RA3-3A1 (Coffman & Weissman, 1981); F4/80 (Austyn & Gordon, 1981); I-A with antibody M5.144 (Bhattacharya et al., 1981); CD3 with antibody 145-2C11 (Leo et al., 1987), panTCRß with antibody H57-597 (Ley et al., 1989); and the following fluorochrome-conjugated reagents from Beckton-Dickenson: Thy-1 was detected with goat anti-mouse Thy 1.2-FITC, CD4 with either antibody GK 1.5-FITC or goat anti-mouse L3T4-PE, CD8 with goat anti-mouse Lyt-2-FITC. Surface Ig was detected with goat anti-mouse IgG+IgM-FITC from Southern Biotech.

tumor with a multiple phenotype (DBX11.10); some of the $B220^+$,$F4/80^{lo}$, IA^+ cells are now expressing Ig on their surface. The last tumor (DB34.4) is a classic mature B cell expressing all four markers. The important point in this analysis is that using the tumors with the multiple phenotypes allows one to assemble a differentiation tree of cell types. Earlier work with *myc* transgenic mice has indicated that B lineage cells retain a limited ability to differentiate (Langdon, et al., 1986; Vaux et al., 1987) and that clonally related but mixed preB/B phenotype tumors could arise (Harris, et al., 1988) These results illustrate the assumption we have made here that inappropriate *myc* expression can transform cells in transition from one differentiative step to the next.

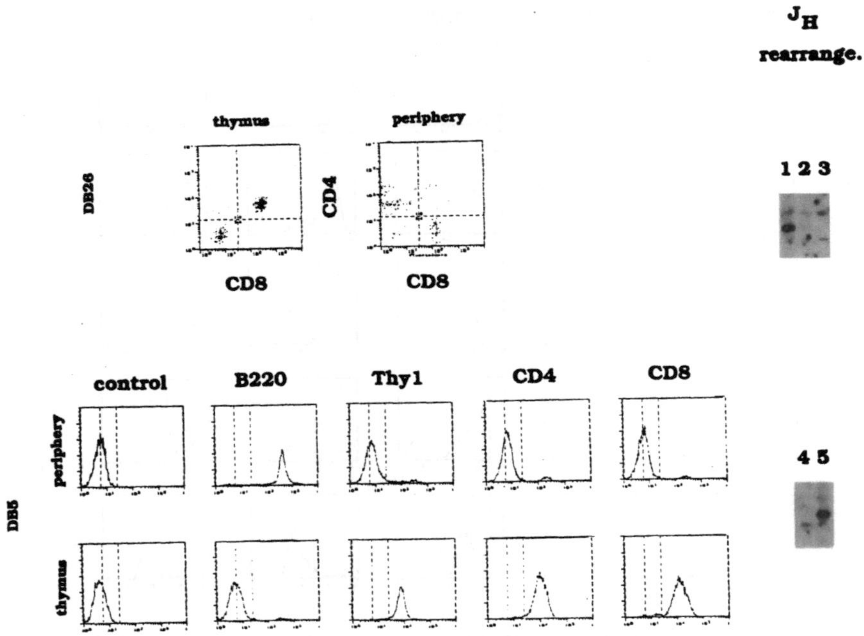

Figure 5. Clonal relationships among multiple phenotype tumors. FACS analysis reagents and profiles are as described for Fig. 4. DB26 cells were stained with both L3T4-PE (vertical axis; CD4) and Lyt-2-FITC (horizontal axis; CD8). At the right are shown results from Southern blot analysis of EcoR1 digested DNA (2.5μg per lane) from peripheral (lymph node or spleen) tumor cells, lanes 2 and 4, or from thymic tumor cells, lanes 3 and 5, probed with a 1.4 kb BamH1-Xba1 fragment spanning J_{H3-4} derived from plasmid j_{11}(Lang et al, 1982). Lane 1 contains a germline control.

One caveat to this approach is that it is important to establish a clonal relationship between the two types of cells; one way to do this is to examine the pattern of antigen receptor rearrangements. DB26 contained a T lineage tumor that was $B220^-$ and $Thy1^+$. In the thymus the cells were mainly CD4,CD8 double positive but in the peripheral lymphoid organs they were either single CD4 or CD8 positive (Fig. 5). The question is are these cells clonally related? If we examine J_H rearrangements by Southern Blot analysis we find that in each case the cells contain the same rearrangement pattern, with both J_H alleles rearranged; therefore they are clonally related. Establishing clonal relationships using this kind of analysis is important for confirming lineage relationships of the two cell types in multiple tumors. On the other hand mouse DB5 had a CD4,CD8 double positive tumor in the thymus but had a $B220^+$, $Thy1^-$ preB tumor in the periphery. Southern analysis of J_H in these two cell types reveals that the thymic tumor contains only germline heavy chain alleles while the peripheral cells have rearrangements in both alleles. Therefore, in this case it is not possible to

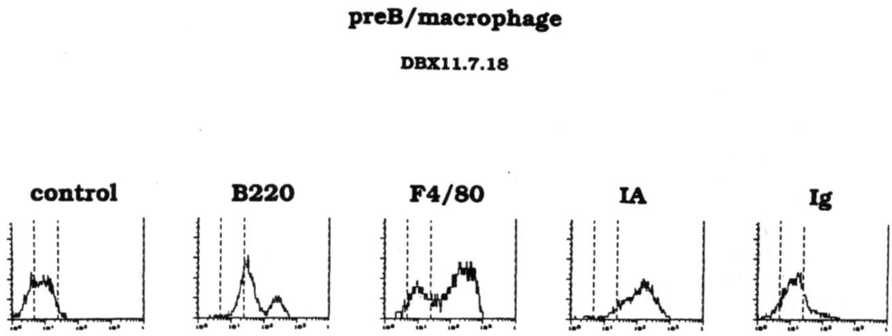

Figure 6. Mixed tumor with two cell types connecting macrophage and B lineages. FACS analysis and reagents are as described for Fig. 4.

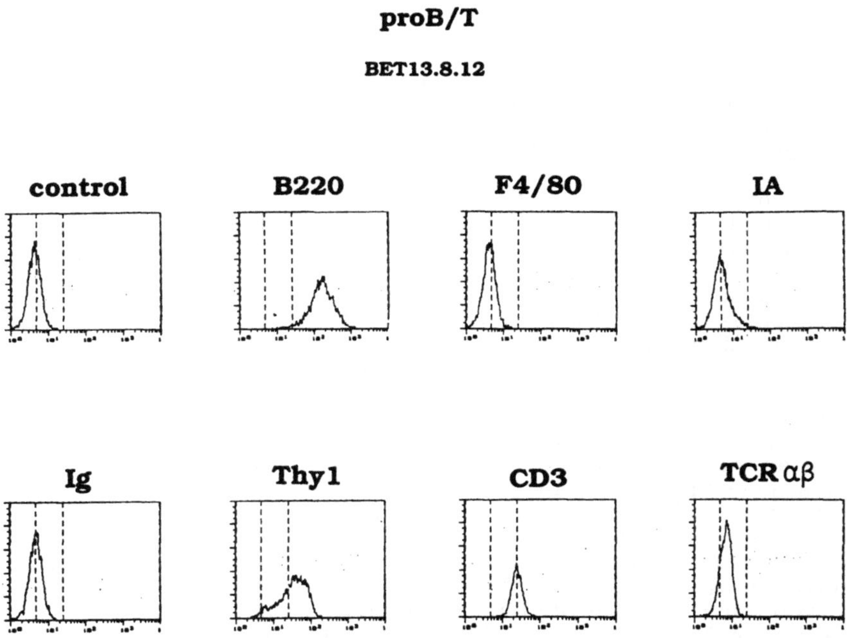

Figure 7. Single tumor population displaying both early T and B lineage markers. FACS analysis and reagents are as described for Fig. 4.

establish a clonal relationship between the two cell types. Another interesting possibility is that these two types of tumors arose independently due to independent contributing mutations in the same or different oncogenes.

Finally, a couple of interesting tumors have come out of this analysis. In mouse DBX11.7.18 the tumor cells were 40% F4/80^{lo} and 60% F4/80^{hi} (Fig. 6); a similar pattern was seen with Mac-1 staining and most of the cells were esterase positive by histochemical staining (data not shown) suggesting that the F4/80^{hi} cells were macrophages. A reciprocal pattern is seen with B220-40%hi and 60% lo; but all cells are IA^{+}. This result supports the idea that there is a direct relationship between the preB and macrophage lineages as seen in tumors (Bauer, et al., 1986) and normal spleen and bone marrow cells (Holmes and Morse,1984).

BET13.8.12 tumor cells showed staining with B220 but were negative for F4/80, IA, and Ig and thus resembled early B lineage cells except that they also stained with Thy1 and CD3, two T lineage markers (Fig. 7). Since these cells were TCR $\alpha\beta$ negative they may represent a kind of lymphoid precursor cell. This type of cell may help bridge the gap between the B and T cell lineages. CD3^{+} but TCR^{-} thymomas have recently been described but no B220 staining was indicated (Ley, et al., 1989).

DISCUSSION

Our observations of a wider variety of tumor types and a different pattern of RNA expression in DB and DBX mice as compared to IGH mice supports the idea that NF-μNR binding site deletions can alter the expression pattern of the transgene and thus alter the lymphoid lineage cells transformed. Although the mechanism by which the NF-μNR sites function remains unclear these mice present an opportunity to study the effects of NF-μNR mutations *in vivo* in a chromosomal context and may increase our understanding of transcriptional regulation in general and of the IgH enhancer in particular.

Since earlier and multiple tumors arise these mice also present a model system for examining early lymphoid-myeloid lineage relationships. Based on the tumors analyzed one can assemble a differentiation tree as depicted in Figure 8. Since cells of the B, T, and macrophage types can be connected in terms of cell surface marker expression and sometimes antigen receptor gene rearrangements, this branch of hemopoiesis has been termed the Antigen Responsive Lineages. The earliest cell detected may be this B220^{+} cell which is negative for all the other markers. As these cells differentiate down the B lineage they acquire first low level expression of F4/80 and then IA before they put Ig on their surface. At some point during this progression the cells can apparently decide to down regulate or change from B220 to another Ly-5 form and upregulate macrophage markers like F4/80 in order to branch off to the macrophage lineage. On the other hand, if the cells acquire expression of Thy1 and CD3 before F4/80 they are funnelled into the T cell lineage followed by the expression of CD4, CD8, and T cell receptor. The findings in our *myc* transgenic mice contrast in some respects with studies of presumptive normal hemopoietic precursors which are Thy-1+ but B220- (Spangrude et al, 1988). However, Ly-5 is observed on precursors of both myeloid and lymphoid lineages and mature, lineage-specific forms arise by alternative splicing (reviewed in Holmes & Morse, 1988) so that our results may reflect early activation of the Ly-5 gene in a multilineage precursor.

ANTIGEN RESPONSIVE LINEAGES

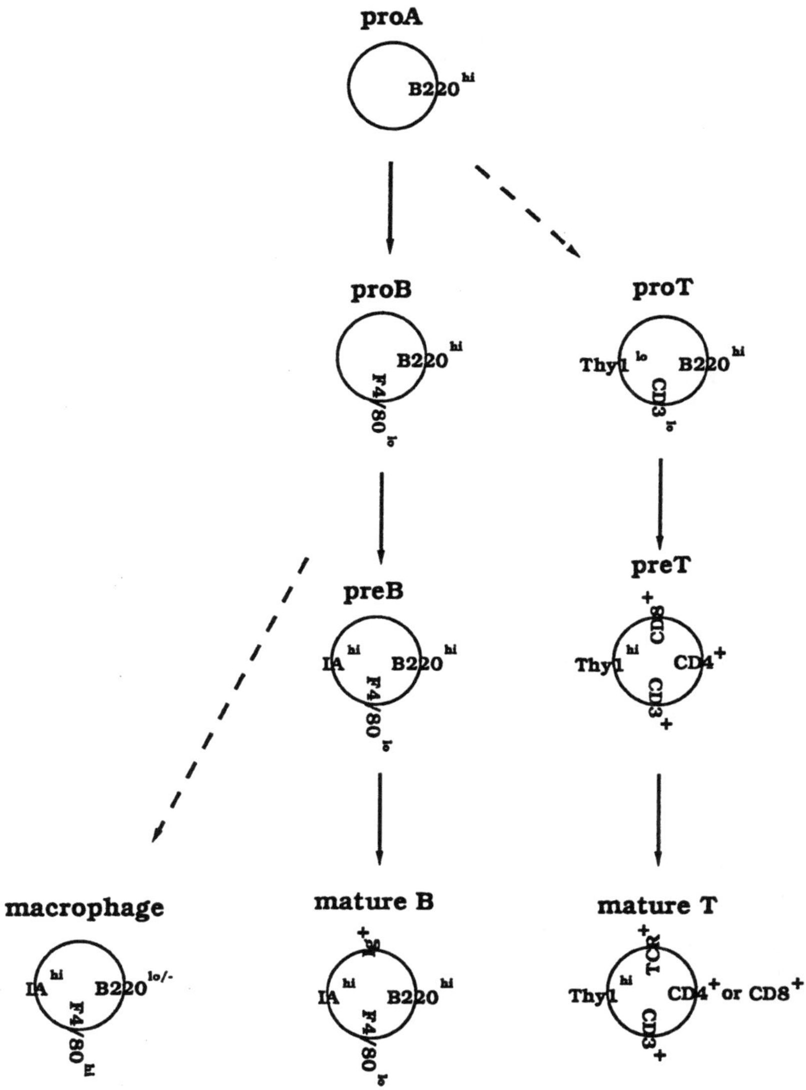

Figure 8. Differentiation of Antigen Responsive Lineages. Arrows represent connections between different tumor phenotypes appearing within an individual animal. Dashed arrows represent putative relationships not yet seen as multiple tumors.

Another use of these mice may be to investigate the possible secondary mutations contributing to the final tumors. This will be particularly interesting in mice which have tumors from different lineages and where the secondary mutational events may be similar or different. These possibilities are now under investigation and may give useful insight into the multistep process of neoplastic transformation.

ACKNOWLEGEMENTS

We thank Urs Muller for generation of the transgenic mice and D. Lenig and Heidi Baechtold for excellent technical assistance. The Basel Institute for Immunology was founded and is supported by F. Hoffmann-La Roche Ltd. Co., Basel Switzerland

REFERENCES

Adams JM, Harris AW, Pinkert CA, Corcoran LM, Alexander WS, Cory S, Palmiter RD, Brinster RL (1985) The c-myc oncogene driven by immunoglobulin enhancers induces lymphoid malignancy in transgenic mice. Nature 318:533-538

Alexander WS, Schrader JW; Adams JM (1987) Expression of the c-*myc* oncogene under control of an immunoglobulin enhancer in Eμ-*myc* transgenic mice. Molec. Cell. Biol. 7: 1436-1444

Alonso S, Minty A, Bourlet Y, Buckingham M (1986) Comparison of three actin-coding sequences in the mouse; evolutionary relationships between the actin genes of warm-blooded vertebrates. J Molec. Evol. 23: 11-22

Austyn JM, Gordon S (1981) F4/80, a monoclonal antibody directed specifically against the mouse macrophage. Eur. J. Immunol 11:805-815

Bauer SR, Holmes KL, Morse HC III, Potter M (1986) Clonal relationship of the lymphoblastic cell line P388 to the macrophage cell line P388D1 as evidenced by immunoglobulin gene rearrangements and expression of cell surface antigens. J Immunol. 136: 4695-4699

Bhattacharya A, Dorf ME, Springer T (1981) A shared alloantigenic determinant on Ia antigens encoded by the I-A and I-E subregions: evidence for I region gene duplication. J. Immunol. 127: 2488-2495

Cleveland JL, Morse HC III, Rapp UR (1988) *Myc* oncogenes and tumor induction. ISI Atlas of Science: Biochemistry/1988: 93-100

Coffman R and Weissman IL (1981) B220: a B cell-specific member of the T200 glycoprotein family. Nature 289: 681-683

Harris AW, Pinkert CA, Crawford M, Langdon WY, Brinster RL, Adams JM (1988) The Eμ-myc transgenic mouse. A model for high-incidence spontaneous lymphoma and leukemia of early B cells. J. Exp. Med. 167: 353-371

Holmes KL, Morse HC III (1988) Murine hematopoietic cell surface antigen expression. Immunol Today 9: 344-350

Krimpenfort P, de Jong R, Uematsu Y, Dembic Z, Ryser S, von Beohmer H, Steinmetz M, Berns A. (1988) Transcription of T cell receptor β-chain is controlled by a downstream regulatory element. EMBO J. 7: 745-750

Kudo A, Bauer SR, Melchers F (1989) Structure, Control of expression and putative function of the preB cell-specific genes VpreB and λ5 *in* F. Melchers et. al (Eds.) Progress in Immunology, Vol. VII. Springer-Verlag Berlin

Lang RB, Stanton LW, Marcu KB (1982) On immunoglobulin heavy chain switching: two γ2b genes are rearranged via switch sequences in MPC-11 cells but only one is expressed. Nuc. Acids Res. 2: 611-630

Langdon WY, Harris AW, Cory S, Adams JM (1986) The c-*myc* oncogene perturbs B lymphocyte development in Eμ-*myc* transgenic mice. Cell 47:11-18

Leo O, Foo M, Sachs DH, Samelson LE, Bluestone JA (1987) Identification of a monoclonal antibody specific for a murine T3 polypeptide. Proc. Natl. Acad. Sci USA 84: 1374-1378

Ley SC, Tan KN, Kubo R, Sy MS, Terhorst C (1989) Surface expression of CD3 in the absence of T cell receptor (TcR): evidence for sorting of partial TcR/CD3 complexes in a post-endoplasmic reticulum compartment. Eur J. Immunol 19: 2309-2317

Scheuermann RH, and Chen U. (1989) A developmental-specific factor binds to suppressor sites flanking the immunoglobulin heavy-chain enhancer. Genes Devel. 3: 1255-1266

Schmidt EV, Pattengale PK, Weir L, Leder P(1988) Transgenic mice bearing the human c-*myc* gene activated by an immunoglobulin enhancer: a pre-B-cell lymphoma model. Proc. Natl. Acad. Sci. USA 85: 6047-6051

Spangrude GJ, Heimfeld S, Weissman IL (1988) Purification and characterization of mouse hematopoietic stem cells. Science 241: 58-62

Stanton LW, Watt R, Marcu KB (1983) Translocation, breakage and truncated transcripts of the c-*myc* oncogene in murine plasmacytomas. Nature 303: 401-406

Vaux DL, Adams JM, Alexander WS, Pike B (1987) Immunologic competence of B cells subjected to constitutive c-*myc* oncogene expression in immunoglobulin heavy chain enhancer *myc* transgenic mice. J. Immunol. 139: 3854-3860

Yukawa K, Kikutani H, Inomoto T, Uehira M, Bin SH, Akagi K, Yamamura KI, Kishimoto T (1989) Strain dependency of B and T lymphoma development in immunoglobulin heavy chain enhancer (Eμ)-myc transgenic mice. J. Exp. Med. 170: 711-726

Isolation of Normal and Tumor-Specific *Pvt*-1 cDNA Clones

K. HUPPI, D. SIWARSKI, R.M. SKURLA,JR., J. GOODNIGHT AND J.F. MUSHINSKI

Laboratory of Genetics, National Cancer Institute, Bethesda, MD20892

One of the first examples of activation of a proto-oncogene by direct chromosomal translocation was identified in lymphoid malignancies with the aberrant association between c-*myc* and antigen receptor genes, i.e. immunoglobulin (Ig) or T- cell receptor (TCR) (reviewed in Cory 1986). More recently, a number of genes i.e. *Bcl*-2 (Cleary et al. 1986; Tsujimoto and Croce 1986), *Bcl*-3 (Ohno et al. 1990), *Lyl*-1 (Cleary et al. 1988), *Ttg* (McGuire et al. 1989) *E2A* (Nourse et al. 1990) and *Il*-3 (Grimaldi and Meeker 1989), have also been found associated with Ig or TCR in hematopoietic malignancies as a result of aberrant chromosomal translocation. Nearly 80% of mouse plasmacytomas contain a rcpt(12;15) chromosomal translocation which juxtapositions c-*myc* and Ig heavy chain genes and ultimately results in deregulation of c-myc gene expression (reviewed in Potter et al. 1984). Another 10-20% of mouse plasma cell tumors exhibit a rcpt (6;15) chromosomal translocation which results in the aberrant association of an Ig light chain gene with a region located more than 100kb distal to c-*myc*. Nevertheless, northern blot analysis of these tumors reveals similar levels of c-*myc* deregulation despite the enormous distance between this cluster of breakpoints, termed *Pvt*-1, and c-*myc* (Mushinski 1988). *Pvt*-1 is consistently found 100-300kb distal to c-*myc* in rat, mouse and humans and, therefore, represents a conserved linkage group among these species (Fig. 1). In studies of rat thymomas, a region of common integration sites (*Mis*-1, *Mlvi*-1) for the Moloney murine leukemia virus (Mo-MuLV) were discovered which proved to be equivalent to mouse *Pvt*-1 (Tsichlis et al. 1983; Villeneuve et al. 1986; Koehne et al. 1989). In fact, Mo-MuLV or MCF proviral integration sites in the region of *Pvt*-1 have been found in 25-50% of rat thymomas (Lemay and Jolicoeur 1984) or murine T lymphomas (Graham et al. 1985). The translocation in the t(2;8) Burkitt's lymphoma, JBL2, juxtaposes an Ig light chain to a region homologous to mouse *Pvt*-1 (Graham and

Current Topics in Microbiology and Immunology, Vol. 166

Adams 1986). While many Burkitt's lymphomas with t(2;8) or t(8;22) translocations contain chromosomal breakpoints which map 3' of c-*myc*, most appear to cluster closer to c-*myc* than *Pvt*-1 (Henglein et al. 1989)(Fig. 1).

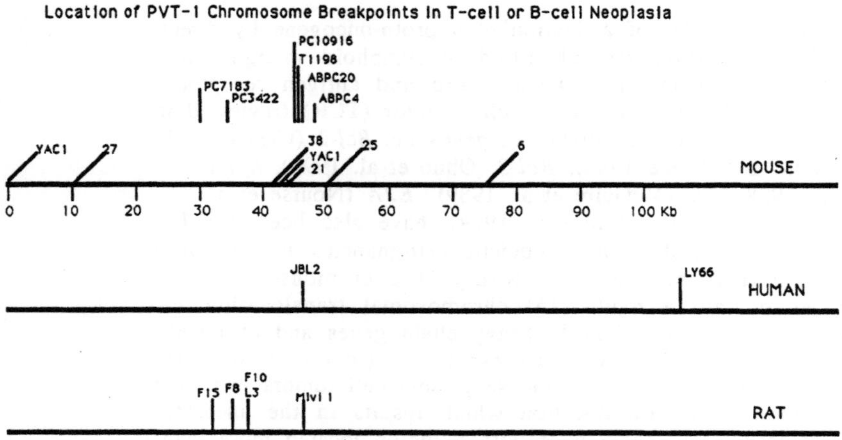

Fig. 1- Schematic comparison of Pvt-1 in mouse, rat and human. The Pvt-1 loci of the respective species are oriented with respect to the observation that the JBL2 translocation breakpoint in humans (Graham and Adams 1986) and the *Mlvi*-1/*Mis*-1 proviral integration sites in rats (Villeneuve et al. 1986; Koehne et al. 1989) are homologous to the major breakpoint cluster of *Pvt*-1 in mouse. Chromosome breakpoints in plasmacytomas (PC7183, PC3422, PC10916, TEPC1198, ABPC20, and ABPC4) are listed above their respective locations in the mouse (Graham et al. 1984; M. Shapiro 1988). Retroviral integration sites for Mo-MuLV-induced (Yac-1) and MCF-induced AKR (21,38,25,6,27) T lymphomas are shown for the mouse (Graham et al. 1985). Two breakpoints resulting from t(2;8) translocations in Burkitt's lymphomas, JBL2 and Ly-66 (Henglein et al. 1989) and proviral integration sites for rat thymomas, F15,F8, F10, L3,and *Mlvi*-1(Graham et al. 1985; Koehne et al. 1989) are depicted in their respective positions.

Since many viral integration sites are associated with regions of transcriptional activity (Kirsch et al. 1985; Yancopoulos and Alt 1986), it seems plausible that the region of *Pvt*-1 could be transcriptionally active as well. Attempts have been made to detect a *Pvt*-1-specific transcript, but with limited success. Recently, cDNA clones specific to the region of *PVT* have been identified in human tumors or placenta (Shtivelman et al. 1989; Shtivelman and Bishop 1990). Two exons identified in these clones extend from a region 57kb downstream of c-*myc* to 7kb just upstream of the JBL2 breakpoint. Thus, the human *PVT* transcript may include several differentially spliced transcripts encoded in a very large region of nearly 200kb.

In order to examine the mouse for evidence of similar *Pvt*-1 transcripts, we turned to a mouse splenic cDNA library as the most likely place to find transcripts associated with lymphoid malignancies. We systematically screened 500,000 recombinants of the cDNA library with genomic DNA probes spanning approximately 20kb of the major breakpoint cluster in mouse plasmacytomas (Fig. 2). From this survey, we identified a single positive clone, Pvt-1-1, with the probe Pvt-a (Cory et al. 1985). Pvt-1-1 contained an insert of 1.4kb which on Southern analysis, hybridized to a single 18kb BamH1 fragment in liver (Fig. 2). We confirmed that Pvt-1-1 maps specifically to the region of Pvt-1 by examination of DNA from several rcpt(6;15) mouse plasmacytomas, which display BamH1 fragments in addition to the 18kb germline fragment. Further DNA mapping and sequencing established that Pvt-1-1 lies within the major breakpoint cluster defined in the mouse and is approximately 2kb proximal to the breakpoint originally identified in the NZB plasmacytoma PC10916 (Cory 1986).

Northern analysis with Pvt-1-1 revealed ca. 14kb transcripts in normal tissues as well as in several plasmacytomas (Mushinski et al. 1990) indicating that the 1.4kb cDNA insert is only a fraction of the entire Pvt-1 transcript. Pvt-1-1 also lacks a poly(A) tail prohibiting the precise location of the sequence with regard to the full-length transcript. Comparison of the Pvt-1-1 sequence with sequences in the GENBANK shows no significant homology with any known sequences including those recently identified as human PVT.

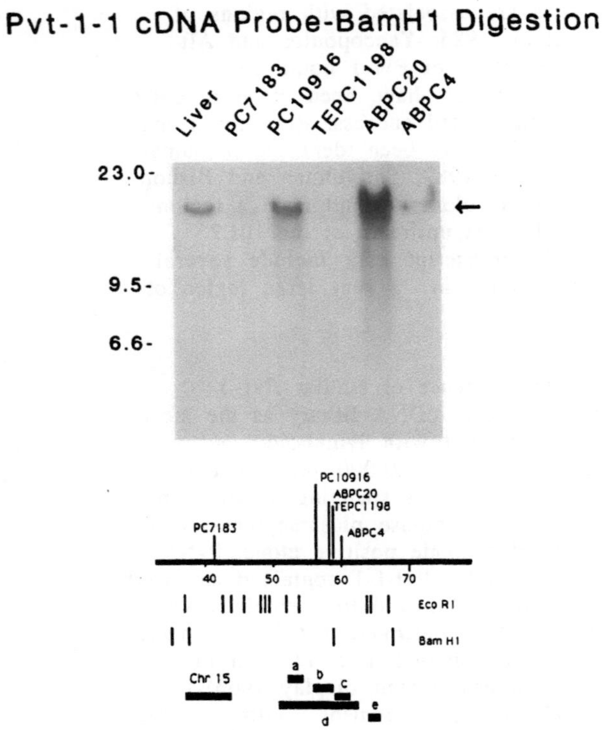

Figure 2- Rearrangements to the mouse Pvt-1 locus are detected by the Pvt-1-1 probe.
Upper panel: BamH1-digested genomic DNA from mouse liver, or plasmacytomas PC7183, PC10916, TEPC1198, ABPC20 or ABPC4 was hybridized to radiolabeled Pvt-1-1 probe. The germline non-rearranged 18kb BamH1 fragment is highlighted (arrow).
Lower panel: Restriction map of mouse Pvt-1. Chromosomal breakpoints are shown with respect to appropriate EcoR1 or BamH1 restriction sites. Below are listed the locations of genomic probes used to screen for mouse cDNA (Cory et al., 1985, M. Shapiro,1988).

Kozak consensus start sites leading to open reading frames (ORFs) of 101 and 104 amino acids. Additional sequencing will be necessary to define further sequence motifs and/or homology.

Since most rcpt(6;15) mouse plasmacytomas contain a breakpoint within the Pvt-1 region, we examined RNA from these tumors for possible alterations of the Pvt-1-1 transcript. It was found that most rcpt (6;15) tumors display a series of truncated Pvt-1-1 transcripts, suggesting that the normal transcript has been interrupted (Mushinski et al. 1990). It was also likely that many of these transcripts may contain sequences derived from the other translocated segment, ie. Ig-κ from chromosome 6. Hybridization of the RNA blot with a Cκ probe would not answer this question, because all of the tumors examined express Ig-κ. Therefore, we constructed a cDNA library from one rcpt (6;15) tumor, ABPC20, which contains four distinct Pvt-1-1 transcripts of 0.9kb, 1.2kb, 7.6kb, and 10.5kb (Fig. 3). Five positive clones were isolated, all of which contained a 1.2kb insert. Sequencing analysis revealed all five clones to be essentially identical (except different sized poly(A) tails) and to contain sequences derived from Pvt-1-1, Ig-Jκ and Ig-Cκ (Fig. 3). Specifically, 363bp of Pvt-1-1 (Pvt-1a) are found at the 5' end juxtapositioned to sequences located 5' of the Jκ4 gene segment. Although the recombination to Jκ4 is imprecise (the V-J joining heptamer motif has not been deleted), a proper splice has taken place between Jκ4 and the Cκ coding region which follows in its entirety. However, no obvious ORF exists which contains both Pvt-1-1 and Ig-κ sequences in the cDNAs, due primarily to the imprecise join to Jκ4. Further cloning and sequence studies are underway to establish the nature of the additional Pvt-1 transcripts in ABPC20. We wish to determine whether these transcripts also contain hybrids of Pvt-1 and Ig-κ and ultimately, whether any of these chimeras are capable of encoding a protein. Similar PVT/Ig-κ chimeras have been identified in a human t(2;8) Burkitt's lymphoma, also without an obvious ORF (Shtivelman and Bishop, 1990).

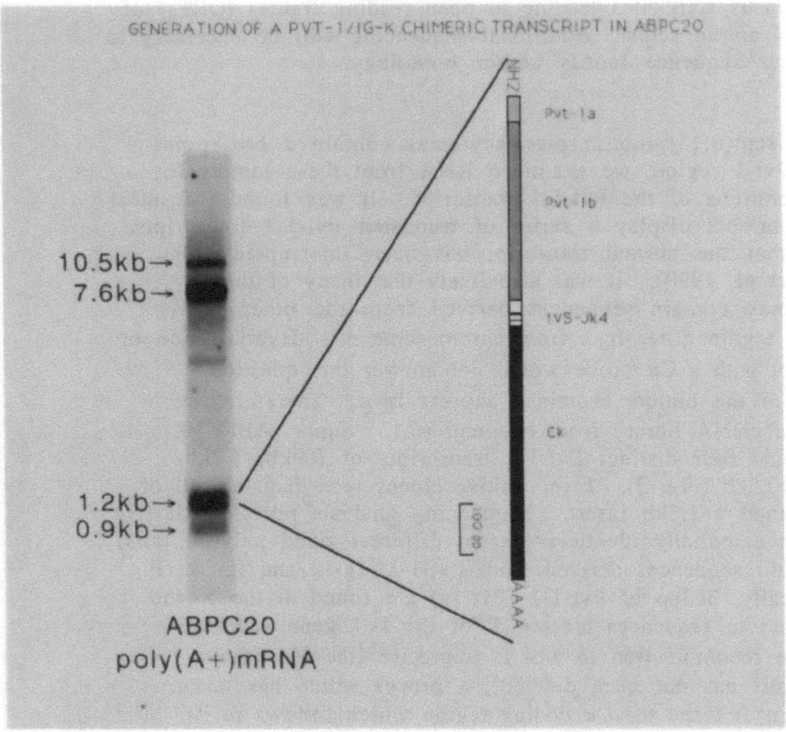

Figure 3 Identification of Pvt-1/Ig-κ chimeric transcripts in the mouse plasmacytoma ABPC20.
Left panel: Northern hybridization of Pvt-1-1 to 5μg of poly(A+) mRNA from the rcpt(6;15) plasmacytoma ABPC20. Size estimates of the major Pvt-1 transcripts are indicated.
Right panel: A schematic diagram of the sequences comprising the cDNA clones isolated from the 1.2kb Pvt-1 transcript. Pvt-1a represents sequences identical to the normal cDNA clone, Pvt-1-1. The region designated Pvt-1b represents sequences derived from another exon of Pvt-1. The IVS-Jκ4 and Cκ regions refer to Ig-κ sequences.

In conclusion, the finding of *Pvt*-1-specific transcripts in normal mouse tissues facilitates new avenues of research regarding the relationship of *Pvt*-1 to c-*myc*. The presence of a *Pvt*-1 transcript correlates well with the general observation that translocation breakpoints or viral integration sites frequently correspond with transcribed genes (Kirsch et al. 1985; Yancopoulos and Alt 1986). Previous dogma, in the absence of evidence of *Pvt*-1 transcripts, suggested that perturbation of chromosome 15 by translocation or viral integration, even at distant sites such as *Pvt*-1, would be sufficiently close to deregulate c-*myc*. Direct comparisons can now be made regarding possible interaction between expressed c-*myc* and *Pvt*-1 genes. For example, one might hypothesize: 1). that the *Pvt*-1 gene product may regulate the transcription or translation of c-*myc* directly; 2). *Pvt*-1 and c-*myc* interaction at the protein level may be necessary to generate a biologically active molecule; or.3). the identification of transcriptional activity at *Pvt*-1 may merely reflect a vestigial gene or sequence which is no longer viable. The fact that *Pvt*-1 is expressed only at certain stages of B-cell development and in higher levels in mouse plasmacytomas (Mushinski et al. 1990) is evidence against the latter hypothesis. Nevertheless, the relationship between human *PVT* and mouse Pvt-1-1 transcripts must be clarified as well as the relationship of *Pvt*-1 to B-cell or T-cell neoplasias. For example, there may be additional Pvt-1 transcriptional units which may be expressed exclusively in T cells. What will be perhaps the most interesting task will be to establish the function of *Pvt*-1 in normal tissues as well as in B-cell or T-cell neoplasias.

REFERENCES

Cleary M, Smith SD, Sklar J (1986) Cloning and Structural Analysis of cDNAs for Bcl-2 and a hybrid *bcl*-2/immunoglobulin transcript resulting from the t(14;18) translocation. Cell 47:19-28

Cleary M L, Mellentin JD, Spies J, Smith SD (1988) Chromosomal translocation involving the beta T-cell receptor gene in acute leukemia. J Exp Med 167: 682-687

Cory S, Graham M,.Webb E, Corcoran L, Adams JM (1985) Variant (6;15) translocations in murine plasmacytomas involve a chromosome 15 locus at least 72kb from the c-*myc* oncogene. EMBO J 4: 675-681

Cory S (1986) Activation of Cellular Oncogenes in Hemopoietic Cells by Chromosome Translocation. Adv Can Res 47: 189-234

Graham M, Adams JM, Cory S (1985) Murine T lymphomas with retroviral inserts in the chromosome 15 locus for plasmacytoma variant translocations. Nature 314: 740-743

Graham M, Adams JM (1986) Chromosome 8 breakpoint far 3' of the c-*myc* oncogene in a Burkitt's lymphoma 2;8 variant translocation is equivalent to the murine *pvt*-1 locus. EMBO J 5: 2845-2851

Grimaldi J, Meeker TC (1989) The t(5;14) chromosomal translocation in a case of acute lymphocytic leukemia joins the interleukin-3 gene to the immunoglobulin heavy chain gene. Blood 73: 2081-2085

Henglein B, Synovzik H, Groitl P, Bornkamm GW, Hartl P, Lipp M (1989) Three breakpoints of variant t(2;8) translocations in Burkitt's lymphoma cells fall within a region 140 kilobases distal from c-*myc*. Mol Cell Biol 9: 2105-2113

Kirsch IR, Brown JA, Lawrence J, Korsmeyer SJ, Morton CC (1985) Translocations that highlight chromosomal regions of differentiated activity. Can Genet Cytogenet 18: 159-171

Koehne CF, Lazo PA, Alves K, Lee J S, Tsichlis PN, O'Donnell P V (1989) The *mlvi*-1 locus involved in the induction of rat T-cell lymphomas and the *pvt*-1/*Mis*-1 locus are identical. J Virol 63: 2366-2369

McGuire EA, Hockett RD, Pollock KM, Bartholdi MF, O'Brien SJ, Korsmeyer SJ (1989) The t(11;14)(p15;q11) in a T-cell acute lymphoblastic leukemia cell line activates multiple transcripts including *ttg*-1, a gene encoding a potential zinc finger protein. Mol Cell Biol 9: 2124-2132

Nourse J, Mellentin JD, Galili N, Wilkinson J, Stanbridge E , Smith SD, Cleary ML (1990) Chromosomal translocation t(1;19) results in synthesis of a homeobox fusion mRNA that codes for a potential chimeric transcription factor Cell 60: 535-545

Mushinski JF (1988) c-myc oncogene activation and chromosomal translocations in BALB/c plasmacytomas. in Cellular Oncogene Activation G Klein (ed) Pp181-222 Marcel Dekker Inc New York

Mushinski JF, Skurla RM Jr, Goodnight J, Siwarski D, Huppi K Expression of c-*myc* and *Pvt*-1. Curr Top Micro Immunol (this volume)

Ohno H, Takimoto G, McKeithan W(1990) The candidate proto-oncogene *bcl*-3 is related to genes implicated in cell lineage determination and cell cycle control. Cell 60: 991-997

Potter M , Wiener F, Mushinski JF (1984) Recent Developments in plasmacytomagenesis in mice. Adv Viral Oncology 4: 139-162

Shapiro M (1988) Mechanisms of recombination at the murine kappa light chain locus. PhD thesis Univ of Penna
Shtivelman E, Henglein B, Groitl P, Lipp M, Bishop JM (1989) Identification of a human transcription unit affected by the variant chromosomal translocations 2;8 and 8;22 of Burkitt lymphoma. Proc Natl Acad Sci USA 86: 3257-3260
Shtivelman E, Bishop JM (1990) Effects of translocations on transcription from *PVT*. Mol Cell Biol 10: 1835-1839
Tsichlis P N, Lohse MA, Szpirer C, Szpirer J, Levan G (1985) Cellular DNA regions involved in the induction of rat thymic lymphomas (*mlvi*-1,*mlvi*-2,*mlvi*-3 and c-*myc*) represent independent loci as determined by their chromosomal map. J Virol 56: 938-942
Tsujimoto Y, Croce CM (1986) Analysis of the structure, transcripts and protein products of bcl-2, the gene involved in human follicular lymphoma. Proc Natl Acad Sci (USA) 83: 5214-5218
Villeneuve L, Rassart E, Jolicoeur P, Graham M, Adams JM (1986) Proviral integration site *Mis*-1 in rat thymomas corresponds to the pvt-1 translocation breakpoint in murine plasmacytomas Mol Cell.Biol 6: 1834-1837
Yancopoulos GD, Alt FW (1986) Regulation of the assembly and expression of variable-region genes. Ann Rev Immunol 4: 339-368.

Expression of c-*myc* and *Pvt*-1

J. F. MUSHINSKI, R. M. SKURLA, JR., J. GOODNIGHT, D. SIWARSKI, AND K. HUPPI

Laboratory of Genetics, National Cancer Institute, Bethesda, MD 20892

Deregulated expression of c-*myc* secondary to chromosomal alterations is a constant feature of mouse plasmacytomas, rat immunocytomas and human Burkitt lymphomas (Mushinski, 1988). One feature of deregulated expression of c-*myc* is that all plasmacytomas have high levels of c-*myc* transcripts, no matter whether their chromosome 15 was interrupted by a rcpt(12;15) translocation, by a rcpt(6;15) translocation, or by interstitial deletion within chromosome 15.

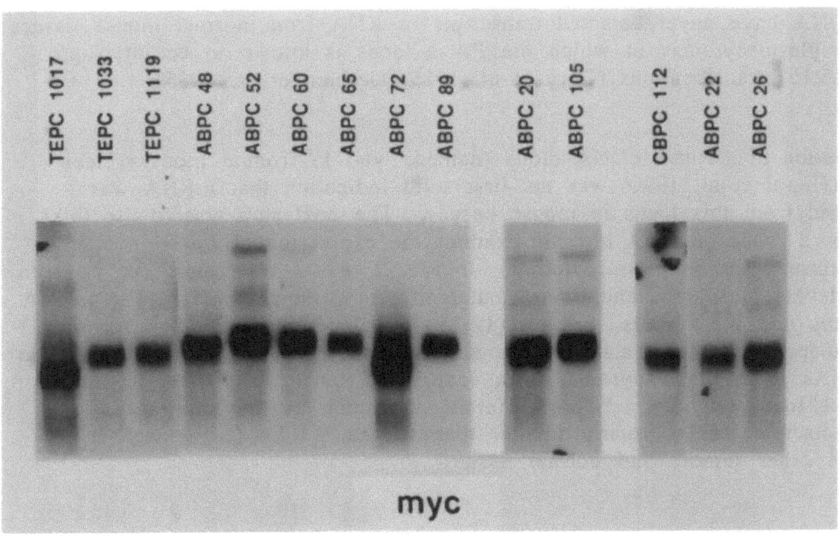

Fig. 1. 5 μg of Poly(A)+ RNA from each of the indicated tumors were electrophoresed on a 1% agarose/formaldehyde gel, blotted onto a nylon membrane, hybridized with ^{32}P-labeled probe for mouse c-*myc* cDNA (Stanton et al. 1983) and exposed to X-ray film for 18 hours. Tumors with rcpt(12;15) translocations are: TEPC 1017, TEPC 1033, TEPC 1119, ABPC 48, ABPC 52, ABPC 60, ABPC 65, ABPC 72 and ABPC 89. "Variant" plasmacytomas with rcpt(6;15) translocations are: ABPC 20 and ABPC 105, and those with interstitial chromosome 15 deletions are: CBPC 112, ABPC 22 and ABPC 26.

Current Topics in Microbiology and Immunology, Vol. 166

The central nature of the *myc* aberration in this form of tumorigenesis makes it important to understand the molecular mechanisms behind c-*myc* deregulation. The rcpt(12;15) chromosomal translocations, which occur in about 75% of mouse plasmacytomas, interrupt the c-*myc* locus in such a way as to separate the coding exons (exons 2 and 3) from the principal regulatory regions in and around exon 1. However, no satisfying mechanism exists to explain how deregulation of c-*myc* expression occurs in rcpt(6;15) translocations, which occur in ca. 20% of mouse plasmacytomas and which contain a breakpoint 100-300 kb downstream of the c-*myc* exons (Webb et al. 1984; Cory et al. 1985; Graham et al. 1986; Huppi et al. 1990). The breakpoints of five of the seven well-studied rcpt(6;15) plasmacytomas have been mapped (Cory et al. 1985) to a region of mouse chromosome 15 called *Pvt*-1 (for plasmacytoma variant translocation). A similar region has been found to be involved in human B-cell tumors (Graham and Adams 1986; Henglein et al. 1989; Shtivelman and Bishop 1989; Shtivelman et al. 1989). One plausible explanation for these findings is that this conserved region may actually encode a protein product which interacts directly, or indirectly, with the c-*myc* gene. This hypothesis would require that this region of the mouse chromosome be transcriptionally active, but probes from a broad stretch of *Pvt*-1 DNA have never detected transcripts in RNA from normal mouse tissues or from plasmacytomas in which the *Pvt*-1 locus is known to be interrupted by rcpt(6;15) translocations (Cory et al. 1985; Graham et al. 1985).

The isolation of a *Pvt*-1 cDNA clone (named Pvt-1-1) from a mouse spleen library (Huppi et al. 1990) was the first solid indication that mRNA was transcribed from this locus in mouse tissues. The next step was to use this cDNA as a probe on RNA blots to examine the expression of *Pvt*-1 in other normal tissues and in plasmacytomas. In Fig. 2, *Pvt*-1 transcripts can be seen in low levels in thymus and, more abundantly, in plasmacytomas. The mRNA in thymus and the tumors with rcpt(12;15) translocations takes the form of high molecular weight (ca.14 kb) transcripts with a rather diffuse appearance. The RNAs from plasmacytomas with rcpt(6;15) translocations which interrupt the Pvt-1 locus (right panel) are different, inasmuch as they appear as several discrete bands, usually smaller than the ca. 14 kb *Pvt*-1 transcripts found in other tumors and normal tissues.

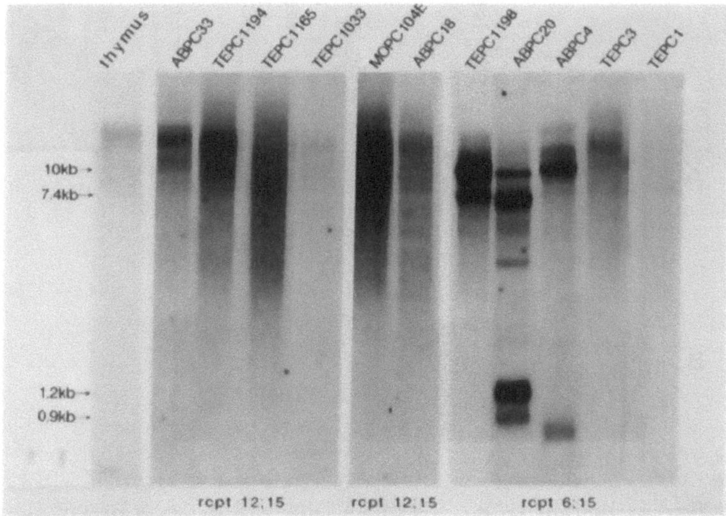

Fig. 2. RNA blot of 5 µg of Poly(A)+ RNA from rcpt(12;15) plasmacytomas (center panels), from rcpt(6;15) plasmacytomas (right panel) and normal thymus (20 µg, left panel) hybridized with ^{32}P-labeled Pvt-1-1 probe. The blot was exposed to X-ray film for 3 days (one day for the TEPC1198 and 14 days for the thymus lane). RNA sizes are indicated in the margin.

It is not clear if the diffuse character of the ca.14 kb mRNA in the *Pvt*-1 RNA blots was due to degradation during preparation of this unusually large mRNA or if it was a true reflection of the heterogeneous nature of these transcripts. Such a microheterogeneity could be due to multiple 5' transcription initiation sites, multiple 3' termination sites, different lengths of Poly(A) tail or differential splicing of multiple exons. To examine this question, Pvt-1-1 was recloned into pGEM-3Z, and single-stranded ^{32}P-labeled riboprobe copies of both strands were synthesized. These riboprobes were hybridized to normal and tumor RNA, digested with ribonuclease, and the protected RNAs examined by electrophoresis on a sequencing gel. The autoradiogram of one such study is shown in Fig. 3.

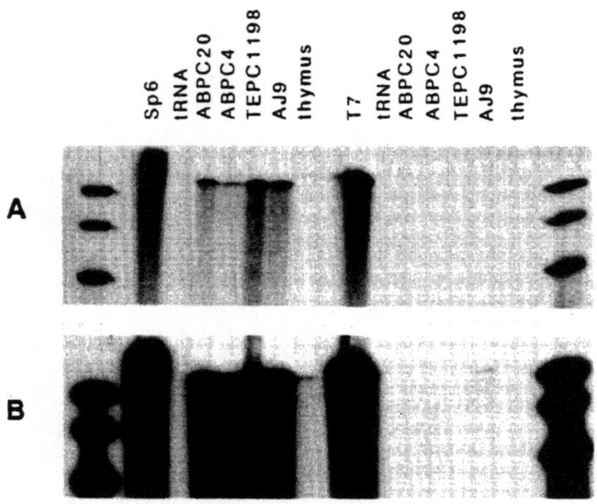

Fig. 3. RNAase protection study performed with 5µg Poly(A)+ RNA from the indicated sources and riboprobe transcripts representing both strands of the Pvt-1-1 cDNA clone. The electrophoregram was exposed to X-ray film for 3 days in panel A and for 10 days in panel B in order to show the thymus band. Positions of end-labeled DNA standards (1353, 1078 and 853 nucleotides) are shown on the left. The size difference between the untreated transcripts and the protected bands is due to multiple cloning site sequences (vector) included in the labeled RNA.

The products from the SP6 promoter were protected, but those from the T7 promoter were not. This establishes the translational orientation of the Pvt-1-1 cDNA sequence (Huppi et al. submitted). Note that normal and tumor RNAs protect the entire length of the 1.4 kb Pvt-1-1 insert; so if there is differential splicing of these transcripts, the Pvt-1-1 sequence is uninterrupted by this process. This and other such RNAase protection experiments also demonstrated that some of the rcpt(6;15) plasmacytomas contain 20-40 times as much *Pvt*-1 mRNA as normal thymus, and some contain even more.

In order to study the effect of B lymphocyte maturation on *Pvt*-1 expression, we hybridized Pvt-1-1 to a blot containing mRNAs from tumors and tumor lines that are frozen in different stages of B-lymphocytic differentiation (Mushinski et al. 1987). The results are shown in Fig. 4.

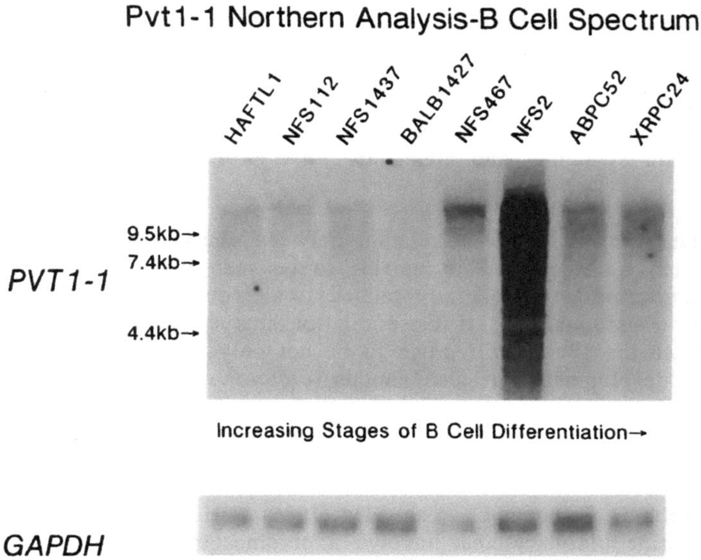

Fig. 4. RNA blot of 5 μg of Poly(A)+ RNA from the indicated lines hybridized to ^{32}P-labeled Pvt-1-1 probe and to a probe for glyceraldehyde phosphate dehydrogenase (GAPDH), a "housekeeping" gene to control for equal amounts of mRNA in each lane.

Pvt-1 steady state RNA levels are low in Pro-B (HAFTL-1) and Pre-B (NFS 112, NFS 1437 and BALB 1427) cell lines. Higher levels of *Pvt*-1 RNA are found in small B lymphocytes (NFS 467, NFS 2) and plasmacytomas (ABPC 52, XRPC 24). These last 4 tumors have rearranged their Ig κ or λ light chains (Mushinski et al. 1987), whereas the less differentiated lines have not. Thus transcription of *Pvt*-1 seems maximal at stages of B-cell development when the light chain genes are being genetically rearranged or being expressed abundantly. This finding of simultaneous transcriptional activation of *Pvt*-1 and normal V_κ-J_κ or V_λ-J_λ recombination offers a plausible explanation for the DNA recombination involving these two chromosomes. What is more, the "variant" translocations involving *Pvt*-1 and J_κ may indicate precisely when transformation occurs during B-cell ontogeny. That is, plasmacytomas or Burkitt lymphomas may be immortalized by the "variant" translocations which happen during light chain V-J assembly. Human follicular cell lymphomas, on the other hand, are transformed by chromosomal translocations between the *bcl*-2 and the J_H chromosomes (Tsujimoto et al. 1985), which could only occur at an earlier stage, namely, during simultaneous $V_H D_H J_H$ assembly and abundant expression of the *bcl* -2 gene (Graninger et al. 1987; Gurfinkel et al. 1987). Similarly, rcpt(12;15)-containing plasmacytomas are immortalized at a later stage, characterized by simultaneous heavy chain switching and abundant c-*myc* expression.

Although our data show a direct correlation between aberrant *Pvt*-1 transcripts and rcpt(6;15) translocations, it remains to be seen whether such alterations in *Pvt*-1 transcription are causally related to deregulation of c-*myc* and/or malignant transformation. If these relationships can be established, it is possible that *Pvt*-1 will prove to be not only an interesting new marker in B-cell development but also another proto-oncogene.

REFERENCES

Cory S, Graham M, Webb E, Corcoran L, Adams JM (1985) Variant (6:15) translocations in murine plasmacytomas involve a chromosome 15 locus at least 72 kb from the c-*myc* oncogene. EMBO J 4: 675-681

Graninger WB, Seta M, Bautain B, Goldman P, Korsmeyer S (1987) Expression of Bcl-2 and Bcl-2-Ig fusion transcripts in normal and neoplastic cells. J Clin Invest 80: 1512-1515

Gurfinkel N, Unger T, Givol D, Mushinski JF (1987) Expression of the *bcl* -2 gene in mouse B lymphocytic cell lines is differentiation stage specific. Eur J Immunol 17: 567-570

Henglein B, Synovzil H, Groitl P, Bornkamm GW, Hartl P, Lipp M (1989) Three breakpoints of variant t(2;8) translocations in Burkitt's lymphoma cells fall within a region 140 kilobases distal from c-*myc*. Mol Cell Bio 9: 2105-2113

Huppi K, Siwarski D, Skurla RMJr, Mushinski JF (1990) Isolation of normal and tumor-specific Pvt-1 cDNA clones. Curr Top Micro Immunol (this volume)

Graham M, Adams JM (1986) Chromosome 8 breakpoint far 3' of the c-*myc* oncogene in a Burkitt's lymphoma 2;8 variant translocation is equivalent to the murine PVT-1 locus. EMBO J 5: 2845-2851

Mushinski JF, Davidson WF, Morse HC III (1987) Activations of cellular oncogenes in human and mouse leukemia-lymphomas: spontaneous and induced oncogene expression in murine B lymphocytic neoplasms. Cancer Invest 5: 345-368

Mushinski, JF (1988) c-*myc* oncogene activation and chromosomal translocations in BALB/c plasmacytomas. *in* Cellular Oncogene Activation, Klein G (ed) Pp. 181-222 Marcel Dekker, Inc., New York

Shtivelman E Bishop JM (1989) The PVT gene frequently amplifies with MYC in tumor cells. Mol Cell Biol 9: 1148-1154

Shtivelman E et al. (1989) Identification of human transcription unit affected by the variant chromosomal translocations 2;8 and 8;22 of Burkitt lymphomas. Proc Nat Acad Sci (USA) 86: 3257-3261

Shtivelman E Bishop JM (1990) Effects of Translocations on Transcription from PVT. Mol Cell Biol 10: 1835-1839

Stanton, LW, Watt R, Marcu KB (1983) Translocation, breakage and truncated transcripts of c-*myc* oncogene in murine plasmacytomas. Nature 303: 401-406

Tsujimoto Y, Cossman J, Jaffe E, Croce CM (1985) Involvement of the *bcl*-2 gene in human follicular lymphoma. Science 228: 1440-1443

Webb E, Adams JM, Cory S (1984) Variant (6;15) translocation in a murine plasmacytoma occurs near an immunoglobulin κ gene but far from the *myc* oncogene. Nature 312: 777-779

Mutational Analysis of the Carboxy-Terminal Casein Kinase II Phosphorylation Site in Human c-*myc*

Alasdair J. Street[1,*], Elizabeth Blackwood[1,2], Bernhard Lüscher[1], and Robert N. Eisenman[1]

[1]Division of Basic Sciences, Fred Hutchinson Cancer Research Center, 1124 Columbia Street, Seattle WA 98104; [2]Department of Pathology, University of Washington School of Medicine, Seattle WA 98195.

SUMMARY

Myc proteins are phosphorylated within two critical regions by casein kinase II (CKII): the central acidic domain and a carboxy-terminal region bordering the basic region-helix-loop-helix segment. In order to test whether the carboxy-terminal phosphorylation site was functionally important we introduced three types of mutations into this region. Two of the mutations would be expected to prevent phosphorylation and minimize negative charge while the third introduced a permanent negative charge. The Myc CKII site mutants were cloned into a retroviral vector and were shown to be efficiently expressed in several different cell types. In one mutant we directly demonstrated loss of the phosphorylation site. When the Myc mutants were used in a cooperative transformation assay of Rat-1 cells with the *bcr-abl* oncogene we were unable to detect a significant difference in transformation efficiency between wild-type Myc and any of the mutants. While the CKII site is non-functional in this assay, the high levels of Myc produced may have overridden potential CKII regulation.

INTRODUCTION

A large body of evidence has accumulated linking oncogene and proto-oncogene functions with mitogenic signal transduction pathways. In this scheme, oncoproteins which are predominantly localized to the cell nucleus, such as Myc, Myb, Fos, Jun, ErbA, Ski, and Ets, can be thought of as transducers of mitogenic signals to the nucleus (see Eisenman 1989 for review). The inducted oncoprotein itself may constitute the signal and indeed Myc, Myb, and Jun are rapidly expressed following mitogenic stimulation of quiescent cells. However for proteins such as Myc and Myb, which are also synthesized continually at a basal level throughout the cell cycle, function could potentially be modulated by other means, including regulation of alternative translational initiation sites, alternative splicing, and post-translational modification. Among the latter, phosphorylation has been shown to be an important means for regulating protein function (Hunter 1987). Indeed all the nuclear oncoproteins are known to be phosphorylated.

* Present address: Dept. of Biochemistry, Medical Sciences Institute, University of Dundee, Dundee, DD1 4HN Scotland

Current Topics in Microbiology and Immunology, Vol. 166

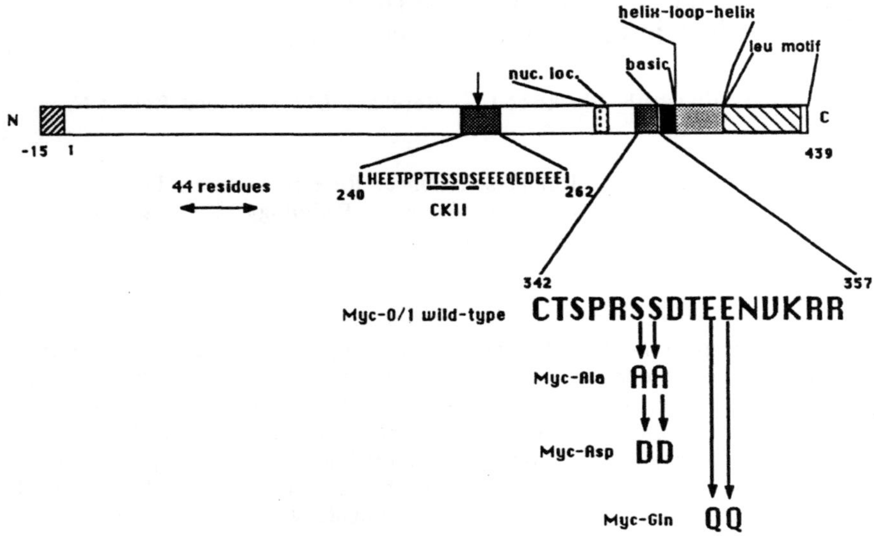

Fig. 1 CKII phosphorylation sites in the human c-myc proteins. A diagram showing major landmarks in human c-myc including the nuclear localization sequence (nuc.loc.), the basic region and the helix-loop-helix and leucine zipper domains. The single letter amino acid code is used to show the sequence of the two CKII sites. Shown in large letters are the mutations studied in this paper.

We had previously determined that human c-Myc proteins are phosphorylated at two major sites by the enzyme casein kinase II (CKII) (Lüscher et al. 1989). We had initially demonstrated that Myc protein in an immunocomplex was a good substrate for CKII phosphorylation and that the regions phosphorylated in this *in vitro* assay corresponded to regions phosphorylated *in vivo*. Using synthetic peptides corresponding to putative CKII consensus sequences within Myc we were able to map the actual phosphorylation sites. Figure 1 shows the location of the two major CKII phosphorylation sites in c-Myc. Both sites had been previously shown to be biologically important in studies using deletion mutants which spanned these regions (Heaney et al. 1986; Stone et al. 1987). We decided to test whether point mutations at one of these phosphorylation sites would affect Myc function. The C-terminal phosphorylation site, located at the border of the basic region and helix-loop-helix/zipper domain, is mutated in several avian retroviruses that have transduced *myc* and is less complicated in terms of its phosphorylation pattern than the central acidic domain (Lüscher et al. 1989). We therefore chose this site as the target for systematic mutagenesis.

Mutagenesis of the CKII Phosphorylation Site in Human c-Myc

The amino acid sequence comprising the carboxy-terminal CKII phosphorylation site in human c-Myc is shown in Fig. 1. Based on the CKII consensus sequence only the first of the two serines (residue 347) would be predicted to be phosphorylated by CKII (Kuenzel et al. 1987). However if the threonine at position 350 were phosphorylated by another kinase, the second serine (residue 348) would also become a target for CKII phosphorylation. Consequently, both serine residues were targeted to ensure complete removal of the site. The following mutations were generated: (i) The serines were changed to alanines (Myc-ala) to prevent phosphorylation at these residues. (ii) The serines were mutated to aspartates (Myc-asp) to establish a "constitutive" negative charge at this site. There are now several examples indicating that aspartic acid can at least partially substitute for phosphorylated residues in terms of enzyme activity and protein-protein interactions (Thorsness and Koshland 1987; Marcus et al. 1988). (iii) The glutamic acid residues at positions 351 and 352 were mutated to glutamines (Myc-gln). Since these glutamic acids constitute the acidic recognition signals for the CKII consensus, their substitution by glutamines would be expected to prevent phosphorylation of the two upstream serines.

The mutations were introduced into the human c-*myc* 0/1 cDNA clone by oligonucleotide-directed *in vitro* mutagenesis. This cDNA, originally isolated from an HL60 cDNA library by D. Bentley, encodes the two c-Myc proteins both *in vitro* and *in vivo* (Hann et al. 1988; see below). Recent nucleotide sequencing of the 0/1 c-*myc* cDNA in our laboratory has confirmed that it has the wild-type c-*myc* coding capacity with the exception of a single silent nucleotide polymorphism (A.J.S. unpublished). The 3' regions of the three CKII site mutant cDNAs were sequenced to confirm the mutagenesis. A summary of the mutants is depicted in Fig. 1.

Expression of c-myc CKII-Site Mutants

The mutant cDNAs and the wild-type 0/1 cDNA were inserted into a murine retroviral vector. The retroviral vector system utilized in this study contains a MuLV LTR, which acts as promoter for the inserted c-*myc* cDNA, and an SV40 promoter which drives expression of a neomycin phosphotransferase gene (LXSN; Miller and Rosman 1989). The vector DNAs were introduced into retrovirus packaging cell lines (psi2 and PA317) and helper virus-free amphotropic viral stocks used to infect Rat-1 cells.

To assay for proper expression of the wild-type and mutant c-Myc proteins, bulk populations of infected cells, selected for resistance to neomycin, were labelled with ^{35}S-methionine and detergent lysates assayed for Myc by radioimmunoprecipitation. The immunoprecipitates were prepared using affinity purified anti-Myc antibodies raised against synthetic peptides corresponding to the carboxy-terminal twelve amino acids of human or murine c-Myc (Hann and

Eisenman 1984). The restricted specificities of such antibodies allow us to distinguish endogenous rat and mouse c-Myc from the introduced human c-Myc proteins. Figure 2 shows an SDS-PAGE analysis of the immunoprecipitates from the different infected cell lines. The anti-human Myc antiserum precipitates no detectable proteins in the molecular weight range expected for Myc from Rat-1 cells infected with the vector lacking a c-*myc* insert. In contrast the antiserum clearly precipitates the p64, and low levels of the p67, human Myc proteins from Rat-1 cells infected with the 0/1 (wild-type), and Myc-ala mutant vectors. Both Rat-1 and PA317 cells infected with the Myc-asp mutant vector synthesize Myc proteins with slightly lower electrophoretic mobility (fig. 2). In addition the PA317 cell clones expressing the Myc-gln mutants produce a p64 doublet. These altered electrophoretic forms of Myc are likely to reflect alterations in charge and/or tertiary structure due to the introduction of the point mutations. These data demonstrate that the retroviral vectors can generate significant levels of wild-type and mutant proteins in infected cells. When we compared the levels of human c-Myc produced with the level of endogenous rat c-Myc we found that the exogenous c-Myc expression was approximately ten fold higher than the endogenous levels (data not shown).

In order to determine whether the mutations actually blocked phosphorylation at the carboxy-terminus of Myc we labelled infected 3T3 cells expressing the Myc-asp mutant and wild-type 0/1 Myc with ^{32}Pi, immunoprecipitated the Myc proteins and prepared two-dimensional tryptic phosphopeptide maps. In a previous study we had demonstrated that the carboxy-terminal CKII phosphorylation site produced two distinctive tryptic phosphopeptides (Lüscher et al. 1989). Our mapping analysis showed that the 0/1 Myc contained the full complement of phosphopeptides while the Myc-asp protein contained all the phosphopeptides with the exception of the two diagnostic carboxy-terminal CKII peptides (data not shown). This confirms that mutations in this CKII region can abolish phosphorylation of the carboxy-terminus of Myc.

Co-Transformation Assays of CKII-Site Mutants

At present there are few direct assays for c-Myc function. Cooperative transformation of primary cells by *myc* and activated *ras* has been frequently used to study mutations in human c-*myc*. However in initial studies using baby rat kidney cells we encountered rather inefficient transformation with *myc* and *ras* (P. Whyte, personal communication) and we therefore utilized a system developed by Witte's laboratory in which v-*myc* and *bcr-abl* were demonstrated to cooperatively transform Rat-1 cells (Lugo and Witte 1989). In collaborative experiments we have shown that c-*myc* will also transform Rat-1 cells together with *bcr-abl* (E. Blackwood, T. Lugo, O. Witte and R. Eisenman; manuscript in preparation) and we have used this transformation system to assay the CKII-site mutants for transforming activity.

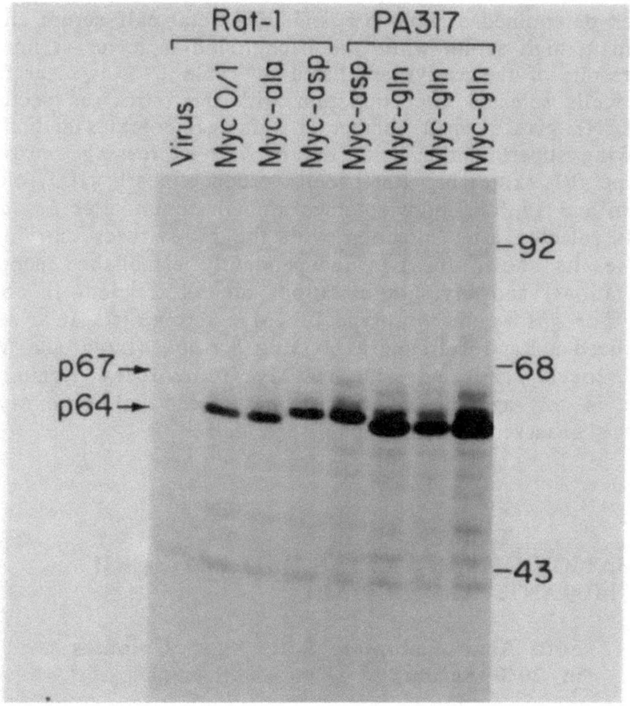

Fig. 2 Myc proteins of human c-*myc* wild-type and CKII site mutants expressed in mouse and Rat-1 cells. The wild-type (0/1) and mutated *c-myc* cDNAs (see Fig. 1 for sequences and designations) were inserted into the LXSN murine retrovirus vector and viral stocks derived in the psi2 and PA317 packaging cell lines. Infected cells were labelled with ^{35}S-methionine, lysed and immunoprecipitated with an anti-Myc C-terminal peptide serum (see Lüscher et al., 1989 for methods). The washed immunoprecipitates were fractionated by SDS-PAGE. An autoradiogram of the dried gel is shown.

We sought to evaluate our Myc-CKII site mutants in a manner which would circumvent the effects of differences in titer between the stocks of *myc* retrovirus mutants. We did this by initially establishing neomycin resistant clones of Rat-1 cells after infection with the mutant *myc* retroviruses. Each non-producer cell population contains a known amount of Myc protein and can then be superinfected with a retroviral vector expressing the P210 *bcr-abl* protein. In each case the doubly infected populations were assayed for transformed colony formation in soft agar.

Colony formation was determined at both 5% and 20% fetal calf serum since larger colonies tend to form at high serum while the background is lower at low serum concentrations. The results of the assays are listed in Table 1. It is clear from the table that the Rat-1 cells selected after infection with the retrovirus vector lacking the c-*myc* insert (LXSN), give a small number of soft agar colonies at both serum concentrations following superinfection with the *bcr-abl* expressing retrovirus. By contrast, the wild-type 0/1 expressing Rat-1 cells produce nearly 100 fold more colonies at low serum and 15 fold more colonies at high serum after *bcr-abl* infection. Rat-1 cells selected after infection with the retroviruses carrying the Myc-ala, Myc-asp (results shown for two independently established neomycin-resistant Rat-1 populations) and Myc-gln mutations are as efficient in cooperative transformation with *bcr-abl* as the wild-type 0/1 Myc retrovirus. In a second type of assay we utilized a Rat-1 cell line expressing *bcr-abl* (Lugo and Witte 1989) which we subsequently infected with our Myc retroviruses. Again no significant difference in soft agar colony formation between wild-type Myc and CKII site mutants was observed (A.J.S. unpublished data).

TABLE 1
CO-TRANSFORMATION OF RAT-1 CELLS WITH MYC CKII MUTANTS AND BCR-ABL

Host for bcr-abl infection	Soft Agar Colonies in 20% Serum*	Soft Agar Colonies in 5% Serum@
Rat-1/LXSN	5	5
Rat-1/ 0/1	77	470
Rat-1/Myc-Ala	48	469
Rat-1/Myc-Asp-1	58	592
Rat-1/Myc-Asp-2	79	509
Rat-1/Myc-Gln	46	474

* Colonies per 500 cells plated. @Colonies per 10,000 cells plated.

CONCLUSIONS

A number of facts originally suggested that the carboxy-terminal casein kinase II phosphorylation site in c-Myc was likely to be of functional importance: (i) the CKII consensus site is evolutionarily conserved in all c-Myc proteins from toads to humans, as well as in all sequenced N-myc and L-Myc-encoded proteins; (ii) the phosphorylated serine(s) lie at the border of the basic region-helix-loop-helix-leucine zipper domain, regions known to be critically important in the function of numerous transcriptional regulatory proteins; (iii) a deletion mutant spanning this region in Myc was inactive in a *ras* co-transformation assay; (iv) several avian acute transforming retroviruses contain *myc* genes which have mutations within the CKII region; and (v) other work from this laboratory has demonstrated that the CKII sites in the c-Myb (Lüscher et al. 1990) and HPV E7 (Firzlaff et al 1989; J. Firzlaff manuscript in preparation) proteins are likely to be critical regulators of function. However our test of three different carboxy-terminal human c-Myc CKII site mutants revealed no significant differences in their ability to transform Rat-1 cells cooperatively with the *bcr-abl* oncogene.

The Myc-ala and Myc-gln mutations would be predicted to significantly decrease negative charge in this region as well as to block CKII phosphorylation (see Fig. 1). The Myc-asp mutation would be expected to replace a phosphoacceptor site with a contitutive negative charge. The loss of the carboxy-terminal CKII phosphopeptides in the Myc-asp mutant and alterations in electrophoretic behavior in the Myc-asp and Myc-gln mutants further attest to the fact that altered proteins were generated.

The lack of a significant effect of the CKII site mutants on co-transformation may have several explanations. One possibility of course is that the CKII phosphorylation at this site is simply functionally irrelevant. However it is also possible that in this transformation assay the relatively high levels of Myc produced, or the presence of *bcr-abl*, allowed the mutated Myc to bypass whatever regulatory role the CKII phosphorylation might have. Another consideration is that we do not know the stoichiometry of phosphorylation at the carboxy-terminal site. If CKII phosphorylation acted as a negative regulator of function, as it appears to do for c-Myb (Lüscher et al. 1990), and the over-produced wild-type Myc was relatively underphosphorylated in the co-transformation assay than one would expect little difference between it and the Myc-ala and Myc-gln mutants. In addition we have little notion of the extent to which the asp substitution will actually mimic a constitutively phosphorylated state. In this regard it is worth noting that an asp substitution at a phosphorylation site in isocitrate dehydrogenase was only 50% as effective as the phosphoamino acid (Thorsness and Koshland 1987). Thus several possibilities may explain the lack of effect on transformation observed in this study. It will be important to determine the stoichiometry of phosphorylation by CKII and to test the CKII phosphorylation site mutants in other assays for Myc function such as autoregulation, direct transformation of immortalized B cells, and downregulation of integrins.

ACKNOWLEDGEMENTS

We thank Tracy Lugo and Owen Witte for the *bcr-acl* construct and for the co-transformation assay procedure. We also thank Jan Smith for assistance with the manuscript. B.B. was supported by the Molecular Training Program in Cancer

Research, CA09437. This research was funded by grant R01 CA20252 from the National Cancer Institute to R.E.

REFERENCES

Eisenman RN (1989) Nuclear Oncogenes, Chapter 7 in Oncogenes and the Molecular Origins of Cancer, pp. 175-221, Cold Spring Harbor, NY.

Firzlaff JM, Galloway DA, Eisenman RN, Luscher B (1989) The E7 protein of human papillomavirus type 16 is phosphorylated by casein kinase II. The New Biologist 1:44-53

Hann SR, Eisenman RN (1984) Proteins encoded by the human *c-myc* oncogene: differential expression in neoplastic cells. Mol Cell Biol 4:2486-2497

Hann SR, King MW, Bentley DL, Anderson CW, Eisenman RN (1988) A non-AUG translational initiation in c-*myc* exon 1 generates an N-terminally distinct protein whose synthesis is disrupted in Burkitt's lymphomas. Cell 52:185-195.

Heaney ML, Pierce J, Parsons JT (1986) Site-directed mutagenesis of the *gag-myc* gene of avian myelocytomatosis virus 29: biological activity and intracellular localization of structurally altered proteins. J Virol 60:167-176.

Hunter T (1987) A thousand and one kinases. Cell 50:823-829.

Kuenzel EA, Mulligan JA, Sommercorn J, Krebs EG (1987) Substrate specificity determinants for casein kinase II as deduced from studies with synthetic peptides. J Biol Chem 262:9136-9140.

Lugo T, Witte ON (1989) The BCR-ABL oncogene transforms Rat-1 cells and cooperates with v-myc. Mol Cell Biol 9:1263-1270.

Lüscher B, Kuenzel EA, Krebs EG, Eisenman RN (1989) Myc oncoproteins are phosphorylated by casein kinase II. EMBO J 8:1111-1119.

Marcus F, Rittenhouse J, Moberly L, Hiller E, Rogers DT (1988) Yeast (Saccharomyces cerevisiae) fructose-1, 6-biphosphatase: Properties of phospho and dephospho forms and of two mutants in which serine 11 has been changed by site-directed mutagenesis. J Biol Chem, 263:6058-6062.

Miller AD, Rosman GJ (1989) Improved retroviral vectors for gene transfer and expression. BioTechniques 7:980-988.

Stone J, de Lange T, Ramsay G, Jakobovits E, Bishop JM, Varmus H, Lee W (1987) Definition of regions in human *c-myc* that are involved in transformation and nuclear localization. Mol Cell Biol 7:1697-1709.

Thorsness PE, Koshland DE (1987) Inactivation of isocitrate dehydrogenase by phosphorylation is mediated by the negative charge of the phosphate. J Biol Chem 262:10422-10425.

The Transforming Activity of PP59C-*MYC* is Weaker Than That of v-*myc*

E. H. Humphries and E. J. Filardo.

Department of Microbiology, University of Texas Southwestern Medical Center, 5323 Harry Hines Blvd., Dallas, Texas 75235-9048

INTRODUCTION

We and others have speculated that elevated expression of c-*myc* plays an important role in the establishment of the transformed follicle, a preneoplastic lesion essential for the development of the ALV-induced bursal-dependent lymphoma (Baba and Humphries 1985, Clurman and Hayward 1989, Neiman *et al* 1985). The discovery of base substitutions and missense mutations in some tumor c-*myc* alleles raises the possibility that qualitative as well as quantitative alterations within the chicken c-*myc* proto-oncogene may play a role in the genesis and/or progression of bursal lymphomas (Westaway *et al* 1984). In order to assess the contribution that deregulated expression of an unaltered c-*myc* allele makes during ALV lymphomagenesis (Hayward *et al* 1981), we have constructed a recombinant avian retrovirus, SNVc-*myc*, that encodes the second and third exons of normal chicken c-*myc* and expresses only the smaller of the two c-*myc* proto-oncogene products, pp62c-*myc* and pp59c-*myc* (Hann *et al* 1988). Ultimately, this virus will enable the effects of specific mutations within the context of wild-type pp59c-*myc* to be assessed.

In this paper, we report the biological characterization of the SNVc-*myc* retrovirus, EF168, that expresses the pp59c-*myc* proto-oncogene. Our results demonstrate that the transforming activity of elevated levels of pp59c-*myc* can be distinguished in several assays from the transformation potential of pp110*gag*v-*myc*, a naturally occurring mutant allele derived from c-*myc* and found in the avian retrovirus MC29 (Bister *et al* 1977). Relative to pp59c-*myc*, pp110*gag*v-*myc* is a structurally altered protein that contains substitutions, deletions and missense mutations (Ramsay *et al* 1982). This result indicates that mutations within the c-*myc* gene product are capable of potentiating its transforming activity and, therefore, producing mutant *myc* alleles the expression of which result in biological consequences that are distinct from those produced by elevated expression of wild-type c-*myc*.

RESULTS

A Recombinant Spleen-Necrosis Retrovirus that Expresses the Proto-Oncogene, pp59c-*myc*.

The spleen necrosis virus was chosen as the virus-vector in which to construct a recombinant avian retrovirus expressing the proto-oncogene pp59c-*myc* (SNVc-*myc*). The SNVc-*myc* plasmid, pEF168, was

Current Topics in Microbiology and Immunology, Vol. 166

derived from pSW272, a molecular clone of a recombinant spleen necrosis virus genome containing the herpes simplex virus thymidine kinase gene (Watanabe and Temin 1983), as summarized in Figure 1. This construct contains a 2.2 kb insert between the SNV splice donor sequence and the cellular c-*myc* splice acceptor site. EF168 virus generated from pEF168 expresses the herpes simplex virus thymidine kinase gene from a genomic length 5.3 kb mRNA transcript and the chicken c-*myc* proto-oncogene from a subgenomic 2.6 kb mRNA transcript.

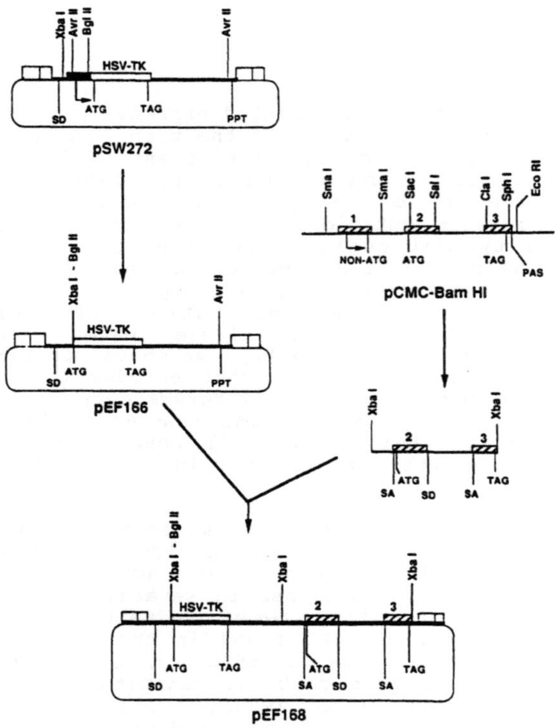

Fig. 1. Construction of a recombinant spleen necrosis virus genome containing the chicken c-*myc* proto-oncogene. Solid bars indicate SNV sequences, large open boxes denote LTR sequences, hatched boxes indicate exons from c-*myc*, dotted boxes designate the herpes simplex virus thymidine kinase coding sequences, and the black box with an arrow indicates herpes simplex virus thymidine kinase promoter sequences. Thin lines represent pBR322 related sequences. pCMC-*Bam* HI contains a 9.4 kb *Bam* HI fragment encompassing the chicken c-*myc* proto-oncogene inserted into the *Bam* HI site of pBR322 (Vennstrom *et al* 1982). SD = splice donor, SA = splice acceptor, ATG = translation initiation codon, TAG = translation termination codon, PAS = polyadenylation signal, NON-ATG = a non-ATG translation initiation codon, PPT = polypurine tract.

An uncloned stock of EF168 was generated by cotransfection of quail embryo fibroblasts with pEF168 and pSW253, an infectious molecular clone of REV-A helper virus DNA (Chen *et al* 1981). The pSW253/pEF168 transfected culture exhibited a morphology that was indistinguishable from quail embryo fibroblasts transformed by the v-*myc* containing avian leukemia viruses MC29 and MH2. Virus harvested from the pSW253/pEF168 transformed culture was used to establish an infected clone from which stock EF168 virus was harvested. This clone, clone 28, contained a single EF168 provirus. The genetic structure of the EF168 provirus in quail embryo fibroblast clone 28 was determined by Southern hybridization analysis. In order to assess the structural integrity of the c-*myc* allele within EF168 proviral DNA, restriction endonucleases were selected which would confirm the presence of sites within the c-*myc* proto-oncogene as well as document the appropriate splicing of intron sequences between exons 2 and 3. The RNA transcripts produced by EF168 in total cellular RNA in clone 28 were examined by Northern analysis. Two transcripts, 5.3 kb and 2.6 kb, that are the appropriate sizes for genomic and subgenomic RNAs from which the 971 bp intron has been spliced were synthesized by EF168. We isolated the EF168-28 provirus as a molecular clone in order to examine its structure and to determine the DNA sequence of it c-*myc* allele (Fig. 2). Southern analysis indicated the presence of a

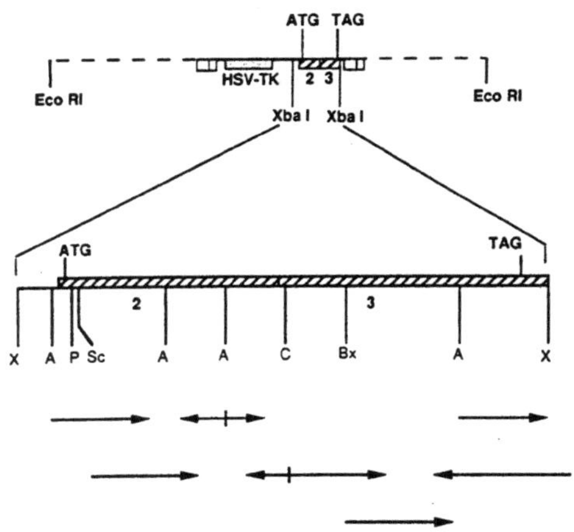

Fig. 2. Molecular cloning and sequencing strategy of the EF168-28 provirus. The 12 kb *Eco* RI insert, originally cloned into λEMBL-4 and subsequently subcloned into pUC19 to form p28.5 is shown. From within p28.5, the 1.6 kb *Xba* I fragment containing the entire coding region for exons 2 and 3 of c-*myc* was subcloned into pUC19. The 1.6 kb fragment was then subcloned as a series of fragments into M13mp18 and M13mp19 for sequence analysis. The sequence of the entire coding region of the chicken c-*myc* allele carried by the EF168-28 provirus was generated from overlapping subclones as indicated. Restriction sites are indicated as follows: *Ava* I (A), *Bst* XI (Bx), *Cla* I (C), *Pst* I (P), *Sac* I (Sc). Arrows indicate the strand that was sequenced.

single EF168 provirus within a 12 kb *Eco* RI fragment in quail embryo clone 28 DNA.

An enriched genomic library was prepared in phage λEMBL-4 from sucrose gradient size-fractionated, *Eco* RI-digested EF168-28 DNA. A single phage clone, λ28.5, containing both c-*myc* and SNV LTR sequences was isolated, subcloned into pUC19, and designated p28.5 (Fig. 2). Cotransfection of quail embryo fibroblasts with p28.5 and pSW253 generated a virus complex with transforming activity equivalent to that produced by cotransfection with pEF168 and pSW253. Sequence analysis of the EF168 c-*myc* allele demonstrated that the splice acceptor for exon two was intact, the splice donor from exon 2 was joined precisely to the splice acceptor of exon 3, and that the *Xba* I site which demarcates the 3′ end of exon 3 was conserved. The 1248 bp open reading frame which encodes the 416 amino acid pp59c-*myc* gene product was preserved within EF168-28 and, when compared to the previously published sequence for this chicken c-*myc* allele (Watson *et al* 1983), contained neither insertions, deletions nor missense mutations.

pp59c-*myc* Possesses Weaker Transforming Activity than V-*myc*.

Three transformation-associated parameters indicate that elevated expression of pp59c-*myc* produced cell lines similar to those which result from expression of v-*myc*. Transformed quail lines established by EF168 were morphologically indistinguishable from cloned lines transformed by the v-*myc* encoding avian retroviruses, MC29 and MH2. Further, transformed clones such as EF168, clone 28, formed colonies in semi-solid growth medium of similar size and with approximately equal efficiency when compared with those formed by MC29. Finally, EF168-28, MC29 and MH2 infection of quail embryo fibroblasts results in the establishment of immortalized cell lines at approximately equivalent efficiencies. These results indicate that elevated expression of pp59c-*myc* is able to sustain a transformed phenotype in quail embryo fibroblasts that is indistinguishable from that produced by either MC29 or MH2.

In contrast to this analysis, however, three biological assays that measure the initiation of transformation of quail cells demonstrated that it was possible to distinguish between the biological effects of pp59c-*myc* and v-*myc* expression. Using cell-free virus, EF168-28, MC29 and MH2 were compared in a standard focus assay using fibroblasts from quail embryos (Table 1). In replicate assays, EF168-28 foci were smaller, more diffuse, and less refractile than those generated by MC29 and MH2. Furthermore, the ability of EF168 to transform quail peripheral blood macrophage cultures differed dramatically from that of MC29 and MH2. While both MC29 and MH2 induced foci with titers consistent with their ability to initiate transformation of quail embryo fibroblasts, EF168 showed no demonstrable activity in this assay. Finally, the ability of newly infected fibroblasts to establish colonies in soft agar was also examined (Table 2). EF168, clone 28, and a clone of MC29 transformed quail produced approximately equal titers of infectious virus (Table 1). Virus from these clones was used to infect quail fibroblasts. Twelve to sixteen hours later, the cells were recovered and plated in soft agar. These results demonstrate that EF168-28 is several thousand-fold less efficient than MC29 in establishing anchorage-independent colony formation by newly infected cells.

Table 1. Transformation titers by focus formation on quail embryo fibroblasts and adherent peripheral blood cell cultures.

	QEF	QPBC
MC29	4×10^4	8×10^2
MH2	5×10^4	2×10^3
EF168-28	2×10^4	<5

All QEF and QPBC assays are presented as focus-forming units/ml and were done in triplicate.

Table 2. Efficiency of colony formation of newly infected quail embryo fibroblasts in semi-solid growth medium.

Virus	Colonies/ 3×10^6 cells[a]
Mock	0
EF168	2
MC29	14,700

[a] Cells were infected at an moi of approximately 0.01. Twelve to sixteen hours post-infection, quail monolayers were recovered by trypsinization and seeded in soft agar medium into ten 60 mm petri dishes. Colony formation efficiencies for newly-infected quail embryo fibroblasts were established from quadruplicate assays.

DISCUSSION

We report the construction of an avian retrovirus, EF168, that expresses the pp59c-*myc* product of the chicken c-*myc* locus. As deduced from its DNA sequence, pp59c-*myc* contained neither deletions, insertions nor missense mutations. The virus is designed to synthesize pp59c-*myc* from a spliced subgenomic RNA in the fashion that MH2 synthesizes pp57*gag*v-*myc* except that pp59c-*myc* is translated from the AUG codon in c-*myc* exon 2 that is 3′ of the SNV splice donor, does not contain *gag* sequences and is, therefore, not a fusion protein. The relative abundance of the EF168 subgenomic RNA species appears equal to that expressed by MH2 transformed quail embryo fibroblasts and is elevated several hundred to a thousand-fold above the steady-state levels of c-*myc* RNA transcribed from the endogenous c-*myc* locus in uninfected quail embryo fibroblasts. This result provides direct evidence that an unaltered version of the pp59c-*myc* chicken proto-oncogene has transforming potential for quail embryo fibroblasts when its expression is regulated by the transcriptional control elements within a retroviral LTR. In contrast, at least 3 assays distinguish between the initiation of transformation established by pp59c-*myc* and that of v-*myc*. The morphological changes associated with focus formation, the ability to establish transformation of adherent peripheral blood cells and the growth of newly infected cells in soft agar all illustrate that c-*myc* is less transforming than v-*myc*. Our analysis suggests, therefore, that specific mutations within c-*myc* alter the biological consequences of transformation with pp59c-*myc*.

Previous studies have emphasized the transforming capacity of c-*myc* and its similarity to v-*myc* (Adams *et al* 1985, Baumbach *et al* 1986, Martin *et al* 1986, Zhou and Duesberg 1988). Our results document three different assays in which the ability of pp59c-*myc* to induce *in vitro* transformation is, when compared with v-*myc*, significantly diminished or not detected. This reduced ability to establish transformation is, at least for fibroblasts, only a transitory effect. This and other observations (Dotto *et al* 1985, LaRocca *et al* 1989, Stoker *et al* 1966, Weiss 1970) suggests that surrounding normal tissue may exert an anti-oncogenic effect that inhibits the initial phenotypic consequences of elevated expression of the c-*myc* proto-oncogene. Such an inhibitory effect does not appear to influence the initiation of transformation by at least two v-*myc* alleles. The construction of isogenic SNVc-*myc* viruses bearing one or more of the specific mutations found in v-*myc* alleles is presently underway and will make it possible to determine the biological significance of such alterations for both *in vitro* transformation and tumor progression.

ACKNOWLEDGMENTS

We gratefully acknowledge the expert secretarial assistance of Susan B. Campbell for her help in assembling this manuscript. This work was supported by Public Health Service grant CA32295 from the National Institutes of Health.

REFERENCES

Adams JM, Harris AV, Pinkert CA, Corcoran LM, Alexander WS, Cory S, Palmieri RD and Brinster RL (1985) The c-*myc* oncogene driven by immunoglobulin enhancers induced lymphoid malignancy in transgenic mice. Nature 318:533-538

Baba TW and Humphries EH (1985) Formation of a transformed follicle is necessary but not sufficient for development of an avian leukosis virus-induced lymphoma. Proc Natl Acad Sci USA 82:213-216

Baumbach WR, Keath EJ and Cole MD (1986) A mouse c-*myc* retrovirus transforms established fibroblast lines *in vitro* and induces monocyte-macrophage tumors *in vivo*. J Virol 59:276-283

Bister K, Hayman MJ and Vogt PK (1977) Defectiveness of avian myelocytomatosis virus, MC29: Isolation of long term nonproducer cultures and analysis of virus specific polypeptide synthesis. Virology 82:431-448

Chen ISY, Mak TK, O'Rear JJ and Temin HM (1981) Characterization of reticuloendotheliosis virus Strain T DNA and isolation of a novel variant of reticuloendotheliosis virus Strain T by molecular cloning. J Virol 40:800-811

Clurman BE and Hayward WS (1989) Multiple proto-oncogene activations in avian leukosis virus-induced lymphomas: Evidence for stage-specific events. Mol Cell Biol 9:2657-2664

Dotto GP, Parada LF and Weinberg RA (1985) Specific growth response of *ras*-transformed embryo fibroblasts to tumor promoters. Nature 318:472-475.

Hann SR, King MW, Bentley DL, Anderson CW and Eisenmann RN (1988) A non-AUG translational initiation in c-*myc* exon 1 generates a N-terminally distinct protein whose synthesis is disrupted in Burkitt's Lymphomas. Cell 52:185-195

Hayward WS, Neel BG, and Astrin SM (1981) Activation of a cellular *onc* gene by promoter insertion. Nature 290:475-480
La Rocca SA, Grossi M, Falcone G, Alema S and Tato F (1989) Interaction with normal cells suppresses the transformed phenotype of v-*myc*-transformed quail muscle cells. Cell 58:123-131
Martin P, Henry C, Ferre F, Duterque-Coquillaud M, Lagrou C, Ghysdael J, Debuire B, Stehelin D and Saule S (1986) Transformation of quail embryo fibroblasts by a retrovirus carrying a normal human c-*myc* gene. Eur Mol Biol Org Jrnl 5:1529-1533
Neiman PE, Wolf C, Enrietto PJ and Cooper GM (1985) A retroviral *myc* gene induces preneoplastic transformation of lymphocytes in a bursal transplantation assay. Proc Natl Acad Sci USA 82:222-226
Ramsay G, Enrietto PJ, Graf T and Hayman MJ (1982) Recovery of *myc*-specific sequences by a partially transformation defective mutant of avian myelocytomatosis virus, MC29, correlates with the restoration of transforming activity. Proc Natl Acad Sci USA 79:6885-6889
Stoker MGP, Shearer M and O'Neill C (1966) Growth inhibition of polyoma transformed cells by contact with static normal fibroblasts. J Cell Sci 1:297-310
Vennstrom B, Sheiness D, Zabielski J and Bishop JM (1982) Isolation and characterization of c-*myc*, a cellular homologue of the oncogene (v-*myc*) of avian myelocytomatosis virus Strain 29. J Virol 42:773-779
Watanabe S and Temin HM (1983) Construction of a helper cell line for avian reticuloendotheliosis virus cloning vectors. Mol Cell Biol 3:2241-2249
Watson DK, Premkumar Reddy E, Duesberg PH and Papas TS (1983) Nucleotide sequence analysis of the chicken c-*myc* gene reveals homologous and unique coding regions by comparison with the transforming gene of avian myelocytomatosis virus MC29, Δ*gag-myc**. Proc Natl Acad Sci USA 80:2146-2150
Weiss RA (1970) The influence of normal cells on the proliferation of tumor cells in culture. Exp Cell Res 63:1-18
Westaway D, Payne G and Varmus HE (1984) Proviral deletions and oncogene base-substitutions in insertionally mutagenized c-*myc* alleles may contribute to the progression of avian bursal tumors. Proc Natl Acad Sci USA 81:843-847
Zhou R-P and Duesberg PH (1988) *Myc* proto-oncogene linked to retroviral promoter but not to enhancer, transforms embryo cells. Proc Natl Acad Sci USA 85:2924-2928

Recombination of the c-*myc* Gene with IgH Enhancer-Sμ Sequences in a Murine Plasmacytoma (DCPC 21) Without Visible Chromosomal Translocations

S. Ohno[1], S. Migita[1], and S. Murakami[2]

[1]Department of Molecular Immunology and [2]Division of Biophysics, Cancer Research Institute, Kanazawa University, 13-1 Takara-machi, Kanazawa 920, Japan

INTRODUCTION

Murine plasmacytomas (MPCs) consistently exhibit chromosomal translocations involving chromosome 15 (t(12;15) or t(6;15)) (Ohno et al. 1979, 1984, 1988). These translocations of chromosomes juxtapose the proto-oncogene c-myc to the one of Ig gene loci (Cory 1986). The juxtaposition of c-myc and the Ig gene strongly suggests that they play a critical genomic role in the development of MPCs probably by deregulating the c-myc gene expression.

During the induction experiment of MPCs in BALB/c mice, we came across one MPC (DCPC 21) that lacks any known chromosomal translocations. If the recombination of c-myc and the Ig gene caused by a chromosomal translocation is critical in the plasmacytomagenesis in mice, it is interesting and valuable to clarify the construction of the c-myc gene in this translocation-negative MPC DCPC 21, whether the c-myc shows the rearrangement without visible chromosomal translocations. We report here that the c-myc gene has moved to the IgC_H gene region on chromosome 12 and recombined with IgH enhancer-Sμ sequences.

INDUCTION OF DCPC 21

DCPC 21 was induced by implantation i.p. with a Millipore diffusion chamber (Merwin and Redmon 1963) in a female BALB/cCrSlc mouse. The diagnosis of plasmacytoma was based on the following features: (1) eccentric nuclei with dense nuclear chromatin; (2) a large amount of basophilic cytoplasm; (3) well-developed parallel arrays of rough endoplasmic reticulum and Golgi apparatus; (4) secretion of a large amount of homogeneous immunoglobulin of γ1, κ type; and (5) strong immunofluorescent staining of the cytoplasm (>95%) with anti-IgG_1 antiserum.

The G-banded karyogram of DCPC 21, primary tumor, is shown in Fig. 1. The modal chromosome number was 77. DCPC 21 has maintained normal morphology in G-banded chromosomes after several *in vivo* transplantations. Of the total 21 MPCs developed, other than this tumor, 18 were t(12;15) and 2 were MPCs of t(6;15) type. Therefore, a translocation-negative MPC DCPC 21 seems to be induced "by chance", not specifically induced by implantation i.p. with a diffusion chamber.

Several pairs of chromosomes 15 and 12 of DCPC 21 are shown in Fig. 2 for the comparison of their band patterns. It seems there are no

Current Topics in Microbiology and Immunology, Vol. 166

morphological aberrations between those two chromosomes. No visible structural changes have also been observed between chromosomes 15 and 6 Therefore, we found DCPC 21 a translocation-negative MPC.

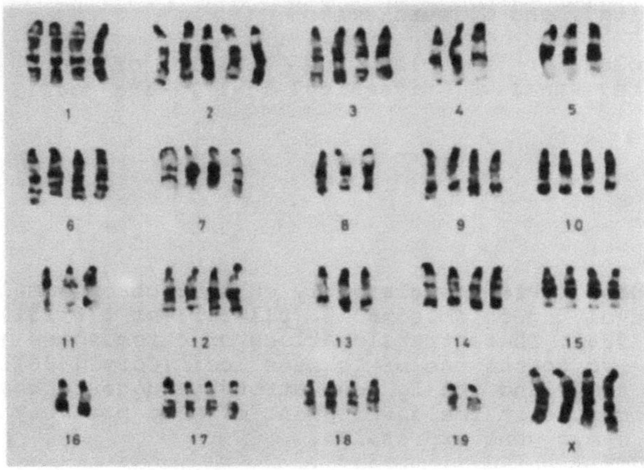

G-banded karyotype of a near-tetraploid plasmacytoma DCPC 21(g. 0). Neither t(6;15) nor t(12;15).

Fig. 1. G-banded karyotype of a near-tetraploid MPC DCPC 21 (primary tumor) showing normal band patterns. Metaphase spreads were prepared from ascitic tumor cells. G-banding was performed using the method of Wang and Fedoroff (1972) with a slight modification. Chromosome identification followed the recommendation of the Committee on Standardized Genetic Nomenclature for Mice (1972). Notice no morphological changes between chromosomes 6, 12, and 15.

IN SITU HYBRIDIZATION OF C-MYC GENE

The c-myc oncogene is normally located on chromosome 15 in mice. To assign the c-myc oncogene to the DCPC 21 chromosomes, in situ hybridization experiment of a [^{3}H]-labelled c-myc probe containing the exon-2,3 (S107) (Taub et al. 1982) was performed essentially as described elsewhere (Harper and Saunders 1981). The distribution of autoradiographic silver grains in 56 metaphase spreads is shown in Fig. 3. Of 441 autoradiographic grains associated with the chromosomes, 59 (13%) and 51 (11%) were located on the chromosomes 15 and 12, respectively. Of the latter grains, 39 (66%) were concentrated within the band D region, mainly involving the D2 of chromosome 15, and 35 (68%) were associated with the chromosomal region corresponding to the 12F1. All the remaining grains were distributed randomly over the whole length of other chromosomes. Statistical evaluation for significance of the grain distribution using a χ^2 test was highly significant (P<<0.001, data not shown). We concluded from the experiment that one half of the total number of c-myc gene moved to the 12F1 where IgH gene are located. This is the same oncogene sift as occurs in the MPCs of th t(12;15) translocation type (Klein 1983). However, neither chromosome 1 nor 12 of DCPC 21 showed any morphological evidence of chromosomal translocations (Figs. 1 and 2).

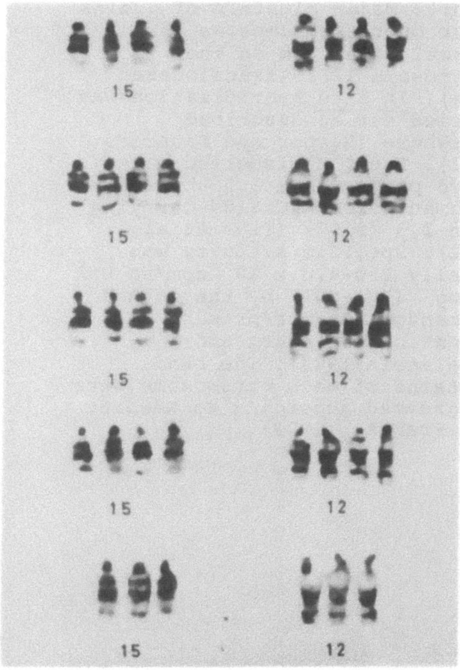

Fig. 2. Comparison of G-banded patterns of chromosomes 15 (left) and 12 (right) of DCPC 21 (primary tumor). The sets of chromosomes 15 and 12 in each row were obtained from the same metaphase spread. All the chromosomes 15 and 12 have a normal morphology.

RESTRICTION ENZYME ANALYSIS OF DCPC 21 DNA

Restriction enzyme analysis of DNA from DCPC 21, using the 5'-c-myc probe (JQ2), showed the presence of two new DNA bands of 0.8 and approx. 3 kb in size by HindIII digestion in addition to the roughly 1.3 kb germline-typed band (Fig. 4). The JQ2 probe is a 630 bp EcoRI/BamHI fragment containing the c-myc 5'-flanking sequence which spans approx. 580 bp, and 119 bp of the c-myc exon-1 (Stanton et al. 1984). Both the 0.8 and 3 kb bands hybridized with the JQ2 indicated the rearranged c-myc sequences that are not found in control BALB/c liver DNA. This result suggested the presence of at least one break (HindIII site) in the c-myc 5'-flanking region $\gtrsim$ 0.8 kb upstream of exon-1.

NUCLEOTIDE SEQUENCING OF THE 5'-FLANKING REGION OF THE REARRANGED DCPC 21 C-MYC

C-myc-Sμ Rearrangement

To define the rearranged c-myc more precisely, we first isolated a clone (Clone 1) containing the 0.8 kb HindIII fragment from a bacteriophage

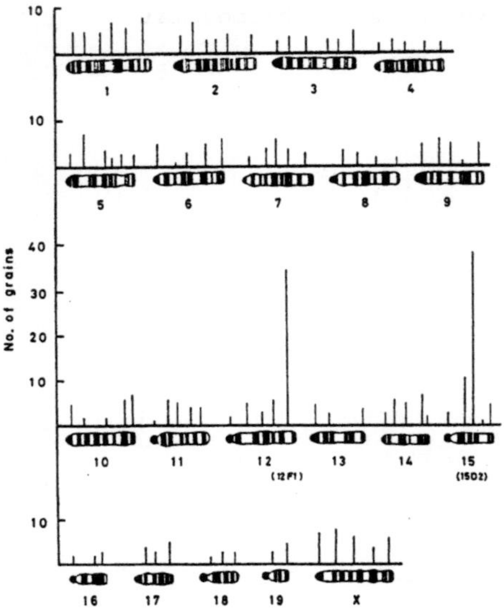

Fig. 3. Distribution of autoradiographic grains on the G-banded 56 metaphase chromosome spreads from DCPC 21 (primary tumor). Major clusters of grains occur on the chromosome 15D2 (normal site) and on the chromosome 12F1 (translocated site). In situ hybridization was carried out as described elsewhere (Harper and Saunders 1981). The $[^3H]$-labelled mouse c-myc probe was a 5.5 kb BamHI fragment from MPC S107 carrying exon-2,3 (S107) (Taub et al. 1982). Specific activity was usually $2.0-4.0 \times 10^7$ cpm/μg DNA using $[^3H]$-dNTP by the method of random hexamer/primer extension (Feinberg and Vogelstein 1983). The band patterns of each chromosome were diagrammed according to Nesbitt and Francke (1973).

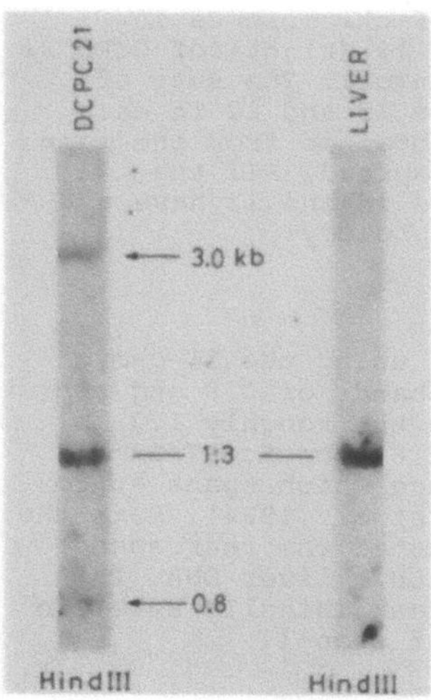

Fig. 4. Genomic Southern blot analysis of HindIII-digested DNA (20 μg) from DCPC 21 using JQ2 as a probe. HindIII-digested BALB/c liver DNA (20 μg) was run as a control on the same 0.8% agarose gel. Size and position of the JQ2 probe is indicated in the restriction map of the murine c-myc gene (Fig. 5b).

library that had been constructed from DCPC 21 and subcloned into M13 vectors. DNA sequence of this region was determined using the dideoxy-chain termination method (Fig. 5a) (Sanger et al. 1977). The DN sequencing data demonstrated that IgSμ region sequence (Sakano et al. 1980) was involved in the rearrangement of the DCPC 21 c-myc gene. Comparison of the sequences near the recombination site indicated by arrows with previously published work (Sakano et al. 1980) revealed tha

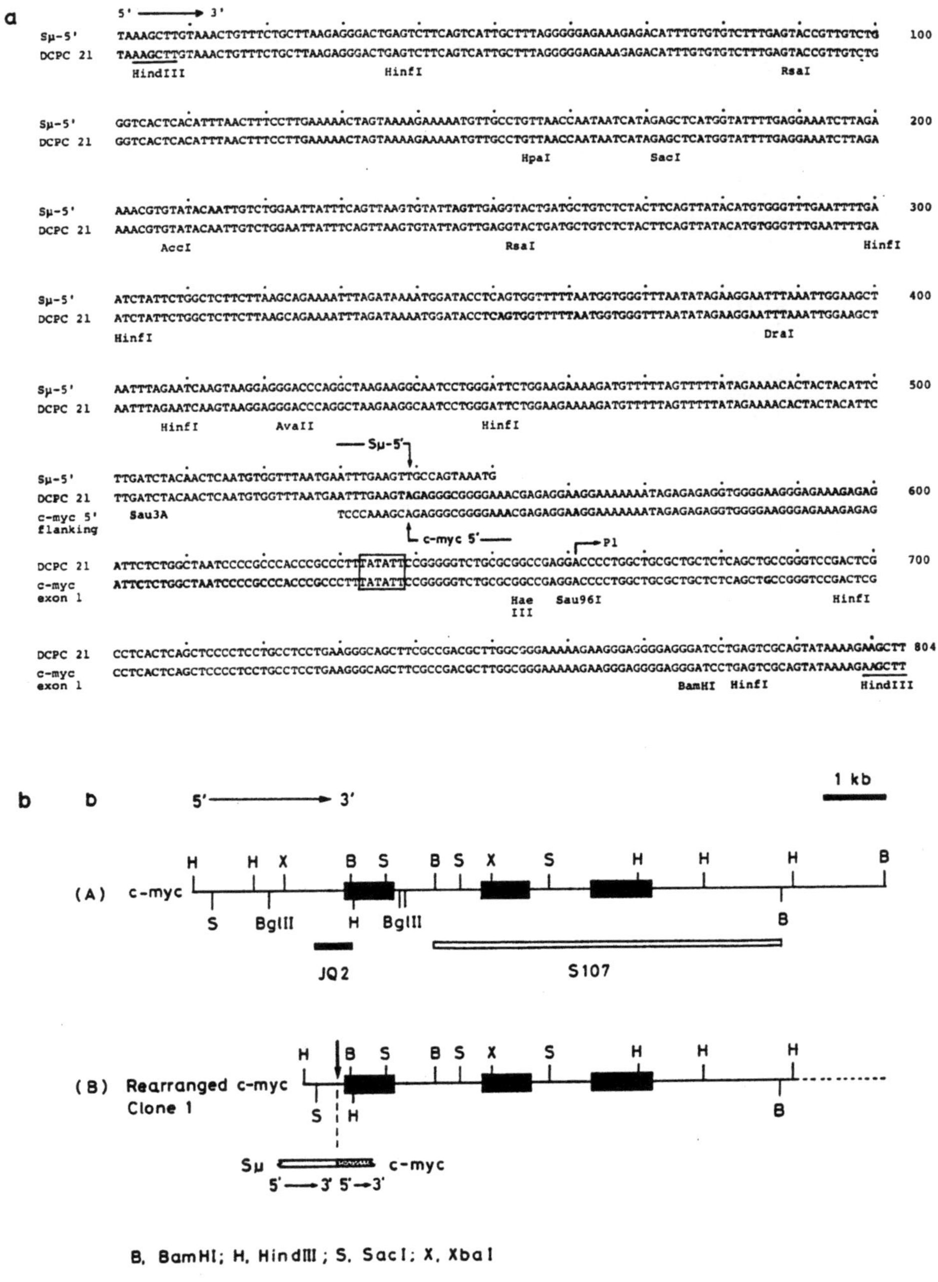

Fig. 5(a) and (b). Nucleotide sequence (a) and the restriction map (b) of the 0.8 kb HindIII fragment indicating the recombination site.

Fig. 5. (a) Nucleotide sequence of the 5'-flanking region of the rearranged DCPC 21 c-myc. DNA sequencing data indicate that the IgSμ region sequence was juxtaposed directly to the c-myc 5'-flanking of DCI 21. Recombination site is indicated by arrows. (b) Restriction map of DCPC 21 c-myc rearrangement and its germline components. (A) Germline c-myc; (B) DCPC 21 c-myc rearrangement. The c-myc probes used, JQ2 and S107, are also shown. Black boxes indicate the c-myc exon-1,2,3, respectively, from left to right. DNA fragments of 10-15 kb, isolated l sucrose gradient fractionation of a partial Sau3A digest of DCPC 21, were ligated to lambda phage vectors (Lambda FIX, STRATAGEGE, La Jolla USA) (Zabarovsky and Allikmets 1986). The ligated DNA was packaged (Gigapack, STRATAGENE, La Jolla, USA) (Rosenberg 1985) and the resulta library of approx. 1 x 10^6 pfu was screened for the rearranged c-myc using the probes S107 (exon-2,3) (Taub et al. 1982) and JQ2 (exon-1) (Stanton et al. 1984). The 0.8 kb HindIII fragment was isolated from tl approx. 15 kb Sau3A insert of the clone (Clone 1) and subcloned into M vectors. DNA sequences were determined using the method of Sanger et a (1977) on the subcloned fragment of HindIII digests which spans approx 0.8 kb upstream of the HindIII site of the exon-1.

the c-myc has broken at 122 bp upstream of the first c-myc promoter an recombined directly with Sμ 5'-flanking in a "head-to-tail" (5' to 3') orientation.

Figure 5b shows a restriction map of DCPC 21 c-myc rearrangement (Clon

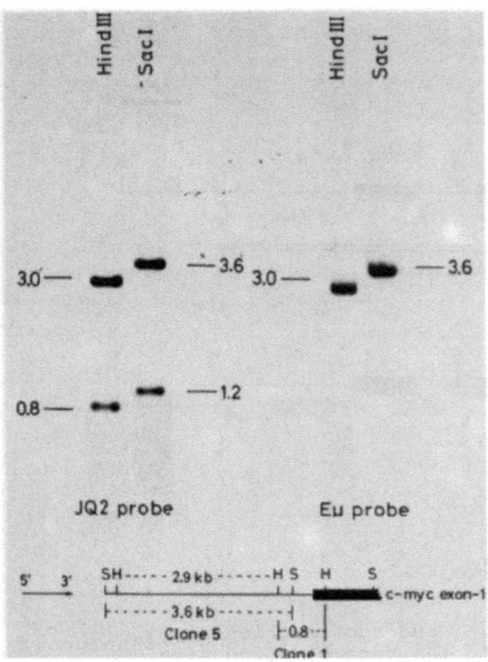

Fig. 6. Southern blot analysis of Clone 5 DNA from DCPC 21. Clone 5 DNA (1 μg) was digested with indicated endonuclease and hybridized either with the JQ2 (left) or the Eμ probe (right). The Eμ probe is a 680 bp XbaI/EcoRI fragment (Gillies et al. 1983). H, HindIII; S, SacI.

1) and its germline components. The results, together with the in situ hybridization experiment (Fig. 3), enabled us to conclude that this recombination event occurred on the chromosome 12 without any visible chromosomal translocations.

Recombination of C-myc with IgH Enhancer-Sμ Sequences

As shown in Fig. 4, the genomic Southern analysis of DCPC 21 DNA using the JQ2 probe demonstrated the presence of the two rearranged c-myc of 0.8 and 3 kb in length, respectively, by HindIII digestion. To define this 3 kb rearranged c-myc, we next isolated a clone containing the 3 kb fragment. For this purpose, a 3.6 kb SacI fragment was isolated from a lambda clone having an approx. 15 kb insert of DCPC 21 DNA prepared by a partial Sau3A digestion. The phage library was screened for the rearranged c-myc using the probes Eμ (IgH enhancer) (Shimizu et al. 1982; Gillies et al. 1983), S107 and JQ2, and subcloned into pUC118 vector.

Figure 6 shows the Southern blot analysis of the clone (Clone 5) containing the 3 kb rearranged fragment. The 3.6 kb SacI fragment as well as the 3 kb HindIII fragment hybridized with both Eμ and JQ2 probes. As expected from a map indicated below, the 3.6 kb SacI fragment encompasses the 3 kb HindIII fragment within it. Deletion mutants carrying the different sizes of the 3.6 kb SacI insert were prepared by

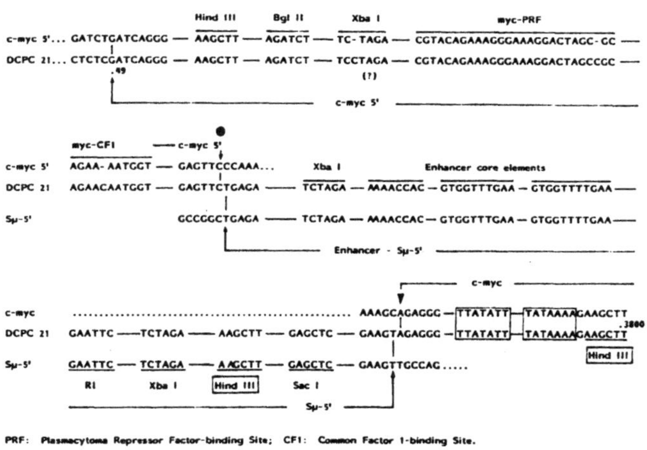

Fig. 7. Nucleotide sequence of the Clone 5 containing the 3.6 kb SacI fragment together with that of the Clone 1 containing the 0.8 kb HindIII fragment. The latter sequence spans between two HindIII sites which are boxed. The recombination sites of the enhancer-Sμ sequences are indicated by a closed circle (5'-boundary of the enhancer-Sμ) and a triangle (3'-boundary of the enhancer-Sμ), respectively. Myc-PRF, myc-CF1 (Kakkis et al. 1989) and enhancer core elements are also shown.

a partial exonuclese III digestion and DNA sequences were determined according to Sanger et al. (1977).

Sequencing data of the 3.6 kb fragment of Clone 5 together with those o the 0.8 kb fragment of Clone 1 (Fig. 5a) are summarized in Fig. 7. For the sake of of convenience, not all the sequences are shown in this figure. The solid lines, however, indicate identities of the nucleotide sequences between the germline c-myc or Sμ and DCPC 21 except 7 bp deletion in DCPC 21 c-myc at the site indicated by a closed triangle (see Fig. 8). From the results presented here, the following conclusion will be drawn: (1) the upstream of the 0.8 kb Sμ-c-myc fragment was IgH enhancer-Sμ sequences; (2) the 5'-flanking of the IgH enhancer-Sμ was again a c-myc 5'-flanking sequence of approx. 1 kb in length. This sequence juxtaposed directly to the enhancer-Sμ in a "head-to-tail" orientation.

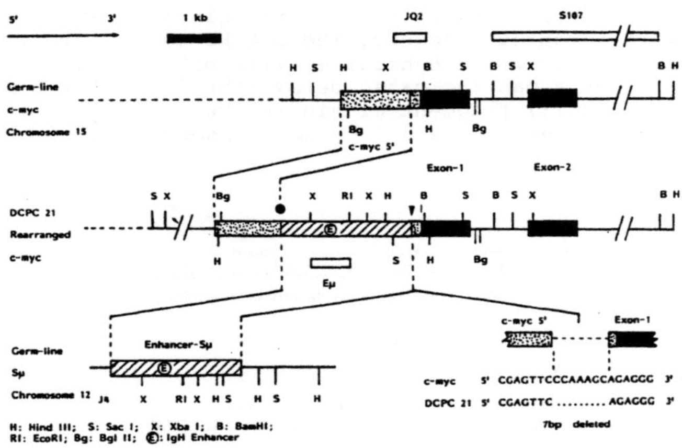

Fig. 8. Restriction map of DCPC 21 c-myc rearrangement obtained from nucleotide sequencing analyses of both the Clone 1 (Fig. 5a) and the Clone 5 (Fig. 7). Germline c-myc and the enhancer-containing Sμ sequences are compared with DCPC 21 c-myc. The recombination sites of the enhancer-Sμ sequences are indicated by a closed circle (5'-boundary of the enhancer-Sμ) and a triangle (3'-boundary of the enhancer-Sμ), respectively. In the lower right corner shows 7 bp deletion in DCPC 21 c-myc at the site of recombination with 3'-boundary of the enhancer-Sμ sequences.

The restriction map of DCPC 21 c-myc rearrangement obtained from nucleotide sequencing analyses of both the Clone 1 and the Clone 5 is shown in Fig. 8. In comparison of the map with previously published one of the germline c-myc and enhancer-Sμ region, we concluded that the rearrangement of the c-myc reflected an insertion of the IgH enhancer-S sequences of approx. 2.3 kb in length into the c-myc region, thus, 5': c-myc 5'-IgH enhancer-Sμ-c-myc 5'-c-myc-exon-1 : 3'.

In situ hybridization experiment (Fig. 3) provided strong support for the conclusion that this recombination event occurred on the chromosome 12 without any visible chromosomal translocations. Indeed, the in situ hybridization of DCPC 21 chromosomes using the Eμ probe revealed that only chromosome 12 was labelled. No specific association of the enhancer sequence with chromosome 15 was found (data not shown). The insert has the same transcriptional orientation as c-myc, in striking contrast to the "head-to-head" alignment of the C_H locus and c-myc gene in every murine MPC of t(12;15) type.

As shwon in Fig. 5a, DCPC 21 has broken at 122 bp upstream of the first c-myc promoter and recombined directly with the Sμ 5'-flanking in a "head-to-tail" fashion. Comparison of the sequences near this recombination site indicated by a closed triangle (Figs. 7 and 8) with the germline c-myc revealed that 7 bp in c-myc 5'-flanking sequence of DCPC 21 were deleted (Fig. 8). Thus, in contrast to the germline c-myc sequence; 5': CGAGTTC CCAAAGC AGAGGG : 3', the rearranged c-myc has the following sequence; 5': CGAGTTC .... AGAGGG : 3' (see Discussion). No c-myc sequence was deleted or duplicated at the recombination site with 5'-boundary of the enhancer-Sμ indicated by a closed circle.

DISCUSSION

Most of the MPCs (>95%) consistently contain the chromosomal translocations involving chromosome 15 and an associated deregulation of c-myc gene transcription, irrespective of the inducing agents (Ohno et al. 1979, 1984, 1988). DCPC 21, however, is an MPC containing no chromosomal translocations (Fig. 1).

We clarified in this paper that DCPC 21 showed c-myc gene rearrangement probably caused by an insertion of the IgH enhancer-Sμ fragment of approx. 2.3 kb in size into the c-myc 5'-flanking region without visible chromosomal translocations. This finding provides strong support for the conclusion that transposition of the c-myc gene to Ig gene regions is a prerequisite genomic event in the development of MPCs. Recently, the c-abl and bcr (breakpoint cluster region) gene rearrangements have been reported in chronic myeloid leukemia patients whose blood cells have no t(9;22) chromosomal translocation (Ph-negative) (Morris et al. 1986; Dreazen et al. 1987).

There must be several alternatives for the c-myc transposition in DCPC 21. Here, we will just show a "simple" model (Fig. 9). The first breaks occur on the chromosome 15 at both sites in the c-myc 5'-flanking region. These breaks cut an approx. 1 kb DNA fragment (a) containing 5'-portion of the c-myc 5'-sequence out of the chromosome 15. On the chromosome 12, the break also occurs between IgH enhancer and J_H gene regions. These breaks would be accompanied by the 1st transposition of the 5'-portion of the c-myc 5'-sequence (a) into the 5'-flanking region of the IgH enhancer sequence. A paracentric inversion must be necessary to occur in this transposition event. Because the IgH enhancer-c-myc 5'-J_H sequences on the inserted allele have the same transcriptional orientation. Next, the 2nd breaks occur on the chromosome 15 at both sites 5' to the c-myc 5'-region and 3' to the c-myc-exon-3. The exact break point downstream of the c-myc-exon-3 has not yet been determined. These breaks again cut the small DNA fragment containing 3'-portion of the c-myc 5'-sequence (b)-c-myc-exon-1,2,3-...3' out of the chromosome 15. Finally, the 2nd transposition of this fragment into the chromosome 12 would again be followed by a paracentric inversion of the fragment. Consequently, there remain no visible structural changes in two chromosomes 12 and 15.

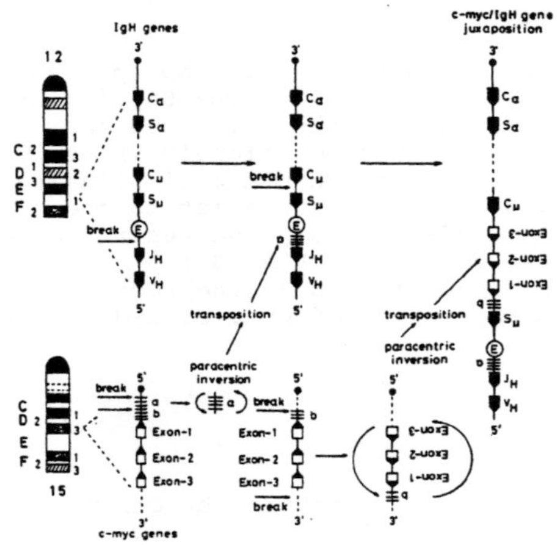

Fig. 9. A "simple" model for the DCPC 21 transposition. In situ hybridization experiments using S107 (Fig. 3) and Eμ probes (data not shown) strongly suggested that this transposition event should be occurred on chromosome 12, not on chromosome 15.

The size of DNA fragment containing the c-myc gene transposed into the chromosome 12 remains unknown at present. Particularly, we do not know how the long sequence between c-myc exon-3 and including PVT-1 affects c-myc transcription. If the c-myc gene transposes into the chromosome 12 with this long sequence 3' of the gene so far upstream tc the IgH enhancer that the enhancer does not have an influence on activating the c-myc gene, it might be necessary that the enhancer sequence moves close to the transposed c-myc gene. We would speculate i this case that the enhancer-Sμ sequences transpose into the c-myc 5'-flanking region as a next event. This could be one of the other alternatives.

Corcoran et al. (1985) reported a complex recombination of a c-myc gene in an MPC ABPC 17 which was induced with Abelson leukemia virus in a pristane-primed BALB/c mouse. The rearrangement reflected an insertion of a 2.3 kb fragment containing IgH enhancer-Sμ sequences, essentially the same size as the insert found in DCPC 21, into the c-myc 5'-flankir region. The complex recombination presented here is reminiscent of thei findings. However, ABPC 17 now turned out to be an invt(12;15) translocation type of chromosomes by F. Wiener (Karolinska Inst., personal communs.).

Based on the gremline J_H-Sμ sequences, the insert comprises 2258 bp, with the enhancer lying approx. 900 bp from its 5'-end (data not shown). Concerning the sequences that might be participating in the recombination are found near the recombination site. There is limited homology adjacent to this site: 5 out of 7 bp in the c-myc 5'-sequence immediately upstream of the insertion site, CGAGTTC, and in the sequence of the insert adjacent to its 5'-boundary, GAAGTTG. At the 3'-boundary of the insert, there is also limited homology, GAAGTTG CCA. Here, 3 bp CCA are identical to a CCA immediately downstream of the insertion site of the c-myc 5' that was deleted in DCPC 21 c-myc rearrangement (see 7 bp deletion in Fig. 8). The GAAGTTG is an essentially homologous sequence to the one of three types of Sμ repeat elements, a YAGGTTG and its analog CAGGTCG (Nikaido et al. 1981; Marcu et al. 1982). The scission 3' to the transposed region may have been made by a switch recombination machinary (Corcoran et al. 1985). Details and further developments will appear elsewhere.

ACKNOWLEDGEMENTS

We express our sincere thanks to Drs M. Potter, J.F. Mushinski, NCI, and K.B. Marcu, State University of New York, for kindly supplying the c-myc probes, S107 and JQ2. We thank Dr T. Honjo, Kyoto University, and JCRB (Japanese Cancer Research Resources Bank), Tokyo, for the Eμ probe. We are also grateful to Drs G. Klein and F. Wiener, Karolinska Institute, for their interest and encouragement. Immunofluorescent staining by Dr H. Tokuyama, Cancer Research Institute Kanazawa University, is greatly appreciated. A part of this work has been published in Oncogene (4: 1513, 1989). We express our gratitude to The Macmillan Press Ltd, for kind permission to use the data presented in the Journal. This work was supported in part by a Grant-in-Aid for Cancer Research from the Ministry of Education, Science and Culture, Japan.

REFERENCES

Committee on Standardized Genetic Nomenclature for Mice (1972) Standard karyotype of the mouse, Mus musculus. J Hered 63: 69-72
Corcoran LM, Cory S, Adams J (1985) Transposition of the immunoglobulin heavy chain enhancer to the myc oncogene in a murine plasmacytoma. Cell 40: 71-79
Cory S (1986) Activation of cellular oncogene in hemopoietic cells by chromosome translocation. Adv Cancer Res 47: 189-234
Dreazen O, Rassool F, Sparkes RS, Klisak I, Goldman JM, Gale RP (1987) Do oncogenes determine clinical features in chronic myeloid leukemia? Lancet 1: 1402-1405
Feinberg AP, Vogelstein B (1983) A technique for radiolabeling DNA restriction endonuclease fragments to high specific activity. Anal Biochem 132: 6-13
Gillies SD, Morrison SL, Oi VT, Tonegawa S (1983) A tissue-specific transcription enhancer element is located in the major intron of a rearranged immunoglobulin heavy chain gene. Cell 33: 717-728
Harper ME, Saunders GF (1981) Localization of single copy DNA sequences on G-banded human chromosomes by in situ hybridization. Chromosoma 83: 431-439
Kakkis E, Riggs KJ, Gillespie W, Calame K (1989) A transcriptional repressor of c-myc. Nature 339: 718-721

Klein G (1983) Specific chromosomal translocations and the genesis of B-cell-derived tumors in mice and men. Cell 32: 311-315
Marcu KB, Lang RB, Stanton LW, Harris LJ (1982) A model for the molecular requirements of immunoglobulin heavy chain class switching. Nature 298: 87-89
Merwin RM, Redmon LW (1963) Induction of plasma cell tumors and sarcoma in mice by diffusion chambers placed in the peritoneal cavity. J Natl Cancer Inst 31: 998-1007
Morris CM, Reeve AE, Fitzgerald PH, Hollings PE, Beard MEJ, Heaton DC (1986) Genomic diversity correlates with clinical variation in Ph'-negative chronic myeloid leukemia. Nature 320: 281-283
Nesbitt MN, Francke U (1973) A system of nomenclature for band patterns of mouse chromosomes. Chromosoma 41: 145-158
Nikaido T, Nakai S, Honjo T (1981) Switch region of immunoglobulin Cμ gene is composed of single tandem repetitive sequences. Nature 292: 845-848
Ohno S, Babonits M, Wiener F, Spira J, Klein G, Potter M (1979) Nonrandom chromosome changes involving the Ig gene-carrying chromosomes 12 and 6 in pristane-induced mouse plasmacytomas. Cell 18: 1001-1007
Ohno S, Migita S, Wiener F, Babonits M, Klein G, Mushinski, JF, Potter M (1984) Chromosomal translocations activating myc sequences and transduction of v-abl are critical events in the rapid induction of plasmacytomas by pristane and Abelson virus. J Exp Med 159: 1762-1777
Ohno S, Takahashi R, Mihara K, Chen H-L, Migita S, Sugiyama T (1988) An atypical t(12;15) translocation of chromosomes found in murine plasmacytomas and activation of the c-myc oncogene. Int J Cancer 41: 595-600
Ohno S, Migita S, Murakami S (1989) C-myc-Sμ rearrangement in a murine plasmacytoma without visible chromosomal translocations. Oncogene 4: 1513-1517
Rosenberg SM (1985) EcoK restriction during in vitro packaging of coliphage lambda DNA. Gene 39: 313-315
Sakano H, Maki R, Kurosawa Y, Roeder W, Tonegawa S (1980) Two types of somatic recombination are necessary for the generation of complete immunoglobulin heavy-chain genes. Nature 286: 676-683
Sanger F, Nicklen S, Coulson AR (1977) DNA sequencing with chain terminating inhibitors. Proc Natl Acad Sci USA 74: 5463-5467
Stanton LW, Yang J-Q, Eckhard LA, Harris LJ, Birshtein BK, Marcu KB (1984) Products of a reciprocal chromosome translocation involving the c-myc gene in a murine plasmacytoma. Proc Natl Acad Sci USA 81: 829-833
Taub R, Kirsch I, Morton C, Lenoir G, Swan D, Tronick S, Aaronson S, Leder P (1982) Translocation of the c-myc gene into the immunoglobulin heavy chain locus in human Burkitt lymphoma and murine plasmacytoma cells. Proc Natl Acad Sci USA 79: 7837-7841
Wang HC, Fedoroff S (1972) Banding in human chromosomes treated with trypsin. Nature (New Biol.) 235: 52-54
Zabarovsky ER, Allikmets RL (1986) An improved technique for the efficient construction of gene libraries by partial filling-in of cohesive ends. Gene 42: 119-123

Downstream Regulatory Elements in the c-*myc* Gene

M. Zajac-Kaye, B. Yu, and N. Ben-Baruch

Medicine Branch, Clinical Oncology Program, National Cancer Institute, National Institute of Health, Maryland 20892

INTRODUCTION

Deregulated expression of the *c-myc* gene appears to play an important role in the development of Burkitt's lymphoma (BL) and certain other lymphoid tumors. Physiologic regulation of the *c-myc* gene is complicated, involving the interplay of distinct transcriptional and post transcriptional mechanisms. There are many regulatory elements that bind to the upstream region of the *c-myc* gene, however, this chapter will focus on the factors that bind to the intron I region of this gene and will discussed the possible role intron I sequence may have in the regulation of the *c-myc* gene.

SOMATIC MUTATIONS CLUSTERED WITHIN THE PROTEIN BINDING SITE IN THE INTRON I OF THE *c-myc* GENE.

In Burkitt's lymphoma, somatic mutations in the activated allele of the *c-myc* gene has been found to cluster near the exon I/intron I boundary. Mutations in exon I sequences have been implied to play a important role in *c-myc* deregulation in BL cells (Pelicci et al 1986, Cesarman et al 1987). However, abnormalities in the 5' region of the intron I sequences in the human *c-myc* gene such as large and small deletions, duplications and point mutations, may be more prevalent than these found in exon I of this gene. Therefore, we examined whether regulatory factors bind to the sequences located in intron I of the human *c-myc* gene and we asked whether mutations identified in BL would interfere with protein binding. We have identified a 20 bp region in intron I of the *c-myc* gene which binds a nuclear protein and we have shown that this protein binding was abolished by point mutations present in a corresponding region in Burkitt's lymphoma *c-myc* DNA (Zajac-Kaye et al. 1988). The protein binding sequence and its location in the human *c-myc* gene is shown in Fig. 1. This 20 bp recognition sequence is frequently mutated in Burkitt's lymphoma *c-myc* DNA. We found that 7 out of 9 BL's (for which the sequence of the *c-myc* intron I were available in the literature) contained mutations clustered within the protein binding site (Fig. 1). Interestingly, a Burkitt's lymphoma cell line, BL2 which was cited in literature as a example of a BL in which *c-myc* is deregulated but " not altered in its putative regulatory regions" (Showe et al. 1987) contained mutations in the 20 bp region of the protein binding site. The number of cases of Burkitt's lymphoma possessing mutations within the binding site for this factor is growing steadily and comprises a large majority of cases.

Current Topics in Microbiology and Immunology, Vol. 166

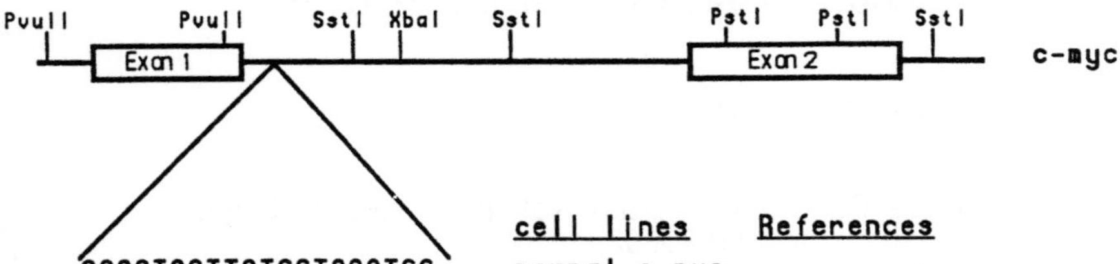

	cell lines	References
AGAGTAGTTATGGTAACTGG	normal c-myc	
------------A-T-----	PA682	Zajac-Kaye et al. 1988
----------------T---	BL2	Showe et al.1987
--------------T-----	KK124	Denny et al. 1985
---A--CC-T------G---	Raji	Rabbitts et al. 1983
---A-----G-TC-------	LY47	Zajac-Kaye et al.1988
-----G--G-C---G-A---	BL3	Morse et al. 1989
-------------A------	pAb.23	LeStrange et al. 1988

Fig. 1. Clustering of somatic mutations in protein binding site in *c-myc* intron I in Burkitt's lymphoma.

PURIFICATION AND CHARACTERIZATION OF THE FACTOR WHICH BINDS TO THE INTRON I OF THE *c-myc* GENE

The protein binding to *myc* intron I was purified from HeLa cell nuclear extracts by cation-exchange chromatography followed by passage over oligonucleotide affinity columns coupled either with the 20bp wild type *c-myc* intron I binding oligonucleotide or with a mutated oligonucleotide derived from the BL *c-myc* DNA (Zajac-Kaye and Levens 1990). The logic behind coupling the mutated oligonucleotide to the affinity column was that the Myc Intron Factor (which specifically binds the wild type (wt) sequence) will not bind strongly to the mutated sequence and thus we could enrich for it in the flow through (FT) of the mutated affinity column. Using South-western blot analysis a 138 kD protein was identified with the *c-myc* wt-oligo probe but was not observed when the mutated bl-oligo probe was used. The wt-oligo column eluate revealed a strong signal for the 138 kD protein and a faint signal for lower molecular weight proteins. After passage through the mutated oligo column, the 138 kD protein was enriched in the FT of the affinity column while the proteins producing the faint bands of lower molecular weight were preferentially retained on the mutated column. Thus, we have identified a specific protein which binds to the 20bp region in intron I of the *c-myc* gene and could not bind to the mutated region present in the corresponding region of BL *c-myc* DNA. We designated this protein MIF for Myc Intron Factor.

To confirm that the 138 kD MIF protein was responsible for the retarded band on the mobility shift assay two dimensional electrophoresis was performed.

The retarded band containing the specific protein:DNA complex and controls were excised from a mobility shift gel and the gel slices were rotated ninety degrees and embedded on the top of an SDS polyacrylamide gel. Following electrophoresis the proteins were electroblotted to nitrocellulose filters and visualized by gold staining. A 138 kD protein was detected only in presence of the *c-myc* specific oligonucleotide probe and it was not detected in presence of a nonspecific oligonucleotide or when oligonucleotides were omitted. This 138 kD protein band detected on the SDS gell aligned exactly under the upper band of the doublet seen with the mobility shift assay indicating that this 138 kD protein is a component of the specific complex generated with the *c-myc* oligonucleotide probe. Thus, using South-western blot analysis or gel retardation assay combined with SDS-PAGE electrophoresis a 138 kD protein was independently identified as the nuclear protein which binds to the 20bp *cis* element in the intron I of the *c-myc* gene (Zajac-Kaye and Levens 1990).

To determine, whether MIF was a phosphoprotein HeLa cells were labeled with ^{32}P orthophosphate and MIF was purified on cation exchange and on wild type and mutated affinity columns. A phosphoprotein in the range of 138 kD behaved chromatographicaly like the MIF protein. It showed affinity for the wt-oligo column, did not bind to the mutated column under similar conditions and was present in the flow through of the mutated affinity column. Phosphoaminoacid analysis indicated that this protein is phosphorylated on serine residues. To test the function of MIF phosphorylation we asked whether changes in phosphorylation state of MIF altered binding with its recognition sequence. Treatment of purified MIF with potato acid phosphatase nearly abolished binding to its recognition sequence while this effect was inhibited by addition of phosphatase inhibitors to the binding reaction. To identify the polypeptide with binding properties altered by phosphatase treatment South-western blot analysis were performed. Purified MIF was treated with phosphatase in the presence or absence of phosphatase inhibitor. Binding to the 138 kD protein detected with the wt *c-myc* oligo probe was again nearly abolished after phosphatase treatment and was protected in the presence of phosphatase inhibitor. These results, demonstrate that the 138 kD MIF is a phosphoprotein and that phosphorylation of MIF is required for binding to the 20bp sequence in intron I of the *c-myc* gene (Zajac-Kaye and Levens 1990).

THE ROLE OF MIF BINDING SEQUENCE IN REGULATION OF TRANSCRIPTION FROM HETEROLOGUOS AND *c-myc* PROMOTERS.

To study the role of MIF protein in regulation of transcription we asked whether MIF can influence expression of SV40 as well as the *c-myc* promoters. The 20 bp synthetic oligonucleotide representing MIF DNA binding region was cloned into a vector (Baker et al 1989) in which transcription of the CAT gene is driven by the SV40 promoter and enhancer. Transient transfection into HeLa cells resulted in 50% reduction of CAT activity by the vector containing the 20 bp element cloned in sense orientation, while no effect was seen when the binding region was cloned in the antisense orientation. RNase protection analysis showed that changes in steady state CAT-RNA levels reflected that observed with CAT enzymatic activity. This results suggests that the 20 bp binding site can act as a weak negative element on the SV40 promoter.

Next, we constructed a series of vectors to examine the effect of MIF binding on the regulation of the promoters in the *c-myc* gene. The MYC/CAT parent vector consisted of 3.2 kb Hind III/Sst I fragment from the *c-myc* gene fused to the CAT reporter gene, so transcription of CAT is now driven by the *c-myc* promoters. This construct included intron I sequence containing the MIF binding region. The MIF binding site was than deleted by removing a 283 bp fragment from the intron I of the *c-myc* gene. A 216 bp region 3' downstream from MIF binding site was also deleted. This vector retained the MIF binding site. The three vectors were transfected into HeLa cells and the transient expression of the CAT gene was compared between each construct. Deletion of the 283 bp MIF containing fragment resulted in 2 fold increase of CAT activity compared to the parent MYC/CAT vector. Surprisingly, when the sequence 3' from MIF were deleted CAT activity was increased 3-4 fold as compared to the parent vector. These results suggest, that another regulatory element is located in intron I, downstream from the MIF binding site.

To examine binding of a nuclear protein to the sequence located downstream from the MIF binding site an exonuclease protection assay was performed. In this assay, two bands were detected which correspond to two protein binding sites. One band represented MIF binding to its recognition sequence and the second band represents another binding sequence located in the region downstream from the MIF binding site. We are in the process of localizing the binding sequence for this factor. In addition, a competition experiment indicated that the 20 bp oligo which represents MIF recognition sequence not only competed binding of MIF but also competed binding of the downstream factor as well. Thus, the effect of MIF binding to its recognition sequence on the control of *c-myc* expression may be dependent on adjacent *cis* elements located in intron I of the *c-myc* gene.

We do not know whether MIF can bind to both sites or whether there is a second factor with the specificity for MIF recognition sequence (Fig. 2). In any event, the interaction between these two different regions and its binding factors may be perturbed in BL due to mutations frequently observed in MIF binding site. In addition, another level of complexity in the regulation of the *c-myc* gene comes with the observation that MIF is a phosphoprotein and that MIF phosphorylation is required for binding to its recognition sequences. The interaction between the intron I *cis* element, its binding factor (MIF), and serine kinases and phosphatases to phosphorylate or dephosphorylate the MIF protein may comprise an important physiologic circuit; alteration of this circuit may be similar to the BL mutations namely reduced MIF binding to DNA. The appropriate balance between serine kinase and phosphatases may play important role in controlling the phosphorylation state of MIF and may also contribute to altering the *c-myc* expression.

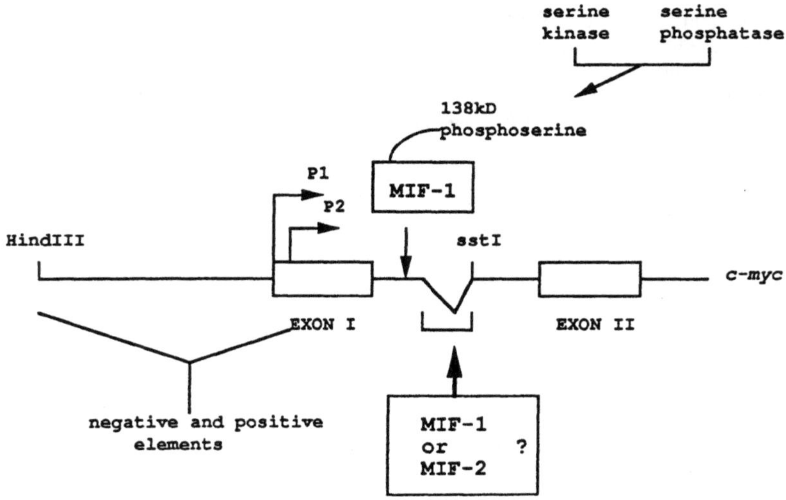

Fig. 2. Intron I elements in the *c-myc* gene

REFERENCES

Baker CC (1989) An improved chloramphenical acetyltransferase expression vector system for mapping transcriptional and post-transcriptional regulatory elements in animal cells. In M-L huhtala and A Vaheri (ed), Recombinant systems in protein expression. Elsevier Science Publishers BV, Amsterdam, the Netherlands.

Cesarman E, Dalla-Favera R, Bentley D, Groudine M (1987) Mutations in the first exon are associated with altered transcription of *c-myc* in Burkitt's lymphoma. Science 238: 1272-1275

Danny CT, Hollis GF, Magrath IT, Kirsch IR (1985) Burkitt lymphoma cell line carrying a variant translocation creates new DNA at the breakpoint and violates the hierarchy of immunoglobulin gene rearrangement. Mol. Cell. Biol. 5: 3199-3207

LeStrange RC (1988) Molecular analysis of the 8;14 translocation in Burkitt's lymphoma and associated changes in the *c-myc* gene. Ph.D. Thesis, Cornell University. Sequence provided by W. Hayward as a personal communication

Morse B, South VJ, Rothberg PG, Astrin SM (1989) Somatic mutation and transcriptional deregulation of *c-myc* in endemic Burkitt's lymphoma disease: heptamer-nonamer recognition mistakes. Mol. Cell. Biol. 9: 74-82

Pelicci P-G, KnowlesII DM, Magrath I, Dalla-Favera R (1986) Chromosomal breakpoints and structural alterations of the *c-myc* locus differ in endemic and sporadic forms of Burkitt lymphoma. Proc. Natl. Acad. Sci. USA 83: 2984-2988

Rabbitts TH, Hamlyn PH, Baer R (1983) Altered nucleotide sequences of translocated *c-myc* gene in Burkitt lymphoma. Nature 306: 760-765
Showe LC, Moore CA, Erikson J, Croce CM (1987) MYC oncogene involved in a t(8;22) chromosome translocation is not altered in its putative regulatory regions. Proc. Natl. Acad. Sci. USA 84: 2824-2828
Zajac-Kaye M, Gelmann EP, Levens D (1988) A point mutation in the *c-myc* locus abolished binding of a nuclear protein. Science 240: 1776-1780
Zajac-Kaye M, Levens D (1990) Phosphorylation-dependent binding of a 138-kDa Myc Intron Factor to a regulatory element in the first intron of the *c-myc* gene. J. Biol. Chem. 265: 4547-4551

DNA Repair in the c-*myc* Locus

E.J. Beecham, J.F. Mushinski, E. Shacter, M. Potter and V.A. Bohr.

Laboratories of Molecular Pharmacology and Genetics, National Cancer Institute, BLdg. 37 Rm 5C25, NIH, Bethesda, MD 20892, USA.

We have studied DNA repair after UV damage in the murine c-myc locus. It appears that a region in B cells upstream of the murine c-myc gene is repaired with a different efficiency in plasmacytoma-resistant DBA/2N mice than in plasmacytoma-susceptible BALB/cAn mice. The region just uptream of c-myc is inefficiently repaired in B lymphoblasts derived from BALB/cAn mice. In contrast, this same region of c-myc is efficiently repaired in B lymphoblasts derived from DBA/2N mice. DNA fragments located in the coding region of c-myc and in another gene, dihydrofolate reductase (DHFR), are repaired with equal efficiency in cells from these two strains of mice. It is possible that repair efficiency of the 5' flank of c-myc may be involved in tumor susceptibility of the mouse strain.

INTRODUCTION

Plasmacytomas can be consistently induced in the inbred mouse strains BALB/cAn and NZB by introducing into the peritoneum agents such as pristane (2,6,10,14-tetramethyl-pentadecane) that cause irritative inflammatory reactions. Most other strains of mice such as DBA/2N are resistant to this form of tumor induction. This suggests that specific genes play a role in plasmacytoma induction. A second important factor is the structural and/or functional state of the c-myc protooncogene. In virtually all of BALB/c plasmacytomas c-myc is structurally and/or functionally altered. The contribution of this genetic alteration to the pathogenesis and progression of plasmacytomas is not clear but its consistent presence suggests that it is essential for neoplastic transformation.

The genetic and molecular mechanisms that limit the carcinogenic effect of pristane in plasmacytoma-resistant mice are not known. The peritoneal granuloma formed by macrophages and other inflammatory cells in pristane-treated mice is a source of many biologically active factors; e.g., reactive oxygen intermediates, growth factors, etc. One hypothesis is that DNA damaging agents produced by inflammatory cells are processed differently in cells from resistant and susceptible mice.

Cells from plasmacytoma-susceptible BALB/cAn mice may not be able to repair DNA damage as effectively as cells from other strains of mice. This would increase the frequency of mutation in BALB/c cells and could potentially lead to a higher incidence of transformation. There is some evidence to support such a hypothesis. "Chromatin repair" studies performed on cells from several inbred strains of mice have shown that plasmacytoma-susceptible BALB/cAn and NZB mice were unable to repair X-ray induced damage as efficiently as several other

Current Topics in Microbiology and Immunology, Vol. 166

plasmacytoma-resistant strains of mice such as DBA/2N (Potter et al. 1988a, Sanford et al. 1986). Furthermore, "chromatin repair" studies performed on a number of BALB/cAn.DBA/2N (C.D2) congenic mice showed that DBA/2N loci Rep-1 on chromosome 1 or Rep-2 on chromosome 4 were able to confer an efficient (DBA-like) repair phenotype in BALB/c congenic mice (Potter et al., 1988b). One of these mice, the C.D2-Fv-1 n/n mouse (chromosome 4) is also partially resistant to plasmacytomagenesis. Thus, the region of DNA surrounding Fv-1 on chromosome 4 carries genes that are involved in controlling resistance to plasmacytomagenesis as well as genes associated with DNA repair. Lin and Ruddle (1981) have described a murine repair gene on chromosome 4 that can restore wild type repair in mouse-xeroderma pigmentosum somatic cell hybrids. The relationship of Rep-2 to this gene has not been established.

The c-myc protooncogene is a target for chromosomal translocation in B-lymphocytic tumors from at least 3 mammalian species. A relevant question concerns how this gene is repaired after DNA damage in normal B cells. It has recently been established that mammalian cells repair damaged DNA in a selective manner. Certain regions of the genome are repaired much more efficiently than other regions or than the genome as a whole; this has been termed preferential DNA repair of active genes (Bohr et al. 1989). Preferential DNA repair or the lack thereof in selected genomic regions may explain some of the differences seen in the carcinogenic response of different tissues or different organisms.

METHODOLOGY FOR GENE SPECIFIC REPAIR EXPERIMENTS

"Chromatin repair" studies measure the average DNA repair in the genome as a whole. Techniques have recently been developed to study damage and repair in genes and other defined sequences (Bohr et al. 1985; Bohr and Okumoto 1988; Bohr et al. 1989). In general this approach can be used to determine repair in any restriction fragment of interest. A requirement is that a strand break be generated at the site of the DNA damage and this is accomplished in different ways for different kinds of damage. Some agents cause strand breaks directly (e.g., ionizing radiation and bleomycin) and in other cases, such as UV damage, DNA lesions are detected with specific endonucleases that cleave the DNA. The frequency of strand breaks in specific restriction fragments is determined through quantitative Southern analysis.

The technique for measuring UV damage and repair will be briefly discussed. Cells are uniformly prelabeled with ^{3}H-thymidine to tag the DNA. After UV irradiation the cells are incubated for repair in the presence of the heavy thymidine analog bromodeoxyuridine (BrdUrd); this allows for separation of (semiconservatively) replicated DNA from the parental DNA. This step is required because the inclusion of replicated (lesion free) DNA in the assay will lead to an overestimate of repair. Genomic DNA is isolated from the cells, digested with restriction endonuclease, and the parental DNA is separated on CsCl gradients. Two equal aliquots of each sample are then removed and one is treated with the E. coli enzyme T4 endonuclease V. This is a UV pyrimidine dimer-specific enzyme which, through a two-step reaction, cleaves the DNA at sites of the most common DNA adduct formed after UV irradiation - the pyrimidine dimer. DNA samples are then electrophoresed on alkaline gels, transfered to support membranes, probed for a gene of interest, and exposed to X-ray film. Autoradiographic bands are quantified by densitometry or directly from the membrane using a (Betagen) blot analyzer. The number of lesions per fragment is calculated from the fraction of fragments free of damage by using the Poisson distribution.

PREFERENTIAL DNA REPAIR IN ACTIVE GENES

Most available results on gene specific repair have been obtained after UV damage to the cells. It was initially shown that repair of UV damage in the essential dihydrofolate reductase gene (DHFR) in Chinese hamster ovary cells (CHO) is much more efficient than the repair of the overall genome (Bohr et al. 1985). In normal repair proficient human cells, the whole genome is repaired after 24 hours. However, essential genes are repaired faster than non-coding sequences in the DNA (Mellon et al. 1986). A preferential DNA repair domain has been described in the CHO DHFR locus: a 60-80 Kb region centered around the 5' end of the gene (Bohr et al. 1986). While the initial level of damage is similar in all fragments within this domain, the repair efficiency differs considerably and is maximal at the 5' end of the gene. The size of this domain is similar to a loop or higher order structure of chromatin and suggests a relation between DNA repair processes and chromatin structure.

Oncogenes, Repair and Transcription

Different genes within the same cell can be repaired with different efficiencies. In a study on the repair of some protooncogenes in mouse cells, it was found that in NIH 3T3 cells the c-abl gene is repaired much more efficiently than the c-mos gene (Madhani et al. 1986). One important difference between these two genes is that the c-abl gene is actively transcribed in these cells whereas the c-mos gene is not. These experiments suggest a correlation between the level of transcription and the efficiency of repair in a given gene. This relationship was further supported by studies on the repair of UV damage in the metallothionein gene in CHO cells (Okumoto and Bohr 1987) and in human cells (Leadon and Snowden 1988). Repair in the metallothionein gene is markedly more efficient when the gene is transcriptionally active than when it is not. These findings suggest that there is some functional association between repair processes and transcription machinery.

The repair experiments presented here were performed to determine the efficiency of repair in the c-myc gene of normal B-lymphocytes from different inbred strains of mice. We are interested in studying how this gene is repaired in these cells, because c-myc mutations appear frequently in B-cell neoplasms.

RESULTS

LPS blasts

LPS blasts were prepared by perfusing the spleens of 8-12 week old untreated mice. Red blood cells were removed by lysis in ammonium chloride (ACK) buffer. Cells were then incubated at 37° for 36-48 hours in culture medium containing 50 μg/ml LPS (lipopolysaccharide). Non-adherent cells were then collected, washed once in buffered saline and irradiated with UV light at a dose of 10 J/m^2.

CHO B-11 Cells.

The technique utilized in these investigations and the type of data they yield are best illustrated by our initial experiment in CHO cells with an amplified DHFR gene. Repair after a standard dose of UV light (20 J/m^2) was studied in a 14 kb fragment encompassing the 5' half of the DHFR gene. The Southern blot and a plot of the repair efficiency in the CHO DHFR gene is shown in Fig 1. The repair analyses of fragments in the mouse c-myc gene in mouse LPS blasts below and in Fig. 2 were obtained from similar experiments.

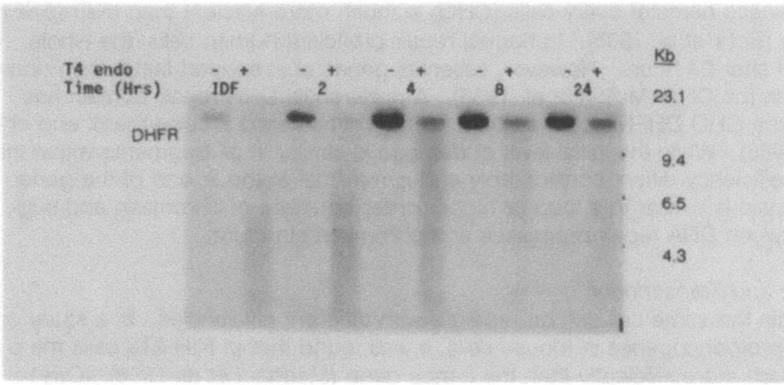

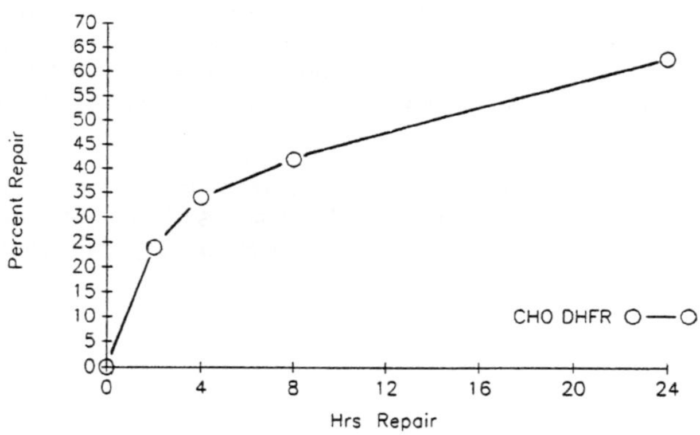

Fig 1. Top: Autoradiogram of a 14 kb KpnI restriction fragment in the 5' end of the DHFR gene in CHO B11 cells after 20 J/m^2 UV irradiation. The - and + signs indicate whether or not the DNA was treated with T4 endonuclease V. IDF, initial dimer frequency; this is damage at the time of irradiation. 2-24 hrs indicate the repair period after damage. Bottom: Plot of repair efficiency. Determinations are from the Southern blot in upper panel.

DNA repair in the 5' region of c-myc

Maps of the mouse c-myc and DHFR genes are shown above the repair curves in Fig. 2. The restriction fragments analyzed for repair are represented by heavy bars under the maps. DNA fragments analyzed for initial damage and repair in these experiments were derived from LPS-stimulated B cells from the spleens of BALB/cAn and DBA/2N mice. DNA repair following UV irradiation at 10 J/m^2 was primarily measured in a 6.5 kb Sac I fragment in the 5' flank of c-myc, and the repair curves are shown in Fig. 2A. B cells from DBA/2N and BALB/cAn mice differ significantly in their ability to repair the 5' region of c-myc. Repair at 12 hrs in plasmacytoma-resistant DBA/2N cells is about 2.5 fold greater than that in plasmacytoma-susceptible BALB/cAn cells. We find that both cell lines sustain similar levels of initial damage, but DBA/2N cells are able to repair damage at a faster rate.

DNA repair in the coding and 3' flanking regions of c-myc

Repair of c-myc was also analyzed in a 13 kb Sac I fragment that spans the third exon and 3' flank of the gene, Fig. 2B. In contrast to the 5' end of c-myc, the rate of repair in the 3' region is similar in both BALB/cAn and DBA/2N B cells. Comparison of repair rates in both the 5' and 3' regions of c-myc also indicates that in DBA/2N cells repair is maximal at the 5' end of the gene. This is analogous to the situation in the DHFR gene of CHO cells, where maximal repair is centered around the 5' end of the coding sequence (Bohr et al. 1986). In contrast, BALB/cAn cells appear to lack rapid repair 5' of the c-myc gene, and both the 5' and 3' regions are repaired with similar efficiencies.

DNA repair in the mouse DHFR gene

The repair analysis of an intragenic 14 kb Sac I fragment from the mouse DHFR gene is shown in Fig. 2C. Repair in this fragment is of similar efficiency in both BALB/cAn and DBA/2N B cells, and it is similar (20-30% repair by 12 h) to that of the 3' c-myc fragment.

Transcription of c-myc in LPS Blasts

Transcription of c-myc was analyzed through northern blots of poly-$(A)^+$ mRNA isolated from both BALB/cAn and DBA/2N cells. The RNA was prepared from untreated spleen cells and from LPS-stimulated cells both before and after UV irradiation. In agreement with previous reports, transcription of c-myc increased significantly when the cells were stimulated with LPS. More importantly the results indicate that both before and after UV irradiation there is no significant difference in the steady state levels of c-myc transcripts in cells derived from BALB/cAn or DBA/2N mice. In addition, we find no detectible steady-state transcripts from the 5' flanking region of c-myc in LPS blasts from either strain (data not shown).

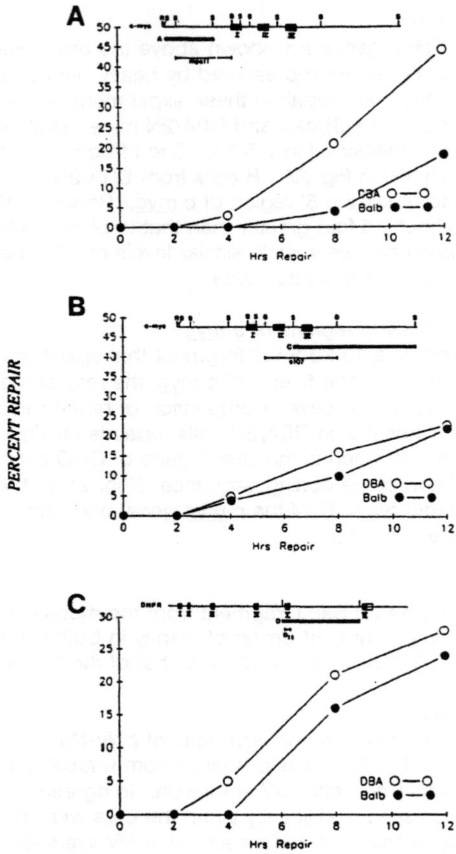

Fig 2. Plots show repair of UV dimers in LPS blasts from DBA/2N (o) and BALB/cAn (●) mice. Data is expressed in percent repair as a function of time after UV irradiation. A: Repair in a 6.5 kb SacI fragment (A) in the 5' flank of the mouse c-myc gene. B: Repair in a 13 kb SacI fragment (C) in the 3'rd exon and 3' flank of c-myc. C: Repair of a 14 kb SacI fragment (D) in the middle of the mouse DHFR gene. Inserts in the graphs show the location of the restriction fragments studied and the probes used.

DISCUSSION

The data presented here indicate that there are different 5' c-*myc* repair phenotypes in BALB/cAn and DBA/2N B lymphoblasts. At a molecular level there are a number of ways that this repair difference could be accounted for. 1) Differences in the transcriptional activity of c-*myc* in B cells from different strains of mice could account for the observed repair differences. This appears unlikely, however, because Northern analysis of c-*myc* levels in BALB/cAn and DBA/2N B lymphoblasts shows no significant differences in the steady state levels of c-*myc* transcripts either before or after UV irradiation. More importantly, the most dramatic differences in repair are seen in the 5' flanking regions, and these regions do not generate sufficient steady-state levels of c-*myc* RNA to be detected at the Northern level. However, we cannot exclude the possibility that there are aberrant transcripts which may explain the repair differences in this region. 2) The repair differences may reflect deficient repair enzymes in BALB/cAn mice. This possibility also appears to be unlikely because repair of the 3' portion of c-*myc* and the DHFR gene are roughly equivalent in both strains. 3) Interstrain differences in the binding of c-*myc* regulatory proteins might block the attachment or activity of repair enzymes in BALB/cAn B-cells. Our preliminary studies on the repair of c-*myc* in primary mouse fibroblasts indicate that in BALB/cAn fibroblasts the 5' end of c-*myc* is repaired more efficiently than in BALB/cAn lymphoblasts. It is possible that the repair efficiency in the 5' flanking region of c-*myc* is affected by the binding of regulatory proteins or enzymes, since it has been demonstrated that there are a number of regulatory elements in this region (Muller et al. 1988). We favor the last hypothesis and are designing experiments to test it. It is evident that there are multiple factors involved in the regulation of c-*myc*. Tissue-specific repair differences in this gene may reflect differential regulation of c-*myc* expression in different tissues.

Central to our interests is the question as to what role c-*myc* repair heterogeneity plays in chromosomal translocations or susceptibility to plasmacytomagenesis. We are currently pursuing this question by analyzing repair of specific genes in a number of C.D2 congenic strains that differ in plasmacytoma susceptibility.

We have found some marked differences in the UV repair efficiency in the 5' flanking region of c-*myc* between two mouse strains. The higher level of repair in this region in the mice which are resistant to tumor formation could link DNA repair events in specific gene regions to the risk of tumor formation. It is reasonable to speculate that DNA repair plays an important role and that these processes may affect the frequency of translocations in a certain DNA region.

In addition, this system is well suited for further studies on features which may regulate the DNA repair in specific sequences as well as the overall genome repair. Preferential DNA repair of genes might be ascribed to the more "open" chromatin structure in actively transcribed genomic regions. The demonstration that DNA repair shows strand specificity toward the transcribed strand (Mellon et al. 1987), however, suggests that repair is directed toward certain genomic regions rather than being dependent solely upon chromatin accessibility. Further studies are needed to examine the relative importance of features such as the local chromatin structure, the primary DNA sequence and DNA regulatory proteins for determining the efficiency and organization of DNA repair.

Mechanism of Negative Feed-back Regulation of c-*myc* Gene Expression in B-Cells and its Inactivation in Tumor Cells

Luigia Lombardi, Francesco Grignani, Lars Sternas, Katarina Cechova, Giorgio Inghirami, and Riccardo Dalla-Favera

Department of Pathology and Cancer Center, College of Physicians & Surgeons, Columbia University, New York, NY 10032, USA

INTRODUCTION

The c-myc proto-oncogene is involved in the control of cellular proliferation and differentiation (for review see Cole 1986). Its deregulated expression, caused by chromosomal translocation (Dalla-Favera 1982, 1983; Taub 1982), amplification (Collins and Groudine 1982; Dalla-Favera 1982) or retroviral insertion (Hayward 1981), is associated with lymphoid and non-lymphoid malignancies in different species. The mechanism regulating normal c-myc gene expression and the cause of its deregulation in tumor cells remains largely unknown. Since normal c-myc alleles are down-regulated in the presence of constitutively expressed c-myc oncogenes in naturally occurring tumors (Leder 1983) as well as in in vitro transformed cells (Rapp 1985; Lombardi 1987) and transgenic mice (Adams 1985), it has been frequently suggested that c-myc expression may be under the control of negative feed-back regulation. However, the actual existence of such a regulatory function remains controversial and its mechanism of action unknown. In an attempt to clarify these issues we have characterized a negative feed-back loop regulating c-myc expression in human EBV-immortalized lymphoblastoid cells and then comprehensively investigated the activity of this mechanism in a number of cell types. Our results provide direct evidence for the existence of a dose-dependent negative feed-back loop and preliminarily define its mechanism. However, we find that this mechanism is inactive in all tumor cell lines tested derived from a variety tissues, suggesting that its inactivation may represent a general regulatory disturbance of transformed cells.

MECHANISM OF C-MYC AUTOREGULATION IN B-CELLS

During the course of studies aimed at investigating the biological activity of c-myc oncogenes introduced into EBV-immortalized human lymphoblasts (LCL) we had previously observed that the expression of c-myc oncogenes constitutively expressed under the control of heterologous promoters, leads to the down-regulation of endogenous c-myc expression (Lombardi 1987), as shown by comparative analysis of endogenous and exogenous c-myc RNA levels. We have used the same system to characterize further the mechanism of autoregulation. Several LCLs were transfected by electroporation using the previously described pHEBoSVmyc2,3 vector (Lombardi 1987) in which a truncated version of the human c-myc gene lacking exon I sequences which has been joined, at the 3' border of the first intron, with the SV40 promoter/enhancer element, allowing for the constitutive expression of the normal 65 kd. c-myc protein product (Hann 1988). In addition, this vector carries the hygromycin-B-resistance gene, which functions as a selectable marker, and the EBV origin of replication which allows its episomal replication in EBV-infected cells (Sugden 1985). In nine stable transfectancts derived from separate LCLs and several subclones the degree of down-modulation of endogenous c-myc RNA and protein was consistently found to directly correlate with the levels of exogenous c-myc RNA and protein (Lombardi 1987; Grignani submitted).

To explore the mechanism, transcriptional or postranscriptional, involved in c-myc autoregulation we studied the levels c-myc gene transcription in transfected LCLs by nuclear run-on transcription analysis. In this case we took advantage again of the fact that sequences from the first exon and 5' portion of the first intron are not present in the pHEBoSVmyc2,3 plasmid and therefore transcription from these sequences is specific for the endogenous genes. In addition, the use of separate cloned fragments to study transcription across the first exon and intron allowed us to examine whether down-regulation of endogenous c-myc expression was controlled at the level of transcript initiation or by block of transcript elongation at the first exon/intron border (Bentley and Groudine 1986). The results (Fig. 1) show that modulation at the level of transcript initiation is responsible for the repression of endogenous c-myc expression.

Current Topics in Microbiology and Immunology, Vol. 166

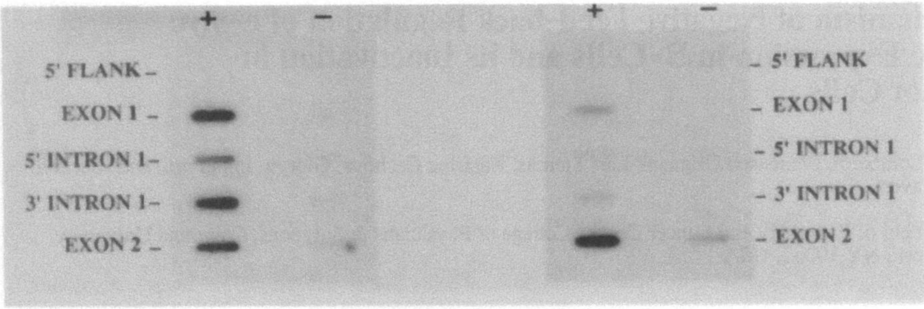

Fig. 1. Analysis of c-myc transcription by the nuclear run-on method in CB33 LCL transfected with pHEBoSV (right panel) or pHEBOSVmyc2,3 (left panel) plasmids, using sense (+) or anti-sense (-) probes representative of 5' flanking, I exon, I intron and II exon sequences. Assay conditions and probes are as previously described (Cesarman 1987).

To define further the specificity and rapidity of c-myc autoregulation, the kinetics of endogenous c-myc expression was studied in LCL transfected with an inducible c-myc gene vector (pHEBoMTmyc2.3, for brevity MTmyc2.3) in which the heavy-metal inducible metallothionein promoter (Stuart 1984) was linked to c-myc coding sequences (II and III exons). This vector, or a control vector (pHEBoMT, for brevity MT) lacking c-myc sequences, was transfected into CB33 LCL and, after antibiotic selection, the transfectants were analyzed for c-myc RNA expression following addition of zinc (Zn) ions to the cell-culture medium. In MTmyc2.3-transfected, but not MT-transfected LCL, Zn-mediated induction of exogenous c-myc expression caused a progressive decrease in the levels of endogenous c-myc RNA detectable 8 hrs after Zn-induction (Fig. 2). Transcriptional down-regulation of the endogenous c-myc gene is already detectable in MTmyc2.3-transfected LCL prior to Zn induction due to leakiness of the MT promoter and further down-regulation of transcription begins to be detectable 8 hrs. after Zn-mediated exogenous c-myc induction (Fig. 3). These data indicate that endogenous c-myc down-regulation in c-myc transformed LCL is not a cell-culture selection- or differentiation-related change, but rather represent a rapid and specific consequence of exogenous c-myc expression consistent with the existence of a negative feed-back regulatory mechanism.

To confirm these results further and to set up an experimental system allowing for further characterization of the autoregulatory mechanism, we tested whether c-myc expressing plasmids could inhibit the initiation of transcription from c-myc promoters and regulatory regions in a transient transfection assay. Towards this end, the pHEBoSV myc2,3 plasmid was used to provide effector molecules which can modulate

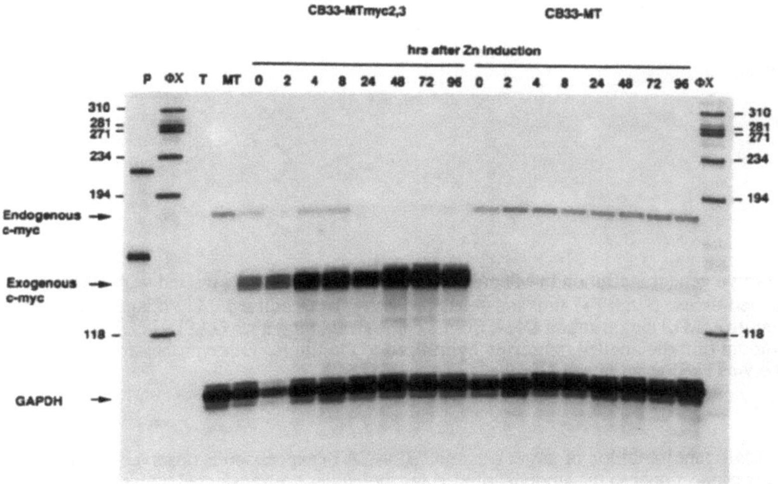

Fig. 2. RNAse protection analysis of the expression of endogenous and exogenous c-myc RNA in CB33 LCL transfected with the pHEBoMT or pHEBoMTmyc 2,3plasmids and treated with 200 μM $ZnSO_4$. A 174 bp probe derived from a Pvu II-Pst I fragment of a human c-myc cDNA clone spanning the first exon-second exon junction was used to discriminate between the endogenous and the exogenous c-myc transcripts, as the first exon is not present in the transfected c-myc alleles. A probe derived from a 120 bp Bal I- Cla I fragment of the human GAPDH gene cDNA was used as a control of the amount of RNA in each lane. P: undigested probes (top band c-myc, bottom band GAPDH; the size of the transcript is increased by the presence of plasmid-derived poly-linker sequences); ΦX: Hae III digested ΦX phage DNA; T: probes digested in the presence of 20μg of tRNA; MT: pHEBoMT-transfected CB33 RNA. Note that endogenous c-myc RNA levels are down-regulated in the MT-myc 2,3 transfected even prior to Zn stimulation due to the leakiness of the MT promoter. A lower amount of RNA was loaded in the 2 hrs. lane of the MTmyc 2,3-transfected CB33 cells.

transcription from target chimeric plasmids. These target plasmids contain 5' c-myc regulatory sequences, including 5' flanking sequences, the two main c-myc promoters and part of the first exon, linked to the chloramphenicol acetyltransferase (CAT) reporter gene (see Fig. 4 C). These plasmids were used in transient co-transfections assays in: i) murine fibroblasts (Balb 3T3) which can be efficiently transfected and display a functional c-myc auto-regulatory mechanism in stable transfection assays (see below) and ii) a Burkitt lymphoma line (P3HR1) chosen as an alternative to LCLs which cannot be used in transient assays due to their relatively low efficiency of transfection. We assumed that P3HR1 cells could be permissive for c-myc autoregulation because the endogenous normal c-myc allele is silent in the presence of the constitutively expressed translocated allele. The results (Fig.4, A-B) show that both in Balb 3T3

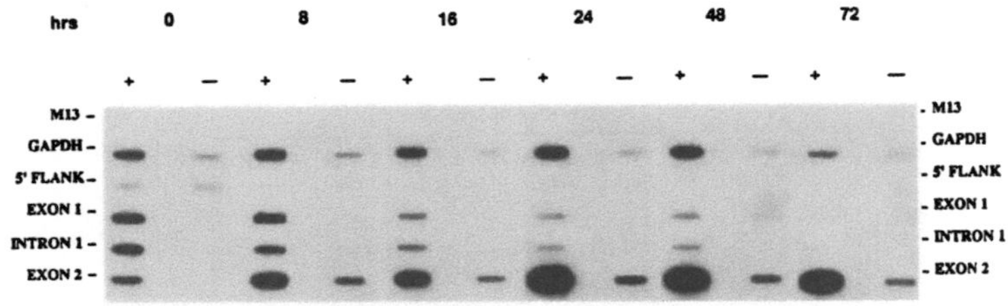

Fig. 3. Analysis of c-myc gene trancription in MTmyc 2,3-transfected CB33 cells treated with 200μM $ZnSO_4$ for the indicated times. Sense (+) and anti-sense probes (-) were utilized. M13 phage DNA was used as a negative control. M13 single strand DNA from a clone containing a rat GAPDH cDNA was used to control for the amount of radiolabelled transcript hybridized to the filters. Other probes are as in fig. 1; the 5' intron 1 probe was used as an intron 1 probe.

and P3HR1 cells a significant inhibition of target plasmid (HPv-CAT) expression is observed and that the degree of inhibition is proportional to the amounts of effector molecules co-transfected. No significant inhibition was observed when the assays were performed using as an effector molecule the pHEBoSVcym2,3 plasmid which is identical to pHEBoSVmyc2,3 except that myc sequences have been inverted into an anti-sense orientation. This indicates that the inhibition of target plasmid (myc/CAT) expression depends upon the presence of c-myc protein and is not simply due to competition for transcriptional factors. Complete inhibition of myc/CAT expression was not achieved in either Balb3T3 nor in P3HR1 cells. This may be due either i)to our inability to achieve sufficiently high effector/target molar ratios because of limitations in the amounts of DNA transfectable or ii) to the fact that necessary cellular factor(s) can be titrated out by the large amounts of target plasmid. A version of the target plasmid (PvPv-CAT, see Fig. 2, C) in which 2 kb. of 5' flanking sequences have been deleted retains the susceptibility to down-regulation, indicating that putative target sequences are contained within the region spanning the two promcters and part of the first exon. Preliminary mapping studies performed in P3HR1 cells indicate that the target region lies <293 bp. 5' to the P1 promoter and therefore does not involve the negative transcriptional regulatory element located 353 bp. 5' to the P1 promoter in the human c-myc locus (Hay 1987) or the one located 290 bp. 5' to the P1 promoter in the murine c-myc locus (Kakkis 1989). The observation that c-myc autoregulation can occur in short term transient assays indicates further that the modulation of c-myc transcription is strictly dependent upon c-myc expression and is not a secondary effect of complex phenotypic changes occurring during c-myc induced transformation.

INACTIVATION OF C-MYC AUTOREGULATION IN TUMOR CELLS

Having characterized the mechanism of c-myc autoregulation in lymphoid cells, we next investigated its activity in other cell types. In particular, we wished to examine whether the discrepancies previously reported on the activity of this mechanism could be explained by differences in the species of origin of the cells , in the tissue derivation or in the growth-phenotype. To test this, we introduced pHEBoSVmyc2,3, MTmyc2.3 or their respective control plasmids in the panel of primary cells and cell lines shown in Table 1 and selected stably transfected derivatives. The presence of c-myc autoregulation was then assayed by analyzing the levels of endogenous (first exon) c-myc transcripts in transfectancts expressing constitutively (pHEBomyc2.3-transfected) or Zn-induced (MTmyc2.3-transfected) exogenous c-myc transcripts by Northern blot hybridization .

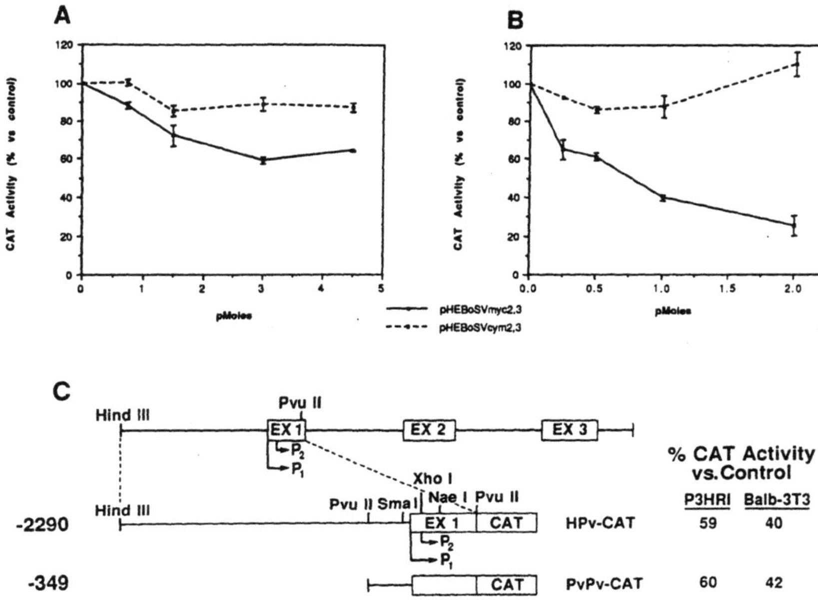

Fig. 4. Analysis of c-myc autoregulation in transient transfection assays. Different amounts (shown in the abscissa) of pHEBoSVmyc2,3 or pHEBoSVcym plasmids were co-transfected into P3HR1 (A) or Balb 3T3 (B) cells with the chimeric pmyc-HPv-CAT plasmid (shown in C) used as a target at 1.5 (P3HR1) or 1.0 (Balb 3T3) pmoles. The pHEBoSVcym plasmid was constructed by inverting the sense of c-myc sequences at the Hind III site of the pHEBoSVmyc2,3 plasmid (5). No expression of sense or anti-sense c-myc RNA transcribed from this plasmid was detectable in transfected cells. At 48 hrs. after transfection, cells were harvested and CAT activity was assayed as described. A plasmid containing the bacterial β-galactosidase gene was also co-transfected with all samples (15 μg.and 2.2 μg in P3HR1 and Balb 3T3 cells, respectively) and β-galactosidase activity in transfected cells was used to normalize CAT values for efficiency of transfection. The values, expressed as % of CAT activity vs. control cells transfected only with the target plasmid, represent the average of at least two experiments, each done in triplicate. Panel C also show the relative activities of the HPv-CAT and PvPV-CAT plasmids when cotransfected with the pSVmyc2.3 (3 pmoles for P3HR1, 1.0 pmoles for Balb 3T3 cells).

Down regulation of endogenous c-myc expression was detected in all non-tumorigenic cells tested including normal primary and immortalized cells of both lymphoid and mesenchymal origin (Table 1). Conversely, all tumorigenic cells derived from a variety of tissues were resistant to down-regulation even in the presence of high ratios of exogenous vs. basal endogenous c-myc gene RNA and protein (not shown). As shown in Fig. 5, no down-regulation of endogenous c-myc expression was detectable in some of the same tumor cell lines even upon Zn-mediated induction of high levels of exogenous c-myc expression driven by transfected pMTmyc2.3 vectors. The absence of c-myc autoregulation correlated with the presence of the tumorigenic phenotype even in cells matched with respect to lineage and stage of differentiation. For instance, compare LCLs and the Bjab cell line, which derives from a B-cell lymphoma and carries structurally and cytogenetically normal c-myc genes. Tumorigenic cells which lack c-myc autoregulation also include three BL cell lines (see Table 1), in which the expression of the endogenous translocated, structurally altered c-myc allele was resistant to down-regulation. In this cases, however, expression of the endogenous normal allele is repressed, suggesting that the lack of down-regulation must

TABLE 1. Loss of c-myc autoregulatory mechanism in tumorigenic cells

CELLS	SPECIE	TISSUE	C-MYC GENE	AUTOREGULATION
Normal				
FS4	human	fibroblasts	normal	+
Immortalized				
LCLs (9 tested)	human	B-lymphoblasts	normal	+
Balb 3T3	mouse	fibroblasts	normal	+
NIH 3T3	mouse	fibroblasts	normal	+
Transformed				
Hela	human	cervical carcinoma	normal	-
U2OS	human	osteosarcoma	normal	-
SK-N-MC	human	neuroblastoma	normal	-
Bjab	human	B-cell lymphoma	normal	-
P3HR1	human	Burkitt lymphoma	translocated	-*
Daudi	human	Burkitt lymphoma	translocated	-*
Ramos	human	Burkitt lymphoma	translocated	-*

* Inactive on the translocated allele (see Fig. 3), but active on normal alleles.

be due to the direct alteration of the c-myc locus rather than to a block in the auto-regulatory loop. We conclude that the c-myc auto-regulatory mechanism functions independently of species and tissue type and is inactivated in tumor cells containing either normal or activated c-myc genes.

CONCLUSIONS

In summary, this study provides evidence for a negative-feed back mechanism regulating c-myc gene expression and shows that this mechanism is inactivated in tumor cells. The existence of such a mechanism had been previously suggested by many studies on the biological activity of the c-myc oncogene in vitro and in transgenic mice (Rapp 1985; Lombardi 1987; Adams 1985). Another study has reported the characterization of a c-myc auto-regulatory activity in murine fibroblasts (Cleveland 1988). Our results extend this characterization by demonstrating that this mechanism is active in a wide spectrum of normal and immortalized cells derived from different tissues. Importantly, our demonstration that c-myc autoregulation is active in short term co-transfection or inducible assays provides more direct evidence for the existence of an auto-regulatory loop by showing its strict dependence upon c-myc gene expression. In addition, these results provide an experimental system which allows the dissection of the effector domains in the c-myc protein and of the target domains within c-myc regulatory sequences and, eventually, the identification of the transcriptional factors mediating these interactions.

The observation that the c-myc auto-regulatory circuit is selectively inactivated in tumor cells provides an interpretation for the discrepancies regarding the existence of the c-myc auto-regulatory mechanism. In general, c-myc autoregulation has been consistently observed when activated c-myc oncogenes are introduced in originally non-tumorigenic cells either in vitro or in vivo (Rapp 1985; Lombardi 1987; Morse 1986; Blasi 1985; Cleveland 1986; Schwartz 1986; Mougneau 1988; Cory 1987). This mechanism was not detected in a limited number of fully-transformed cells studied, typically mouse erythroleukemia cells (Coppola 1986; Dmitrowsky 1986; Prochownik 1986). In addition, the loss of c-myc autoregulation has

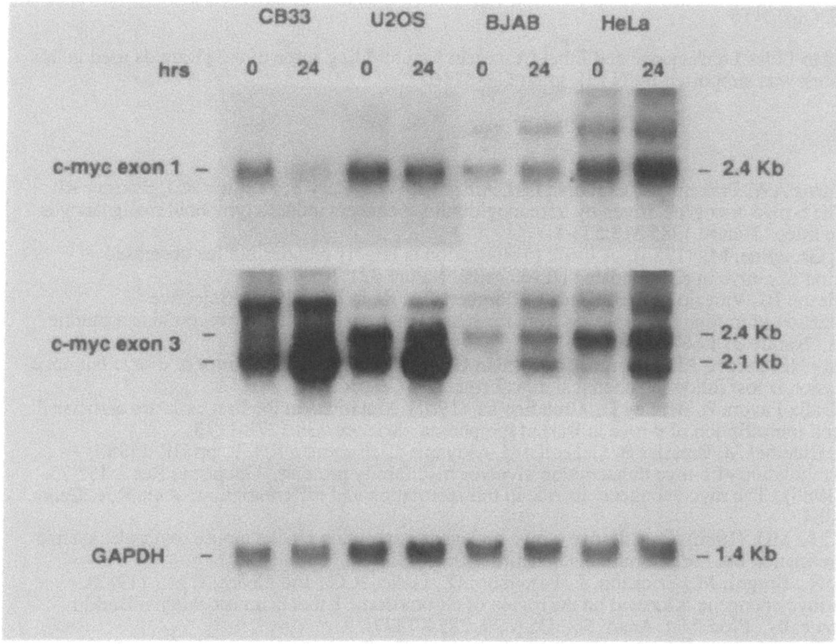

Fig. 5. Northern blot analysis of endogenous and exogenous c-myc expression in the indicated cell lines transfected with the MTmyc 2,3 plasmid. RNA was collected in basal conditions and after 24 hours of culture in the prsence of 200 μM $ZnSO_4$. The filters were sequentially hybridized with radiolabelled a Nae I- Pvu II and a Cla I- Eco RI restriction fragments representative of human c-myc first exon and third exon sequences,respectively, and then with a rat GAPDH gene cDNA probe as a control of the amount of RNA loaded (Lombardi 1987)

been described in long-term cultures of human LCLs (Martinotti 1988) and in murine myeloid cells (Cory 1987) which have progressed in vitro toward a more malignant phenotype. We suggest that the remaining discrepancies are either due to insufficient levels of c-myc protein expression (Stewart 1984; Zerlin 1987) and/or to the in vitro progression of some non-transformed cells, or subclones of them, toward a more malignant phenotype (Zerlin 1987; Freytag 1988; Keath 1984).

The lack of c-myc autoregulation in tumor cells shown here suggests that the inactivation of a regulatory circuit controlling c-myc gene expression may be a general disturbance of transformed cells. This conclusion is consistent with previous reports on the loss of normal cell-cycle regulation of the c-myc gene in a variety of tumor-derived and in vitro-transformed cells, even in the absence of detectable structural alterations of c-myc sequences (Campisi 1984; Erisman 1985, 1989). These observations suggest that c-myc gene deregulation may represent a common end-point due to alterations at various levels within the same auto-regulatory circuit. In particular, a block of auto-regulation can occur : i) at the level of the c-myc locus itself (e.g. by chromosomal translocation in Burkitt lymphoma) in the presence of an otherwise functional auto-regulatory circuit, or ii) by loss, inactivation or disregulation of other genes of the same circuit in tumors carrying normal c-myc alleles. Using the loss of c-myc autoregulation as an assay, it may now be possible to identify which genetic loci are associated with the disruption of this regulatory pathway.

ACKNOWLEDGMENTS

We are grateful to Luisa Lanfrancone and Ethel Cesarman for providing some of the plasmids used in this study. This work was supported by N.I.H. grant CA-37165.

REFERENCES:

Adams JM, Harris AW, Pinkert CA, Corcoran LM, Alexander WS, Cory S, Palmiter RD, Brinster RL (1985) The c-myc oncogene driven by immunoglobulin enhancers induces lymphoid malignancy in transgenic mice. Nature 1985 318:533-8.

Bentley, D.L., Groudine, M. (1986). A block to elongation is largely responsible for decreased transcription of c-myc in differentiated HL-60 cells. Nature 321:702-706.

Blasi E, Mathieson BJ, Varesio L, Cleveland JL, Borchert PA, Rapp UR (1985) Selective immortalization of murine macrophages from fresh bone marrow by a raf/myc recombinant murine retrovirus. Nature 318:667-70.

Campisi J, Gray HE, Pardee AB, Dean M, Sonenshein GE (1984) Cell-cycle control of c-myc but not c-ras expression is lost following chemical transformation. Cell 36:241-7.

Cesarman E, Dalla-Favera R, Bentley D, Groudine M (1987) Mutations in the first exon are associated with altered transcription of c-myc in Burkitt lymphoma. Science 238:1272-1275.

Cleveland JL, Huleihel M, Bressler P, Siebenlist U, Akiyama L, Eisenman RN, Rapp UR (1988) Negative regulation of c-myc transcription involves myc family proteins. Oncogene Res 3:357-75.

Cole, M.D. (1986). The myc oncogene: its role in transformation and differentiation. Ann. Rev. Genet. 20, 361-384.

Coppola JA, Cole MD (1986) Constitutive c-myc oncogene expression blocks mouse erythroleukaemia cell differentiation but not commitment. Nature 320:760-3.

Dalla-Favera, R., Bregni, M., Erickson, J., Patterson, D., Gallo, R.C., and Croce, C.M. (1982). Human c-myc oncogene is located on the region of chromosome 8 that is translocated in Burkitt lymphoma cells. Proc. Nat. Acad. Sci. USA.79, 7824-7827.

Dalla-Favera, R., Gelmann, E.P., Martinotti, S., Franchini, G., Papas,T.S., Gallo, R.C., and Wong-Staal, F. (1982a). Cloning and characterization of different human sequences related to the onc-gene (v-myc) of avian myelocytomatosis virus (MC29). Proc. Nat. Acad. Sci. USA 79, 6497-6501.

Dean M, Cleveland JL, Rapp UR, Ihle JN (1987) Role of myc in the abrogation of IL3 dependence of myeloid FDC-P1 cells. Oncogene Res 1:279-96.

Dmitrovsky E, Kuehl WM, Hollis GF, Kirsch IR, Bender TP, Segal S (1986) Expression of a transfected human c-myc oncogene inhibits differentiation of a mouse erythroleukaemia cell line. Nature 322:748-50.

Erisman MD, Rothberg PG, Diehl RE, Morse CC, Spandorfer JM, Astrin SM (1985) Deregulation of c-myc gene expression in human colon carcinoma is not accompanied by amplification or rearrangement of the gene. Mol Cell Biol 5:1969-76.

Erisman MD, Scott JK, Astrin SM (1989) Evidence that the familial adenomatous polyposis gene is involved in a subset of colon cancers with a complementable defect in c-myc regulation. Proc Natl Acad Sci USA 86:4264-8.

Freytag SO (1988) Enforced expression of the c-myc oncogene inhibits cell differentiation by precluding entry into a distinct predifferentiation state in G0/G1. Mol Cell Biol. 8:1614-24.

Hann, S.R., King M.W., Bentley, D.L., Anderson, C.W., and Eisenmann, R.N. (1988). A non-AUG translational initiation in c-myc exon 1 generates an N-terminally distinct protein whose synthesis is disrupted in Burkitt's lymphomas. Cell 52, 185-195.

Hay N, Takimoto M, Bishop JM (1989) A FOS protein is present in a complex that binds a negative regulator of MYC. Genes Dev 3:293-303.

Hayward WS, Neel BG, Astrin SM (1981) Activation of a cellular onc gene by promoter insertion in ALV-induced lymphoid leukosis. Nature 290:475-480.

Kakkis E, Riggs KJ, Gillespie W, Calame K (1989) A transcriptional repressor of c-myc. Nature 339:718-721.

Keath EJ, Caimi PG, Cole MD. (1984) Fibroblast lines expressing activated c-myc oncogenes are tumorigenic in nude mice and syngeneic animals. Cell 39:339-48.

Langdon WY, Harris AW, Cory S (1989) Acceleration of B-lymphoid tumorigenesis in E mu-myc transgenic mice by v-H-ras and v-raf but not v-abl. Oncogene Res 4:253-8.

Lombardi, L., Newcomb, E.W., and Dalla-Favera, R. (1987). Pathogenesis of Burkitt lymphoma: expression of an activated c-myc oncogene causes the tumorigenic conversion of EBV-infected human B-lymphoblasts. Cell 49, 161-170.
Morse HC, Hartley JW, Fredrickson TN, Yetter RA, Majumdar C, Cleveland JL, Rapp UR (1986) Recombinant murine retroviruses containing avian v-myc induce a wide spectrum of neoplasms in newborn mice. Proc Natl Acad Sci USA 83:6868-72.
Mougneau E, Cerni C, Tillier F, Cuzin F (1988) Tumorigenic transformation of rat FR3T3 fibroblasts carrying an activated myc oncogene requires subsequent mutational events. Oncogene-Res 2:177-88.
Prochownik EV, Kukowska J (1986) Deregulated expression of c-myc by murine erythroleukaemia cells prevents differentiation. Nature 322:848-50.
Rapp UR, Cleveland JL, Brightman K, Scott A, Ihle JN (1985) Abrogation of IL-3 and IL-2 dependence by recombinant murine retroviruses expressing v-myc oncogenes. Nature 317:434-8.
Schwartz RC, Stanton LW, Riley SC, Marcu KB, Witte ON (1986) Synergism of v-myc and v-Ha-ras in the in vitro neoplastic progression of murine lymphoid cells. Mol-Cell-Biol 6:3221-31.
Stewart TA, Pattengale PK, Leder P (1984) Spontaneous mammary adenocarcinomas in transgenic mice that carry and express MTV/myc fusion genes. Cell 38:627-37.
Stuart, G.W., Searle, P.F., Chen H.Y., Brinster, R.L., and Palmiter, R.D. (1984). A 12-base-pair DNA motif that is repeated several times in metallothionein gene promoters confers metal regulation to a heterologous gene. Proc. nat. Acad. Sci. USA 81:7318-7322.
Sugden, W., Marsh, K., and Yates, P. (1985). A vector that replicates as a plasmid and can be efficently selected in B-lymphoblasts transformed by Epstein-Barr Virus. Mol. Cell. Biol. 5, 410-413.
Taub R, Kirsch I, Morton C, Lenoir G, Swan D, Tronick S, Aaronson S, Leder P (1982) Translocation of the c-myc gene into the immunoglobulin heavy chain locus in human Burkitt lymphoma and murine plasmacytoma cells. Proc Natl Acad Sci USA 79:7837-41.
Zerlin M, Julius MA, Cerni C, Marcu KB (1987) Elevated expression of an exogenous c-myc gene is insufficient for transformation and tumorigenic conversion of established fibroblasts. Oncogene 1:19-27.

Moloney Murine Leukemia Virus Integration 1060 Base Pairs 5' of c-*myc* Exon 1 in a Plasmacytoma Without a Chromosomal Translocation

J. SHAUGHNESSY, JR [1,2], K. HUPPI [1], J. F. MUSHINSKI [1], and M. POTTER [1]

[1] Laboratory of Genetics, National Cancer Institute, Bethesda, MD 20892
[2] Department of Zoology, University of Maryland, College Park, MD 20742

INTRODUCTION

Transcriptional activation of cellular oncogenes has been shown to be associated with the integration of replication-competent retroviruses in a variety of tumors. Activation of the c-*myc* oncogene by viral integration within or near the gene has been shown to occur in several tumor systems including bursal B-cell lymphomas of the chicken and murine T-cell lymphomas.
Avian bursal lymphomas are induced by the avian leukosis virus (ALV) and reticuloendotheliosis virus (REV) which integrate within the c-*myc* locus primarily between the untranslated first exon and the coding exons. 95% of proviral integrations are in the same transcriptional orientation as the c-*myc* gene, and expression of c-*myc* is initiated from within one of the viral long terminal repeats. In approximately 90% of MuLV induced murine T-cell lymphomas the proviral integrations occur 0.5 to 1.5 kilobase pairs (kbp) 5' of c-*myc* exon 1, and only 10% occur within exon 1. No integrations have been found within the first intron, the major target of ALV/REV-induced B-cell lymphomas. The orientation of the T lymphoma proviruses, in contrast to the majority of those found in the bursal lymphomas, are in the opposite transcriptional orientation of c-*myc*. Enhancer insertion is thought to be the mechanism of c-*myc* activation in these tumor systems (for a review see Clurman and Hayward 1988). It is puzzling that no c-*myc*-associated viral integrations have been seen in murine B-cell lymphomas or plasmacytomas that were induced by retroviruses.
Over 95% of mouse plasmacytomas contain rcpt(12;15) or rcpt(6;15) chromosomal translocations in which the c-*myc* gene on chromosome 15 is juxtaposed with immunoglobulin heavy or light chain loci on chromosome 12 and 6, respectively (For review see Potter 1990). However, there remains a single example in our collection of plasmacytomas, an Abelson virus/pristane induced tumor (ABPC22), that lacks obvious cytogenetic abnormalities of the D2 region of chromosome 15 (Ohno et al. 1984).

RESULTS

Northern analysis of poly(A)$^+$mRNA isolated from ABPC22 shows that the level of c-*myc* gene transcripts is comparable to other plasmacytomas and that the message size is the expected 2.4 kb indicating that the gene is intact (Mushinski et al. 1990).
Restriction digestion analysis of ABPC22 DNA with Bgl II, Hind III, and Xho I (Fig.1a) using the upstream c-*myc* probe pB/c2.0XbBam1.8 (Fig. 2a) showed a germline and non-germline band, suggesting that one of the c-*myc*-bearing chromosomes had undergone a rearrangement. Digestions with Sac I, Xba I, Bam HI, Kpn I, and Sma I also showed rearrangements (data not shown). From these data we were able to construct a partial restriction map of the rearranged c-*myc* allele (Fig. 2c) which suggested that a 6-7 kbp piece of DNA had been inserted in this locus. The map of the foreign DNA matched the restriction map of the Moloney Murine Leukemia Virus (MoMLV) the helper virus component of the Abelson virus utilized in the tumor induction (Fig. 2b). A unique Xho I site present 1.5 kbp from the 5' LTR in the MoMLV indicated that the virus had integrated in the opposite transcriptional orientation with respect to c-*myc*.

Current Topics in Microbiology and Immunology, Vol. 166

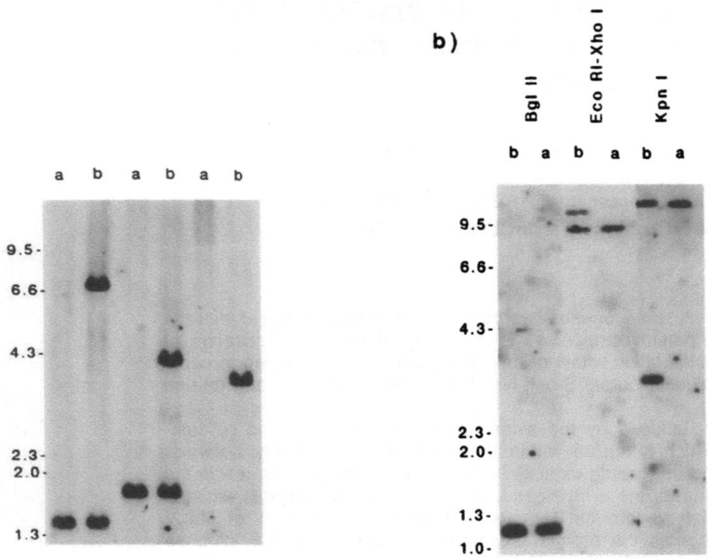

Figure 1. **a**) Southern hybridization of a 5' c-*myc* probe to tumor and genomic DNA. Hind III, Bgl II and Xho I digested DNAs from the plasmacytoma ABPC 22 (lanes a) and liver from mouse inbred strain BALB/cAnPt (lanes b) were hybridized with the probe pB/c2.0XbBam1.8 (Fig 2a). Positions of standard Hind III-digested bacteriophage λ DNA are shown on the left. **b**) Southern hybridization of a c-*myc* probe to tumor and genomic DNA. Bgl II, Eco RI- Xho I, and Kpn I digested DNAs from the plasmacytoma ABPC 22 (lanes a)and liver from mouse inbred strain BALB/cAnPt (lanes b) were hybridized with the probe pB/c2.0XbBg0.8 (Fig. 2c). Positions of standard Hind III-digested bacteriophage λ DNA are shown on the left

The absence of the Xho I site in the Abelson Murine Leukemia Virus (AbMLV) component and the positions of the new Bam HI and Hind III sites in the ABPC22 map indicated that the integrated provirus was the MoMLV and not the AbMLV.

The restriction map indicated that the virus had inserted into the 5' flanking sequences between the Sma I site at -425 bp and the Bgl II site at -1138 bp. In order to determine whether an entire MoMLV genome was present we performed Southern blot analysis using a probe which was specific for c-*myc* sequences 5' of the integration site. Restriction digestions with Kpn I and Xho I-Eco RI (Fig. 1b) using the probe pB/c2.0XbBg0.8 (Fig 2c) revealed rearrangements. A rearranged band was also detected using Sac I, Sma I, and Xba I (data not shown).

The size of the Xho-Eco RI rearranged band was smaller than would be predicted if a complete provirus had integrated and the c-*myc* gene 5' of the breakpoint was unaltered. This indicated that approximately 2-3 kbp of DNA had been deleted from between the Xho I site in the provirus and the Eco RI site normally found 8.0 kbp upstream of c-*myc* exon 1.

Southern blot analysis showed that no rearrangements existed in Bgl II (Fig. 1b) or Hind III digested DNA (data not shown) indicating that the c-*myc* region 5' of the provirus was likely to be germline. From the above data we were able to complete the restriction map of the mutated c-*myc* allele.

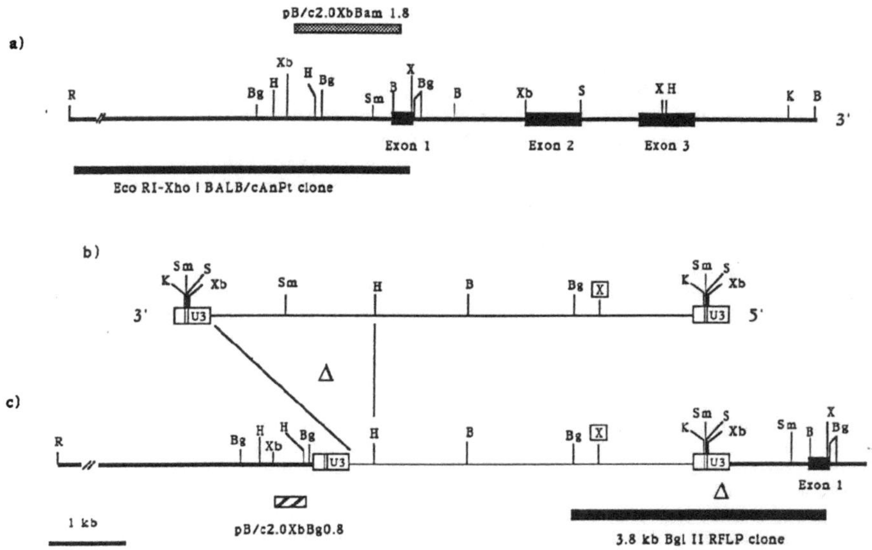

Figure 2. Restriction maps. a) The murine c-*myc* locus. Location of pB/c2.0XbBam1.8 is indicated by a shaded box. The BALB/cAnPt clone used for the sequence comparison is represented by a black box. b) A map of the MoMLV genome. The MoMLV specific Xho I site is boxed. c) The mutated c-*myc* allele from ABPC22. The deleted 72 bp direct repeat and the 2.5 kbp from between the HindIII in the pol gene and the 3' LTR are represented by Δ. The ABPC22 specific 3.8 kbp Bgl II c-*myc* clone and the probe pB/c2.0XbBg0.8 are represented by black and striped boxes respectively. The following abbreviations were used: R=EcoRI, B=BamHI, H=HindIII, Bg=BglII, X=XhoI, K=KpnI, Sm=SmaI, Xb=XbaI, S=SacI

We concluded that both of LTRs of the Moloney provirus were present, that the region of c-*myc* 5' of the insertion site was intact, but that approximately 2.5 kbp of the MoMLV between the single Hind III site and the 3' LTR had been deleted(Fig. 2c)

In order to identify the precise site of the insertion and to analyze the LTR and c-*myc* sequence for somatic mutations, we cloned the rearranged 3.8 kbp Bgl II fragment (Figs. 1a, 2c). Sequencing toward the first exon of c-*myc* from the Xba I site in the LTR revealed that exactly on of the two 72 base pair direct repeats was deleted and replaced by two thymidine residues and that a pair of thymidine residues were deleted from the end of the retroviral LTR.

The sequence of the c-*myc* gene from the Xho I site in exon 1 to the site of the viral integration was compared with similar sequences from an 8.8 kbp Xho I- Eco RI clone derived from a BALB/cAnPt splenic library (Fig. 2a). Except for a duplication of a guanosine-cytosine pair six bases 3' of the break point at -1060, no somatic mutations were detected in the provirus containing c-*myc* allele.

DISCUSSION

In this paper we describe the first example of LTR insertion 5' of the c-*myc* gene in a tumor of the murine B-cell lineage. This result supports the conclusion that perturbation of the c-*myc* gene either by chromosomal translocation or viral integration is a prerequisite event in the development of murine plasmacytomas.
Here we showed that a deleted MoMLV was integrated in the opposite transcriptional orientation 1.1 kbp 5' of c-*myc*. Restriction digestion analysis indicated that approximately 2.5 kbp of the 3' region of the viral genome was deleted, an event that could facilitate tumor progression by allowing an infected cell to escape immune surveillance and/or allow superinfection of the cell which could lead to additional oncogene activation (Goodnow and Hayward, 1987).
A number of *cis*-acting positive and negative control elements have been identified in the exon 1 and 5' flanking region of the c-*myc* locus (Asselin et al. 1989; Remmers et al. 1986; Yang et al. 1986; Nepveu and Marcu 1986; Kakkis et al. 1989). The sequence of the region between the integration site and the end of exon 1 was free of somatic mutations. These results suggest that the placement of an LTR one kilobase 5' of c-*myc* may be sufficient to override the effects of endogenous *cis*-acting negative elements located between the LTR and the gene.
It has been our impression that the steady state level of c-*myc* mRNA in several preperations of RNA from ABPC22 is slightly lower than the levels observed in translocation positive tumors. Recently we discovered two additional tumors, a plasmacytoma (MSVPC524) and a B-cell lymphoma (MSVBL311) with "head to head" viral integrations 2 and1 kbp 5' of c-*myc*, respectively. Northern analysis of poly(A)$^+$ mRNA from these tumors showed that c-*myc* steady state levels are higher than those found in ABPC22. A question arises as to whether the deletion of one of the 72 bp tandem repeats from the LTR of the MoMLV integrated in ABPC22 is resposible for the lower levels of c-*myc* RNA by having reduced the effectivness of the viral enhancer 5' of c-*myc*. Future efforts will be directed toward studying the role of the MoMLV viral LTR in the activation of c-*myc* gene.

REFERENCES

Asselin C, Nepveu A, Marcu KB (1989) Molecular requirements for transcriptional initiation of the murine c-*myc* gene. Oncogene 4: 549-558.

Clurman B, Hayward W (1988) Insertional activation of proto-oncogenes. In Cellular Oncogene Activation. G. Klein ed. pp. 55-94.

Goodnow MM, Hayward WS (1987) 5' long terminal repeats of *myc* associated proviruses appear structurally intact but are functionally mpaired in tumors induced by avian leukosis virus. J Virology 621: 2489-2498.

Kakkis E, Riggs KJ, Gillespie W, Calame K (1989) A transcriptional repressor of c-*myc*. Nature 339: 718-722.

Mushinski JF, Skurla RM Jr, Goodnight J, Siwarski D, Huppi K Expression of c-*myc* and *Pvt*-1. Curr Top Micro Immunol (this volume).

Nepveu A, and Marcu K (1986) Intragenic pausing and anti-sense transcription within the murine c-*myc* locus. EMBO 5: 2859-2865.

Ohno S, Migita S, Wiener F, Babonits, M, Klein G, Mushinski J F, Potter M (1984) Chromosomal translocations activating *myc* sequences and plasmacytomas by pristane and Abelson virus. J Exp Med 159: 1762-777.

Potter M (1990) Neoplastic development in B-lymphocytes Carcinogenesis 11: 1-13.

Remmers E, Yang J, Marcu K (1986) A negative transcriptional control element located upstream of the murine c-*myc* gene. EMBO 5: 899-904.

Yang JQ, Remmers EF, Marcu KB (1986) The first exon of the c-*myc* proto-oncogene contains a novel positive control element. EMBO 5: 3553-3562.

Role of EBV in B-Cell Neoplasia

Cell Phenotype Dependent Down-regulation of MHC Class I Antigens in Burkitt's Lymphoma Cells

M.G. Masucci

Department of Tumor Biology, Karolinska Institute, Box 60 400, 104 01 Stockholm, Sweden

INTRODUCTION

MHC class I antigens are highly polymorphic trans-membrane glycoproteins that play a key role in the regulation of immune responses. Lysis of virus infected or neoplastic cells by cytotoxic T-lymphocytes (CTL) depends on the cognate recognition of processed antigens and MHC class I molecules (Zinkernagel and Doherty 1979). Cell lines derived from Epstein-Barr virus (EBV) carrying Burkitt's lymphoma (BL) are often resistant to lysis by HLA class I antigen restricted EBV-specific CTLs generated by stimulation of lymphocytes from EBV seropositive individuals with autologous virus infected cells (Rooney et al. 1985b, Tosteindottir et al. 1986). Since lymphocytes from BL patients are able to control the proliferation of autologous EBV infected cells in in vitro regression assays (Rooney et al. 1985a), the CTL resistance of the tumor cells was interpreted to result from in vivo immunoselection. The finding that six of the seven EBV encoded antigens that may provide the target for the immune rejection of EBV transformed normal cells are not expressed in the tumors and are up-regulated after in vitro propagation (Rowe DT et al. 1986, Rowe M et al. 1987) was taken as a further evidence for the presence of immunological control mechanisms.

An alternative interpretation has emerged from studies aiming to characterize the phenotype of BL cells and to identify the normal B-cell precursor of the tumor. Biopsy cells from EBV carrying and EBV negative BLs and early passages of the derived cell lines were shown to express a common phenotype. They are positive for CALLA (CD10) and BLA (CD77) and negative for a variety of B-cell activation markers and adhesion molecules that are expressed at high levels on mitogen or antigen activated and EBV transformed B-blasts (Rowe M et al. 1986). Cell lines from EBV negative tumors are relatively stable in vitro while the EBV positive ones tend to drift towards a more "LCL-like" phenotype in parallel with up-regulation of the EBV encoded EBNA 2-6 and LMP (Rowe M et al. 1987). Analysis of MHC class I antigen expression in a panel of BL lines derived from HLA A11 positive individuals has revealed a selective down-regulation of the A11 allele in both EBV carring and EBV negative tumors (Masucci et al. 1987, Masucci et al. 1989). The defect was paralleled by resistance of the tumor cells to HLA-A11 specific and, in the case of EBV carrying tumors, HLA-A11 restricted EBV specific CTLs (Torsteinsdottir et al. 1986, Masucci et al. 1987). The HLA-A11 negative phenotype appeared to be due to a regulatory phenomenon rather than a structural defect because the reactivity with an HLA-A11 specific monoclonal antibody and the sensitivity to A11 specific CTLs were increased after treatment with interferon and tumor necrosis factor (Avila-Carino et al. 1988). Moreover, HLA-A11 was re-expressed in some EBV negative BLs following in vitro EBV

Current Topics in Microbiology and Immunology, Vol. 166

conversion (Masucci et al. 1989). These findings suggest that the aberrant expression of MHC class I antigens, and the parallel resistance of BLs to CTL mediated immune attacks, might be a cell phenotype dependent phenomenon rather that the result of immune selection. The evidence supporting this possibility will be briefly summarized in this chapter.

EXPRESSION OF HLA CLASS I IN PAIRED LCL/BL LINES

Serological and biochemical methods were used to investigate the HLA class I expression of BL lines derived from EBV carrying or EBV negative tumors and in vitro EBV transformed LCLs obtained from the same patients. HLA class I antigens were immunoprecipitated from extracts of metabolically labelled cells using the W6/32 monoclonal antibody that detects fully assembled B_2m/heavy chain complexes. The class I polypeptides were separated by one dimension isoelectric focusing (ID-IEF). This method allows the identification of different class I alleles by virtue of their distinct isoelectric point (Neefjes et al. 1986). Selective down-regulation or losses of HLA-A and -C alleles appeared to be relatively common in BL cells. Discrepancies between the biochemical typing of the LCL and BL lines were observed in nine cases out of thirteen (Table 1).

Table 1. MHC class I antigen expression of BL cells.

Cell line	EBV	phenotypic group[a]	affected allele	IFN sensitivity
BL 28	-	I	A11	+
			Cw7	-
BL 29	+	I/II	Aw69	+
BL 36	+	II	Cw8	-
			Cw4	+
BL 37	+	II	A3	-
			Cw4	+
			Cw5	+
BL 41	-	I	A11	-
			Cw7	-
BL72	+	II	A11	+
WW-1-BL	+	II/III	A11	-
WW-2-BL	+	II	A11	-
			B39	-
Jiyoje M13	+	II	Cw4	+
BL 16	+	III	none	-
BL 18	+	III	none	-
BL 60	+	III	none	-
BL 31	-	I/II	none	-

a Phenotypic characterization was done according to the criteria described by Rowe M et al. 1986. Type I BLs express CD10 and CD77 but do not express CD23 and CD39, type III BLs and LCLs express CD23 and CD29 and do not express CD10 and CD77, type II BL coexpress the two sets of activation/differentiation markers.

Within the limits of resolution allowed by the method, losses of one or more class I heavy chain were demonstrated in six BL lines (BL28, BL36, BL37, BL41, WW-1-BL and WW-2-BL). In six cell lines (BL28, BL29, BL36, BL37, BL72 and Jiyoje M13) certain class I specificites were detected in the autoradiograms but the level of expression appeared to be lower compared to the corresponding LCLs. The extent and selectivity of the down-regulation was estimated by comparing the ratios between the intensities of the B_2m and allele specific heavy chain bands measured by densitometry. The ratios of the down-regulated alleles were in repeated experiments lower in the BL lines compared to LCLs. For some of the cell lines only single alleles within the HLA-A or - C loci were affected (Aw69 in BL29, Cw8 in BL36, A11 in BL72 and WW-1-BL, Cw4 in Jiyoje M13), for others both HLA-A and -C specificities were reduced or absent (A11 and Cw7 in BL28 and BL41; A3,Cw4 and Cw5 in BL37). WW-2-BL was the only BL line that showed a combined loss of one HLA-A and one -B polypeptide.

Gamma-IFN had different effects on the expression of the class I alleles that were down-regulated or lost in BL cells. In general, alleles that were expressed at a lower level in BL than in LCLs were up-regulated by IFN treament, (A11 in BL28 and BL72; Aw69 in BL29; Cw4 in BL36, BL37 and Jijoye M13; Cw5 in BL36), whereas gamma-IFN treatment could not rescue the expression of those alleles that were not detected in the untreated BL cells (Cw7 in BL28 and BL41; Cw8 in BL36; A3 in BL37; A11 in BL41, WW-1-BL and WW-2-BL and B39 in WW-2-BL). An allele-selective effect of gamma-IFN was often observed when more than one allele was affected in the same cell line. In BL28 A11 was up-regulated without effect on Cw7, in BL36 up-regulation of Cw4 was not accompanied by up-regulation of Cw8 and in BL37 both HLA C alleles were upregulated but there was no effect on A3 expression. The variable response to gamma-IFN treatment suggest that different mechanisms may be involved in the down-regulation.

CORRELATION BETWEEN HLA CLASS I EXPRESSION AND CELL PHENOTYPE

BL lines that have been propagated in vitro for a prolonged period of time are heterogeneous with regard to expression of B- cell activation and differentiation markers and, in the case of EBV carrying BLs, viral antigens. EBV negative BL lines, and some of the EBV positive ones, maintain the phenotype of the biopsy cells characterized by high expression of CD10 and CD77, absence of B-cell activation markers and down-regulation of all EBV encoded antigens except EBNA-1 (type I BLs). The majority of the EBV carrying lines tend to progress towards a more "LCL-like" phenotype. They up-regulate EBNA 2-6 and LMP and acquire B-cell activation markers (type II BLs), and may finally loose CD10 and CD77 expression (type III BLs). Allele selective defects of class I antigen expression were documented in nine out of ten BL lines expressing a type I or I/II phenotype (Table 1). The only exception was the EBV negative line BL31. In spite of its high expression of CD10 and CD77 and lack of B-cell activation markers, this line is different from a group I BL. It grows in large clumps and expresses high levels of cell adhesion molecules, cannot be cloned in agarose and is non-tumorigenic in immunosuppressed mice (unpublished observations). Three type II/III or III cell lines resembled the corresponding LCLs also in their MHC class I antigen expression.

The correlation between cell phenotype and expression of class I HLA was further investigated in phenotypically distinct cell clones derived from an early passage of the BL line MUTU (Gregory et al. 1990). Five clones that had retained the original BL biopsy phenotype (type I) expressed the virus-encoded nuclear antigen EBNA-1 but not the other transformation associated proteins EBNA 2-6 and LMP (Figure 1). Two other clones which had acquired an "LCL-like" phenotype (type III) expressed all these proteins and also contained some cells in lytic cycle.

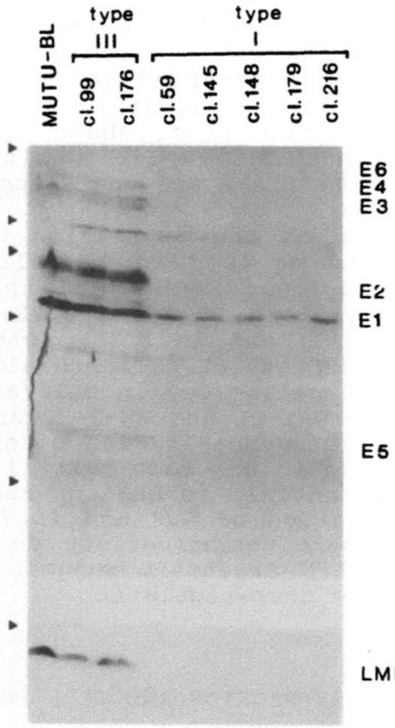

Figure 1. Expression of EBV antigens in phenotypically distinct clones of the BL line MUTU. SDS-polyacrylamide gel electrophoresis and Western blotting were performed as previously described (Masucci et al. 1987). Serum from an healthy EBV seropositive donor (HR: VCA 1:320, EBNA 1:160, EBNA2 1:80) was used for detection of the EBV nuclear antigens. The EBV membrane protein LMP was detected by the S12 monoclonal antibody.

The expression of HLA class I was investigated by ID-IEF after immunoprecipitation of metabolically labelled cell extracts with the W6/32 MoAb. In order to obtain a quantitative as well as qualitative evaluation of class I antigen expression the immunoprecipitations were performed from equal amounts of TCA precipitable material. Type I clones expressed significantly lower levels of all HLA class I alleles compared to the type III clones and the uncloned MUTU line (Figure 2). An EBV transformed LCL derived from the normal B-lymphocytes of the same patient resembled the type III clones (not shown). The total amount of B_2m recovered from type III clones exceeded the amount recovered in type I clones by a factor of three (Table 2). This suggests that, relative to the total protein synthesis, class I antigens are expressed at a lower level in BL lines that maintain the phenotype of the original tumor than in cell lines that progress towards a more "LCL-like" phenotype. Moreover, after normalization of the radioactivity recovered in the HLA A and B heavy chain bands relative to B_2m, HLA A1 was found to be selectively reduced in the two type I clones.

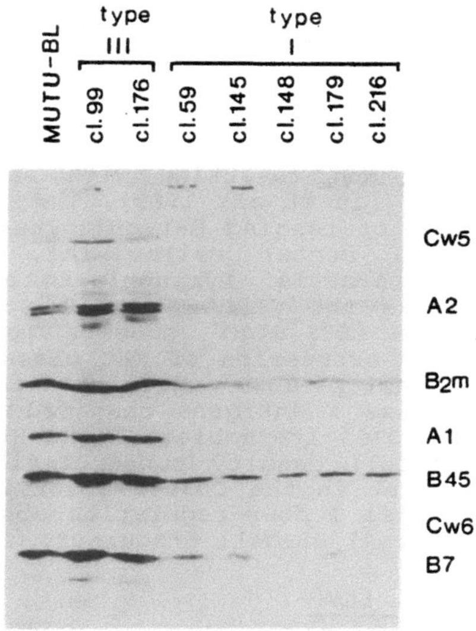

Figure 2. Expression of HLA class I in the MUTU cell lines. Immunoprecipition with the W6/32 MoAb was performed from equal amounts of TCA precipitable material. HLA class I polypeptides were resolved by ID-IEF and detected by autoradiography. Different alleles were identified by comparison with a panel of previously characterized LCLs.

Table 2. The HLA A1 allele is selectively down-regulated in type I clones of the MUTU BL.

	B_2m[a]	B_2m/heavy chain ratio			
		A1	A2	B7	B45
Type I					
cl.148	414	0.36	0.69	0.78	1.28
cl.216	363	0.26	0.86	0.99	1.37
Type III					
cl.176	1130	0.67	0.77	1.03	1.33

a Polypeptides corresponding to the specificities indicated were excised from the gel and the radioactivity associated was determined by liquid spectrometry. Mean of three experiments.

EXPRESSION OF HLA CLASS I IN SUBPOPULATIONS OF NORMAL B-CELLS

A phenotypically similar normal counterpart of the BL cells has been identified in the follicular germinal centers of human tonsils (Gregory et al. 1987). Follicular centroblasts have a high proliferation rate and express CD10 and CD77 but no B-cell activation markers. These cells are thought be the site of somatic mutation and Ig class switch which occurs during affinity maturation of T-cell dependent antibody responses. Tonsil B-cells were separated by Percoll gradient centrifugation and rosetting with anti-CD39 and anti-IgD coated sheep red cells (Liu et al. 1989). The high density subpopulation is mainly composed of resting B-lymphocytes which lack the characteristics of geminal center cells. The low density supopulations contain activated B-cells. Lymphoblasts are enriched in the CD39$^+$SIgD$^+$ subset whereas the CALLA and BLA positive centroblasts are recovered in the CD39$^+$SIgD$^+$ subset. Tonsil B-cell subpopulations differed in their expression of MHC class I antigens (figure 3). The low cell density CD39$^-$ germinal center cells expressed significantly less class I antigens compared to the high cell density resting cells and CD39$^+$ lymphoblasts. The HLA-C alleles were detected in the two low cell density subpopulations but the expression was significantly lower in the CD39$^-$ subset. The results suggest that pattern the HLA class I down-regulation observed in BL cells may be a vestige of the normal BL-cell precursor.

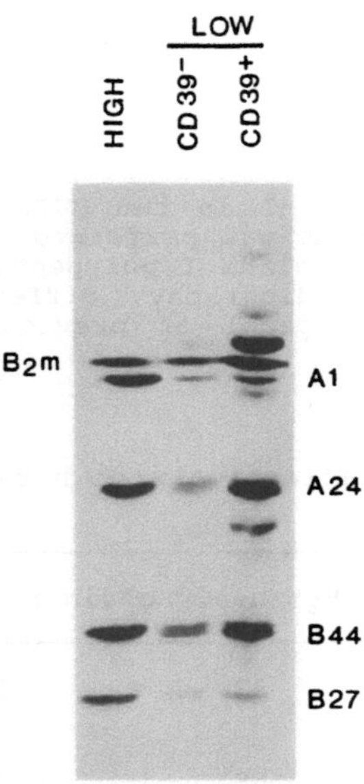

Figure 3. Expression of HLA class I in phenotypically different subpopulations of normal B-cells. Tonsil B-lymphocytes were separated by Percoll gradient centrifugation in high and low density subpopulations. The low density cells were further fractionated by rosetting with anti-CD39 and anti-IgD coated sheep red cells. HLA class I antigens were detected by 1D-IEF after immunoprecipitation with the W6/32 MoAb.

CONCLUSIONS

There are two possibilities to explain the cell phenotype dependent down-regulation of MHC class I antigens in BLs and, by inference, in their phenotypically similar normal counterpart. If progression along the differentiation pathway of B-cells entails a program of changes in gene expression, class I genes may be included in the genes affected because they share certain regulatory features with other genes that are essential for the elaboration of this developmental program. Alternatively, the changes in class I MHC expression may be functionally relevant as they contribute a possible connection between the T and B compartment of the immune system, and because T cells could be critically involved in guiding B cell differentiation. It is noteworthy that, due to their capacity to bind internally processed peptides (Townsend et al. 1986), class I molecules could present not only viral antigens or mutated cellular proteins but also normal proteins which are expressed in a differentiation/activation dependent fashion. The allele-selective down-regulations may be envisaged as the consequence of preferential association of the rejection target expressed in BLs and in their normal counterpart with certain HLA class I specificities.

The deregulated expression of the c-myc oncogene due to translocation in the proximity of Ig coding sequences (Klein et al. 1983) could contribute to the aberrant expression of HLA class I genes in BL cells. High levels of expression of different members of the myc-gene family were shown to correlate with down-regulation of MHC class I antigens in neuroblastoma, small cell lung carcinoma and melanoma (Bernards 1987). Versteeg et al. (1989) have recently shown that transfection of a constitutively activated c-myc construct in human melanoma cell lines preferentially suppresses the expression of alleles within the HLA B locus. The finding that HLA A and C alleles are preferentially suppressed in BLs is compatible with the notion that the oncogene acts on trans-acting cellular factors.

REFERENCES

Avila-Carino J, Torsteinsdottir S, Bejarano MT, Klein G, Klein E, Masucci MG (1988) Combied treatment with interferon (IFN)-gamma and tumor necrosis factor (TNF)-alpha up-regulates the expression of HLA class I determinants in Burkitt lymphoma lines. 117:303-311

Bernards R (1987) Suppression of MHC gene expression in cancer cells TIG 3:298-301

Gregory CD, Rowe M, Rickinson AB (1990) Different Epsetin-Barr virus (EBV) cell interaction in phenotypically distinct clones of a Burkitt lymphoma cell lines. Int.J.Cancer in press

Gregory CD, Tursz T, Edwards C, Tetaud C, Talbot M, Caillou B, Rickinson AB, Lipinski M (1987) Identification of a subset of normal B-cells with a Burkitt lymphoma like phenotype. J.Immunol. 139:313-318

Klein G (1983) Specific chromosomal translocations and the genesis of B cell derived lymphomas in mouse and man. Cell 32:311-315

Liu YJ, Joshua DE, Williams GT, Smith CA, Gordon J, MacLennan ICM (1989) Mechanism of antigen-driven selection in germinal centers. Nature 342:929-931

Masucci MG, Torsteinsdottir S, Colombani J, Brautbar C, Klein E, Klein G (1987) Down-regulation of class I HLA antigens and of the Epstein-barr virus encoded latent membrane protein in Burkitt lymphoma lines. Proc.Natl.Acad.Sci. USA 84:4567-4571

Masucci MG, Stam N, Torsteinsdottir S, Neefjes J, Klein G, Ploegh H (1989) Allele specific down-regulation of MHC class I antigens in Burkitt lymphoma lines. Cell.Immunol. 120:396-400

Neefjes JJ, Breur-Vrisendorp BS, van Seventer GA, Ivanyi P, Ploegh H (1986) An improved biochemical method for the analysis of HLA class I antigens. Definition of new HLA class I subtypes. Hum.Immunol. 16:169-181

Rooney CM, Rickinson AB, Moss DJ, Lenoir GM, Epstein MA (1985a) Cell mediated immunosurveillance mechanisms and the pathogenesis of Burkitt lymphomas. In: Lenoir G, O'Connor G., Olwey CLM (eds) Burkitt lymphoma a human cancer model. IARC Scientific Publications p 249

Rooney CM, Rowe M, Wallace LE, Rickinson AB (1985b) Epstein-Barr virus positive Burkitt lymphoma cells not recognized by virus specific T-cell surveillance. Nature 317:629-631

Rowe DT, Rowe M, Evan GI, Wallace LE, Farrell PJ, Rickinson AB (1986) Restricted expression of EBV latent genes and T-lymphocyte detected membrane antigen in Burkitt lymphoma cells. EMBO J. 5:2599-2607

Rowe M, Rowe DT, Gregory CD, Young LS, Farrel PJ, Rupani H, Rickinson AB (1987) Differences in growth pahenotype reflect novel patterns of Epstein-Barr virus latent gene expression in Burkitt lymphoma cells EMBO J. 6:2743-2751

Rowe M, Rooney CM, Edwards CF, Lenoir GM, Rickinson AB (1986) Epstein-Barr virus status and tumor cell phenotype in sporadic Burkitt lymphoma. Int.J.Cancer 37:357-373

Torsteinsdottir S, Masucci MG, Ehlin-Henriksson B, Brautbar C, Ben Bassat H, Klein G, Klein E (1986) Differentiation dependent sensitivity of human B-cell derived lines to major histocompatibility complex restricted T-cell cytotoxicity. Proc.Natl.Acad.Sci. USA 83:5620-5625

Townsend ARM, Rothbard J, Gotch FM, Badhaum G, Wraith D, McMichael AJ (1986) The epitopes of influenza nucleoprotein re-cognized by cytotoxic T lymphocytes can be defined with short synthetic peptides. Cell 44:959-968

Versteeg R, Kruse-Wolters M, Plomp AC, Van Leeuwen A, Stam NJ, Ploegh HL, Ruiter DJ, Schrier PI (1989) Suppression of class I human hystocompatibility leukocyte antigen by c-myc is locus specific. J.Exp.Med. 170:621-635

Zinkernagel RM, Doherty PC (1979) MHC restricted cytotoxic T-cells, studies on the biological role of polymorphic major transplantation antigens determining T-cell restriction specificity, function and responsiveness. Adv.Immunol. 27:51-63

EBV-Associated B-Cell Lymphomas Following Transfer of Human Peripheral Blood Lymphocytes to Mice with Severe Combined Immune Deficiency

D.E. Mosier, S.M. Baird, M.B. Kirven, R.J. Gulizia, D.B. Wilson, R. Kubayashi[1], G. Picchio, J.L. Garnier, J.L. Sullivan[2] and T.J. Kipps[1]

Division of Immunology, Medical Biology Institute, 11077 North Torrey Pines Road, La Jolla, CA 92037, USA

INTRODUCTION

Mice with severe combined immune deficiency (SCID) accept xenografts of adult human peripheral blood leukocytes (PBL). The transplanted human PBL survive for several months and have been shown to reconstitute human immune function at both the T and B cell level. However, all SCID mice reconstituted with 50 x 10^6 or more PBL from donors with evidence of exposure to Epstein-Barr virus (EBV) develop human B cell lymphomas at 8-16 weeks after PBL engraftment. These tumors involve both lymphatic and non-lymphatic organs, and histologically resemble poorly differentiated immunoblastic lymphomas. The tumors were associated with high levels of human immunoglobulin secretion and serum electrophoresis revealed oligoclonal immunoglobulin banding patterns. Analysis of tumor DNA shows the presence of EBV genomes and oligoclonal patterns of immunoglobulin gene rearrangement. The surface phenotype of the tumor cells is that of mature, activated B lymphocytes. Rare tumors with evidence of *c-myc* oncogene rearrangement are seen. Taken together, these observations suggest an EBV-related proliferation of B lymphocytes leading to the rapid appearance of oligoclonal B cell malignancies following transfer of normal B lymphocytes to SCID mice. The similarity between these tumors and those that develop in immunosuppressed transplant recipients or AIDS patients is striking.

EBV AND HUMAN TUMORS

Epstein-Barr virus (EBV) was first detected in 1964 as virus particles in Burkitt's lymphoma cells (Epstein, 1964). Infection with EBV is found world wide, and is associated with both benign and malignant lymphoproliferative disorders. EBV is a human herpesvirus whose genome has been sequenced; it has a host cell range which is limited to B lymphocytes and certain epithelial cells (Jondal *et al.* 1973; Lemon *et al.* 1977). EBV is maintained in the infected host as a permissive infection of epithelial cells in the oropharynx, and as a latent, non-permissive infection of peripheral B lymphocytes (Sixbey *et al.* 1984; Nonoyama and Pagano 1972). The virus has the capacity to activate and immortalize latently-infected B cells (Pattengale *et al.* 1974; Kirchner *et al.* 1979), so the small number of latent gene products are the primary focus of studies of B cell lymphomagenesis.

---

[1]Research Institute of Scripps Clinic, La Jolla, CA 92037, USA
[2]Department of Pediatrics, University of Massachusetts Medical Center, 55 Lake Avenue N., Worcester, MA 01655, USA

Current Topics in Microbiology and Immunology, Vol. 166

Primary exposure to EBV typically results in asymptomatic infection or in benign infectious mononucleosis (IM) followed by lifelong latent infection. An immunoregulatory mechanism is assumed to exist which prevents the induction and proliferation of EBV-transformed B cells in the host. Abnormal host control of EBV infection occurs in several congenital immunodeficiency diseases (severe combined immunodeficiency, ataxia telangiectasia, Wiscott Aldrich syndrome), in allograft recipients receiving immunosuppressive therapy and in patients with acquired immunodeficiency syndrome (AIDS) (Borzy *et al.* 1979; Saemundsen *et al.* 1981; Perry *et al.* 1980; Penn 1983; Sullivan 1988; Groopman *et al.* 1986). All of these patients demonstrate a high frequency of EBV-induced B-cell lymphomas and lymphoproliferative disorders.

Two malignancies are associated with EBV infection; endemic Burkitt's lymphoma (BL) and nasopharyngeal carcinoma (NPC) (Epstein 1986; Klein 1979a). In the endemic areas of central Africa and New Guinea, 95-98% of BL cases carry the EBV genome. The sporadic form of BL that occurs with lower frequency in other parts of the world has a much lower incidence of EBV expression (Burkitt 1967), but does share the characteristic rearrangement of the *c-myc* oncogene with immunoglobulin heavy (IgH) or light (IgL) chain genes. The role of EBV in the induction of tumors is not clear. The pathogenesis of endemic BL has been proposed to occur in three stages (zurHausen *et al.* 1972; Klein 1979b; Klein 1983):

1. Primary EBV infection, which establishes a pool of latently infected B cells or pre-B cells in the host.

2. Altered T cell regulation leading to a loss of T cell control of EBV-infected cells (Whittle *et al.* 1984) as well as polyclonal B cell activation.

3. Chromosomal translocation which places the *c-myc* oncogene under the control of Ig gene regulatory elements.

In NPC endemic areas, premalignant hyperplastic lesions are often identified. All of these lesions have been found to carry EBV DNA (Mikaelian *et al.* 1989) and many appear to be clonal proliferations. Similar oligoclonal expansions of EBV-positive B lymphocytes have been detected in AIDS patients (Pelicci *et al.* 1986).

These observations suggest a minimal model for EBV-associated B cell lymphomas, which is the activation of latently-infected B lymphocytes due to faulty immunoregulation. The recent observation of B cell lymphomas of human origin in SCID mice transplanted with human PBL (Mosier *et al.* 1988) gives the opportunity to address this hypothesis in more detail. One central issue is whether the lymphomas in SCID mice more closely resemble EBV-induced lymphoblastoid lines (LCL) or Burkitt's lymphomas (BL). The distinction between these transformed B cells (Finke *et al.* 1987; Rowe *et al.* 1987) is summarized in Table 1.

Table 1: Surface phenotype of EBV-transformed lymphoblastoid cell lines (LCL) and Burkitt's lymphoma lines (BL)

		LCL	BL
B cell markers	CD-10 (CALLA)	-	+++
	CD-20 (B1)	+++	+++
	CD-23 (FcRε)	+++	-
	CD-39	+++	-
EBV antigens	EBNA-1	+++	+++
	EBNA-2	+++	-
	LMP	+++	-
cell adhesion molecules	LFA-1	+++	-
	ICAM-1	+++	-
	LFA-3	+++	-
MHC antigens	class I	+++	+ - ++
	class II	+++	+

EBV-ASSOCIATED B CELL LYMPHOMAS IN SCID MICE

SCID mice, which lack T and B lymphocytes because of a autosomal recessive mutation which affects Ig and T cell receptor gene rearrangement (Bosma *et al.* 1983; Schuler *et al.* 1986), are permissive for engraftment of mature human PBL (Mosier *et al.* 1988). We have observed tumors in SCID mice receiving PBL from EBV-VCA seropositive donors (Mosier *et al.* 1988), but not in mice reconstituted with EBV-seronegative donors (see below). The tumors histologically resemble immunoblastic lymphomas, and are malignant since they are (i) invasive, (ii) grow *in vitro*, and (iii) produce lethal secondary tumors when transplanted to SCID recipients. All of the tumors are associated with high levels of monoclonal Ig secretion, although secretory plasma cells make up only a small fraction of the tumor cells. The frequency with which tumors appear is directly proportional to the number of EBV^+ PBL transferred (Fig. 1), which allows the estimate that approximately 1 in 4×10^6 B cells is capable of generating a tumor.

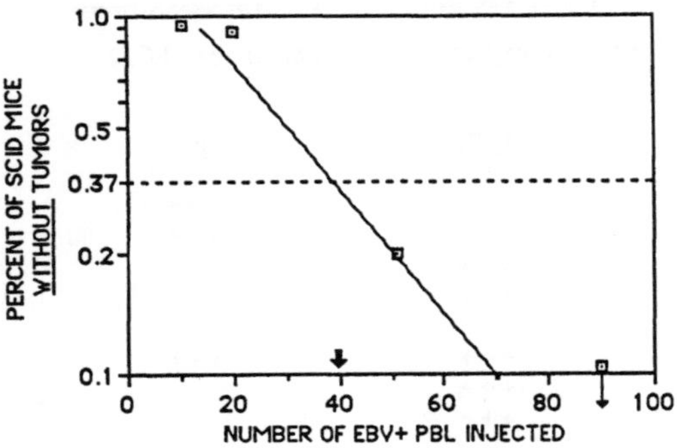

Fig. 1. The relationship of tumor incidence to the number of EBV⁺ PBL injected into SCID mice follows a Poisson distribution. The data indicate approximately one precursor capable of tumor formation is contained in 40 million PBL. The data were compiled from over 100 SCID recipients of PBL from eight different donors.

Since the tumors are B cell lymphomas and B cells represent about 10% of the injected PBL, the data in Fig. 4 would suggest that about 1 in 4 x 10^6 B cells can give rise to a tumor. This must be viewed as a mimimum estimate, however, since the transfer of fewer than 5 x 10^6 PBL fails to reconstitute human immune function in SCID mice; *i.e.*, one must correct for *in vivo* seeding efficiency. It is also probable that pre-malignant precursor numbers differ to some extent between individuals.

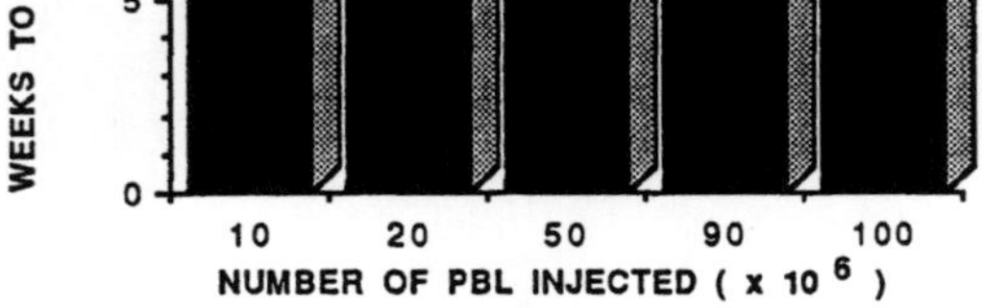

Fig. 2: Relationship between number of PBL injected and time to tumor detection for all mice that developed tumors. At smaller numbers of PBL injected, only a fraction of mice developed tumors (see Fig. 1).

Exposure of the donor to EBV appears to be essential for the development of lymphomas in SCID mice. Fig. 3 illustrates the mortality curve and human immunoglobulin levels in SCID mice receiving PBL from either an EBV+ donor or an EBV- donor. Similar data has been generated using a total of 15 EBV+ donors, 13 of whom gave rise to tumors, and 11 EBV- donors, none of whom gave rise to tumors.

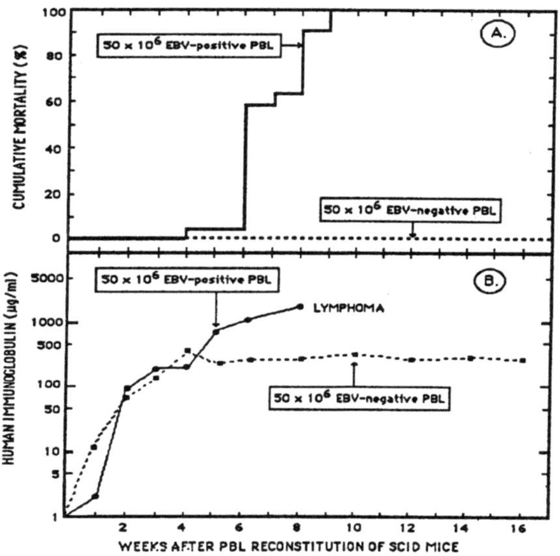

Fig. 3: Mortality of SCID mice transplanted with PBL from EBV-seropositive or seronegative donors (Mosier et al. *1989). Panel A shows the cumulative mortality of two groups of 10 SCID mice each transplanted with PBL at eight weeks of age. Panel B shows the human immunoglobulin (IgM + IgG + IgA) concentration in the two groups of hu-PBL-SCID mice.*

The tumors that appear in SCID mice transplanted with PBL from EBV+ donors are malignant B cell lymphomas by the following criteria:

(i) Tumor-bearing mice have a monoclonal elevation of IgM or IgG. Some mice injected with 50 x 10^6 or more PBL display a biclonal gammopathy with expression of kappa L chain on one Ig and lambda L chain on the other. This finding eliminates class switching within a clonal population as an explanation for multiple J_H rearrangements.

(ii) DNA from all tumors analyzed hybridizes with the EBV Bam W probe for the internal repeat sequence of EBV. Primers specific for the Bam W region also amplify DNA from both tumors and the serum of mice 4 or more weeks following reconstitution with EBV^{+} PBL.

(iii) Staining of tumor cells reveals expression of CD19, CD20, CD23, CD39, LFA-1, and MHC class II antigens, but not CD5 or CD10. This would place their surface phenotype closer to that of EBV-transformed LCL than of BL (see Table 1).

CONCLUSIONS AND DISCUSSION

The transplantation of human PBL to SCID mice represents a new model for studying the effects of EBV on human B cell transformation. Most mice develop tumors with a phenotype consistent with *in vivo* transformation of B cells by EBV, but occasional tumors have *c-myc* rearrangements and may be more closely related to BL. Interestingly, only a few donors seem capable of giving rise to the BL-like tumors. The genesis of these human B cell tumors (or B cell lymphoproliferative disease) in SCID mice is likely to be related to a loss of T cell function that occurs within the first few weeks after transplantation of PBL to SCID recipients. This hypothesis can readily be evaluated in the hu-PBL-SCID model. At a practical level, the transplantation of PBL from potential transplant donors or recipients to SCID mice may predict the frequency with which post-transplant lymphomas will be encountered.

REFERENCES

Borzy MS, Hong K, Horowtiz SD, et al (1979) Fatal lymphoma after transplantation of cultured thymus in children with combined immunodeficiency disease. N Engl J Med 301:565-568

Bosma GC, Custer RP, Bosma MJ (1983) Severe combined immunodeficiency mutation in the mouse. Nature 301:527-530

Burkitt DP (1967) Burkitt's lymphoma outside the known endemic areas of Africa and New Guinea. Int J Cancer 2:562-565

Epstein MA (1986) Historical background: Burkitt's lymphoma and Epstein-Barr virus. In: Burkitt's Lymphoma: A Human Cancer Model, Lyon, IARC, pg 17

Epstein MA, Achong BG (1979) In: The Epstein Barr Virus. Spring-Verlag, Berlin, pp 1-22

Epstein MA, Barr YM, Achong BG (1964) Virus particles in cultured lymphoblasts from Burkitt's lymphoma. Lancet i:702:703

Finke J, Rowe M, Kallin B, Emborg I, et al (1987) Monoclonal and polyclonal antibodies against Epstein Barr virus: Nuclear antigen 5 (EBNA5) detect multiple protein species in Burkitt's lymphoma and lymphoblastoid cell lines. J Virol 61:3870-3878

Groopman JE, Sullivan JL, Mulder C, et al (1986) Pathogenesis of B cell lymphoma patients with AIDS. Blood 67:612-615

Jondal M, Klein G (1973) Surface markers on human B and T lymphocytes. II. Presence of Epstein-Barr virus receptors on B lymphocytes. J Exp Med 138:1365-1378

Kirchner H, Tosato G, Blaese RM, et al (1979) Polyclonal immunoglobulin secretion by human B lymphocytes exposed to Epstein-Varr vius in vitro. J Immunol 122:1310-1313
Klein G (1979a) In: The Epstein-Barr Virus, Springer-Verlag, Berlin, pp 339-350
Klein G (1979b) Lymphoma development in mice and humans: Diversity of initiation is followed by convergent cytogenetic evolution. Proc Natl Acad Sci USA 76:2442-2446
Klein G (1983) Specific chromosomal translocations and the genesis of B-cell-derived tumors in mice and men. Cell 32:311-315
Lemon SM, Hutt LM, Shaw JD, Li J-LH, Pagano JS (1977) Replication of EBV in epithelial cells during infectious mononucleosis. Nature 268:268-270
Mikaelian I, Ellouze R, Dubreuil C, Anavoka T (1989) In: International Conference on Esptein-Barr Virus: The First 25 Years (abstracts). Oxford University, April 9-13
Mosier DE, Gulizia RJ, Baird SA, Wilson DB (1988) Transfer of a functional human immune system to mice with severe combined immunodeficiency. Nature 335:256-259
Mosier DE, Gulizia RJ, Baird SM, Spector SA, Spector D, Kipps TJ, Fox RI, Carson DA, Cooper N, Richman DD, Wilson DB (1989) Studies of HIV infection and the development of Epstein-Barr virus-related B cell lymphomas following transfer of human lymphocytes to mice with severe combined immunodeficiency. In: Current Topics in Microbiology and Immunology (Springer-Verlag, Berlin-Heidelberg), vol 152, pp 195-199.
Nonoyama M, Pagano JS (1972) Nonoyama-separation of Epstein Barr virus DNA from large chromosomal DNA in non-virus producing cells. Nature 238:169-171
Pattengale PK, Smith RW, Berger P (1974) B cell characteristics of human periferal and cord blood lymphocytes transformed by Epstein Barr virus. J Natl Cancer Inst 52:1081-086
Pelicci PG, Knowles DM, Arlin ZA, Wieczorek R, Luciw P, Dina D, Basilico C, Dalla-Favera R (1986) Multiple monoclonal B cell expansions and c-myc oncogene rearrangements in acquired immune deficiency syndrome-related lymphoproliferative disorders. J Exp Med 164:2049-2060
Penn I (1983) Lymphomas complicating organ transplantation. Transpl Proc 15:2790
Perry GS, Spector BD, Schumann LM, Mandel JS (1980) The Wiskett-Aldrich syndrome in the United States and Canada (1892-1979). J Pediatrics 97:72-78
Rowe M, Rowe DT, Gregory CD, Young LS, Farrell PJ, Rupani H, Rickinson AB (1987) Differences in B cell growth phenotype reflect novel patterns of Epstein-Barr virus latent gene expression in Burkitt's lymphoma cells. EMBO J 6:2743-2751
Saemundsen A, Bertel AJ, Henle W, et al (1981) Epstein Barr virus-carrying lymphoma in a patient with ataxia telangiectasia. Brit Med J 282:425-427
Schuler W, Weiler IJ, Schuler A, Phillips RA, Rosenberg N, Mak TW, Kearney JF, Perry RP, Bosma MJ (1986) Rearrangement of antigen receptor genes is defective in mice with severe combined immune deficiency. Cell 46:963-972,
Sixbey JW, Nedrud JG, Raab-Traub N (1984) Epstein Barr virus replication in oropharyngeal epithelial cells. N Engl J Med 310:1225-1230
Sullivan JL (1988) Hematologic consequences of the Epstein Barr virus infection. Sem in Hematol 25:269-279
Whittle HC, Brown J, Marsh K, Greenwood BM, Seidelin P, Tighe H, Wedderburn L (1984) T-cell control of Epstein-Barr virus-infected B cells is lost during P. falciparum malaria. Nature 312:449-450
zurHausen H, Diehl V, Wolf H, Schulte-Holthausen H, Schneider U (1972) Occurence of Epstein-Barr virus genomes in human lymphoblastoid cell lines. Nature 237:189-190

Analysis of Epstein-Barr Virus Gene Expression in Lymphomas Derived from Normal Human B Cells Grafted into SCID Mice

M. Rowe, L.S. Young, and A.B. Rickinson

Department of Cancer Studies, Cancer Research Campaign Laboratories, School of Medical Sciences, University of Birmingham, Birmingham, U.K.

INTRODUCTION

The recent demonstration that C.B-17*scid/scid* mice (hereafter referred to as SCID mice) could be grafted successfully with human haematopoietic cells (Mosier et al., 1988; McCune et al., 1988; Kamel-Reid and Dick, 1988) has opened the way to the use of the SCID-hu mouse as a model for studying human immune-function *in vivo*. However, if SCID-hu mice are generated by intraperitoneal injection of about $5x10^7$ human peripheral blood lymphocytes (PBLs), then consideration must be given to the fact that human B cell lymphomas regularly develop between 2-4 months post-engraftment. The causal role of Epstein-Barr virus (EBV) in the development of these lymphomas has been demonstrated (Mosier et al., 1988). Thus, all of the lymphomas contained EBV DNA and, furthermore, lymphomas did not develop in mice engrafted with PBLs from donors where there was no serological evidence of EBV infection. In this preliminary report we have further characterised these EBV-positive lymphomas in order to determine whether they might be used as a model for EBV-associated malignancies.

EBV-associated diseases: background

Epstein-Barr virus is a herpes virus which is carried as a persistent and life-long infection of epithelial and lymphoid tissues in 80 to 100% of adults in all populations wordwide. As a result of the potent host immune-response that is mounted to EBV (Rickinson, 1986), infection with the virus is usually asymptomatic. However, EBV is the causative agent of infectious mononucleosis and is associated with various malignancies, including nasopharyngeal carcinoma, Burkitt's lymphoma (BL), and lymphomas arising in immunosuppressed post-transplant patients (Henle and Henle, 1985; Cleary et al., 1986).

The pathogenic potential of EBV *in vivo* is illustrated by the ease with which which the virus will infect normal resting B cells *in vitro* and transform them into permanent lymphoblastoid cell lines (LCLs). Such LCLs are generally non-permissive for virus-production but they constitutively express a number of so-called latent genes which encode six separate nuclear antigens (EBNA 1, EBNA 2, EBNA 3a, EBNA 3b, EBNA 3c, and EBNA-LP), a "latent membrane protein" (LMP), and a second less-well characterised membrane-associated "terminal protein" (TP) (reviewed by Knutson and Sugden, 1989). With the development of suitable serological probes, it has been

Current Topics in Microbiology and Immunology, Vol. 166

possible to analyse in some detail the expression of the six EBNAs and of LMP in EBV-associated malignancies. Thus, we recently demonstrated that BL biopsy cells differ from LCLs in that they express detectable levels of EBNA 1 only (Rowe et al., 1987; Gregory et al., 1990). An example of this type of analysis is illustrated by Fig. 1, which shows a Western blot of a BL cell extract and of a cell extract from a normal LCL. The filter was probed with with a polyclonal human serum that detected all six EBNA proteins in the LCL cells, but which detected only EBNA 1 in the BL cells; parallel blots (not shown) probed with a monoclonal antibody to LMP showed that this membrane protein is similarly expressed in the LCL cells but is undetectable in the BL cells. We now refer to the LCL-like pattern of EBV gene-expression as "active latency", and the BL-like pattern of EBV gene-expression as "passive latency". Two other features distinguish BL cells from LCLs. Firstly, BL cells possess characteristic monoclonal chromosome translocations involving the c-*myc* gene on chromosome 8 and one of the immunoglobulin genes on chromosome 14, 2, or 22, (Lenoir and Bornkamm, 1987). Secondly, BL cells downregulate the expression of several B cell "activation" antigens, including the adhesion-molecule family of proteins and instead express markers of germinal centre cells (Rowe and Gregory 1989).

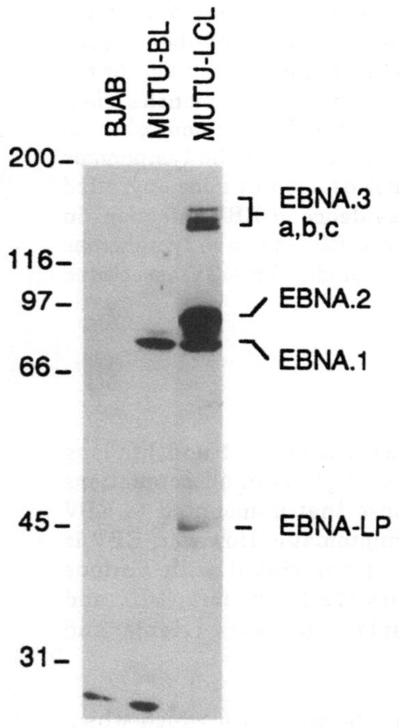

Figure 1. Western blot of SDS-solubilised cell nuclear extracts separated on a 7.5% discontinuous polyacrylamide gel and transferred to a nitrocellulose filter. The filter was probed with a human serum (RS22) which contained antibodies reactive with each of the six EBNAs, and specifically-bound antibodies were detected with ^{125}I–Protein A. BJAB is a reference EBV-negative cell line whilst MUTU-BL and MUTU-LCL are, respectively, tumour cells from a BL patient, and a normal LCL produced by infection of resting B cells *in vitro* with virus derived from the BL tumour cells.

In contrast to BL tumours, B cell lymphomas that arise in post-transplant patients are indistinguishable from LCLs with respect to viral gene expression and cellular phenotype and with respect to the lack of BL-type chromosome translocations (Young et al., 1989; Shapiro et al., 1988). We interpret these observed differences between BL and post-transplant lymphomas as reflecting the importance of the potent anti-viral immune response in controlling persistent EBV infection in healthy individuals. Thus, the clinically-induced immunosuppression in transplant patients allows the outgrowth of EBV-infected B cells which resemble *in vitro* transformed LCLs. However, BL tumours have to evade a relatively intact immune-surveillance system and this is achieved by down-regulating the viral gene-products together with the cellular adhesion molecules that are recognised by cytotoxic T cells. In addition BL tumour cells have been shown to downregulate certain HLA class-I molecules (Torsteinsdottir et al., 1988), which would prevent presentation of processed viral peptides to effector T cells.

EBV-POSITIVE LYMPHOMAS IN SCID-hu MICE

Materials and Methods

Between 40-50 x 10^6 PBLs, isolated from donors with known EBV serological status, were injected intraperitoneally into SCID mice. When animals became unwell they were sacrificed and autopsied.

Tumours and other biopsies were analysed for EBV gene expression in two ways: Western blotting of solubilised tissue, and immunofluorescence on frozen sections. Samples for Western blotting were prepared by homogenising tissues in gel sample buffer at 2.5% weight /volume, and 0.5 mg tissue per track was separated on an SDS-polyacrylamide gel. Control extracts from cultured cell lines were prepared from 10^6 cells. Following transfer to nitrocellulose filters, the resolved samples were probed with EBV-specific antibodies (either selected well-characterised human sera, or monoclonal antibodies) and were detected with ^{125}I-Protein A. Tissues for immunofluorescence were snap-frozen in liquid nitrogen, sectioned, and fixed in cold methanol. Full details of these procedures will be given in a subsequent complete report on this work (Rowe et al, manuscript in preparation). The monoclonal antibodies to LMP (CS.1-4), EBNA 2 (PE2), EBNA-LP (JF186), and the BamZ immediate-early antigen (BZ) were developed by the authors, whilst the anti-VCA monoclonal antibody (V3) was a kind gift of Dr G Pearson.

Results and Discussion

To date, out of 17 SCID mice injected intraperitoneally with PBLs from a total of eight EBV-seropositive donors, 14 (82%) showed signs of illness within 17 weeks, with a mean time for onset of symptoms of 11.9 ± 3.5 weeks. All 14 sick animals showed liver abnormalities, including necrosis and inflammatory lymphoid infiltrates.In addition, between 1 to 5 solid tumours were found, usually in the peritoneal cavity around the stomach, liver, or intestines but also frequently in the thymus. The 3 healthy animals were sacrificed at 19, 23, and 23 weeks and showed no abnormalities at autopsy. Three mice that were injected with PBLs from an EBV-seronegative donor were healthy at 23 weeks and did not produce tumours.

The tumours arising in PBL-engrafted SCID mice were shown to express the same range of latent EBV proteins that are constitutively expressed in LCLs. Thus, Fig. 2 shows a Western blot, probed with a human serum for the detection of EBNAs, of two separate tumours that arose in one mouse injected with human PBLs. Immunofluorescence staining of frozen sections of the tumours with monoclonal antibodies to three of the latent proteins, clearly demonstrated the expression of LMP in addition to EBNA-2 and EBNA-LP in the majority of the tumour cells (Fig. 3).

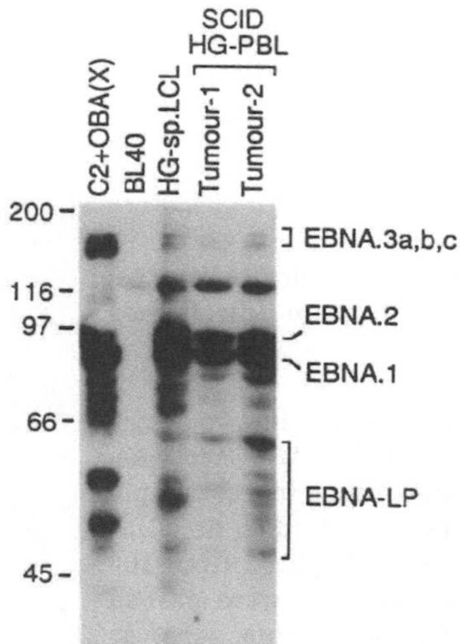

Figure 2. Western blot of two separate tumours arising in a SCID mouse engrafted with PBLs from donor H.G.. The track labelled HG-spLCL contained extracts from a cell line generated by spontaneous outgrowth *in vitro* from PBLs taken from the same donor. C2+OBA[X] is a reference EBV-positive LCL which contains a different isolate of EBV, and BL40 is an EBV-negative reference cell line. The filter was probed with a human serum, PG, which contained antibodies reactive with each of the EBNAs.

One unexpected finding in these studies was that the tumours invariably contained cells in the virus replicative cycle. This is most dramatically illustrated by the Western blots shown in Fig. 4. It will be noted that EBNA-1 was readily detected in two tumour samples, whilst no EBNA-1 was detected in the spleen sample and there were just detectable bands of a possibly degraded EBNA-1 protein in the liver sample (Fig. 4, left blot). The EA[D] and BZ-IEA proteins were readily detected in both tumour samples at levels greatly in excess of those observed in the *in vitro* spontaneous LCL. Furthermore, the sensitivity of the anti-EA[D] human serum probe was such that the viral antigens were unequivocally detected in the liver sample, albeit at lower levels, whilst the spleen sample remained negative (Fig. 4: right panel). Immunofluorescence staining of sections of the SCID tumours with monoclonal antibodies (data not shown) clearly showed a subpopulation of cells that were strongly positive for the BZ-IEA protein and for the viral capsid antigen (VCA). The number of cells in lytic cycle was determined to be as high as 5% when immunofluorescence staining was performed on smears made from cell suspensions prepared by treating the fresh tumour biopsies with a collagenase-based dispersion medium.

Table 1 summarises and compares the observed EBV gene expression in SCID-hu tumours with the gene-expression in BL tumours and in post-transplant lymphomas.

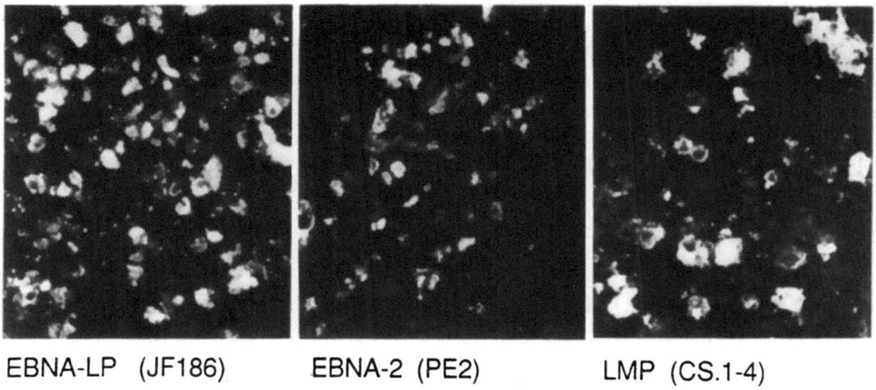

Figure 3. Immunofluorescence photomicrographs of methanol-fixed sections of a thymic tumour that developed in a SCID mouse injected with PBLs from donor M.R.. The three photographs from left to right show staining for EBNA-LP, EBNA-2, and LMP respectively.

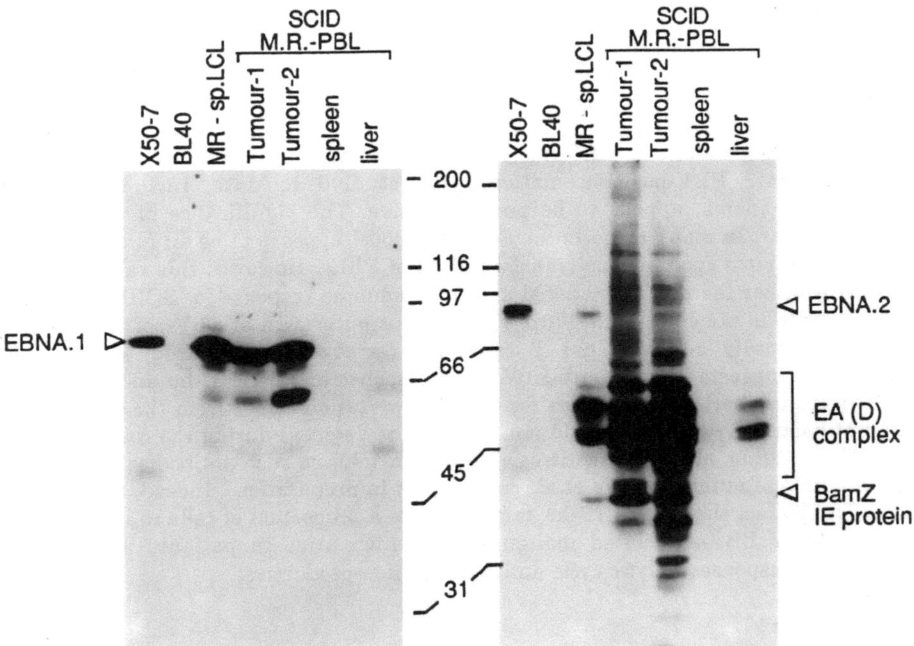

Figure 4. Western blots of two identical filters probed either with the human serum AMo which reacts with EBNA-1 only (left panel), or with the human serum EE which reacts predominantly with the EA[D] and BZ-IEA proteins of the ytic virus-productive cycle, and, to a lesser extent, with the EBNA-2 latent protein (right panel).

With respect to latent gene-expression, the SCID-hu tumours are indistinguishable from post-transplant lymphomas and *in vitro* transformed normal LCLs. This pattern of viral gene-expression is also consistent with the cellular phenotype of the SCID-hu tumours which, from our preliminary data (not shown), is also indistinguishable from post-transplant lymphomas and LCLs, and is distinct from the cellular phenotype of BL tumours. Furthermore, no consistent chromosomal abnormalities have been found in the SCID-hu tumours (Rowe et al., manuscript in preparation).

Table 1: EBV gene-expression in B-lymphocytes

	Latent Proteins		Productive-cycle Proteins	
Cell - type	EBNA.1	EBNA.2,3a, 3b, 3c, -LP	BZ-IEA	VCA
Normal LCL	+	+	–/+	–/+
Burkitt's lymphoma	+	–	–	–
Post-transplant lymphoma	+	+	–	–
SCID-hu lymphoma	+	+	+	+

The distinguishing feature of the SCID-hu lymphomas is that they contain cells in lytic cycle: all other EBV-positive malignancies studied to date, including nasopharyngeal carcinoma, appear to be non-productive. The significance of this observation is unclear. In many respects the SCID-hu model appears to be an *in vivo* equivalent of the *in vitro* spontaneous transformation of PBLs. However, this raises the question of whether there is a functional anti-viral immune response in SCID-hu mice, since when PBLs are cultured *in vitro* spontaneous outgrowth of EBV-positive LCLs does not usually occur unless T cells are removed or inhibited with cyclosporin A. This suggests that the anti-EBV T cell response in our SCID-hu mice is ineffective, absent, or short-lived. Support for this interpretation comes from the fact that whilst antibodies to the viral capsid antigen may be detected during the first 6 weeks after PBL grafting, albeit at low titres of less than 1:40, no such antibodies are detected at the time of autopsy (Rowe et al., manuscript in preparation). Indeed, this may account for the fact that the SCID-hu tumours have a proportion of cells in lytic cycle whereas other EBV-associated malignancies, which arise in patients with normal or elevated responses to lytic cycle antigens, are non-productive.

ACKNOWLEDGEMENTS

This research was supported by grants from the Wellcome Trust and from the Cancer Research Campaign, London. We wish to thank Janis Pearson and Val Nash for invaluable technical assistance, and Lesley Wallace for critically reading the manuscript.

REFERENCES

Cleary ML, Dorfman RF, Sklar J (1986) Failure in immunological control of the virus infection: Post-transplant lymphomas. In: Epstein MA, Achong BG (eds) *The Epstein-Barr Virus: Recent Advances*. William Heinemann Medical Books, London. pp 163-182

Gregory CD, Rowe M, Rickinson AB (1990) Different Epstein-Barr virus (EBV)-B cell interactions in phenotypically distinct clones of a Burkitt lymphoma cell line. *J. Gen. Virol.* (in press)

Henle W, Henle G (1985) Epstein-Barr virus and human malignancies. In: Klein G (ed) *Adv. Viral Oncology* vol. 5. Raven Press, New York. pp 201-238

Kamel-Reid S, Dick JE (1988) Engraftment of immune-deficient mice with human hematopoietic stem cells. *Science* 242: 1706-1709

Knutson JC, Sugden B (1989) Immortalization of B lymphocytes by Epstein-Barr virus: What does the virus contribute to the cell? In: Klein G (ed) *Adv. Viral Oncology* vol. 8. Raven Press, New York. pp 151-172

Lenoir GM, Bornkamm (1987) Burkitt's lymphoma, a human cancer model for the study of the multistep development of cancer: proposal for a new scenario. In: Klein G (ed) *Adv. Viral Oncology* vol. 7. Raven Press, New York. pp 173-206

McCune JM, Namikawa R, Kaneshima H, Schultz LD, Lieberman M, Weissman IL (1988) The SCID-hu mouse: Murine model for the analysis of human hematolymphoid differentiation and function. *Science* 241: 1632-1639

Mosier DE, Gulizia RJ, Baird SM, Wison DB, (1988) Transfer of a functional human immune system to mice with severe combined immunodeficiency. *Nature* 335: 256-259

Rickinson AB (1986) Cellular immunological responses to the virus infection. In: Epstein MA, Achong BG (eds) *The Epstein-Barr Virus: Recent Advances*. William Heinemann Medical Books, London. pp 75-126

Rowe M, Rowe DT, Gregory CD, Young LS, Farrell PJ, Rupani H, Rickinson AB (1987) Differences in B cell growth phenotype reflect novel patterns of Epstein-Barr virus latent gene expression in Burkitt's lymphoma. *EMBO J.* 6: 2743-2751

Rowe M, Gregory C (1989) Epstein-Barr virus and Burkitt's lymphoma. In: Klein G (ed) *Adv. Viral Oncology* vol. 8. Raven Press, New York. pp 237-259

Shapiro RS, McClain K, Frizzera G, Gajl-Peczalska KJ, Kersey JH, Blazer BR, Arthur DC, Patton DF, Greenberg JS, Burke B, Ramsay NKC, McGlave P, Filipovich AH (1988) Epstein-Barr virus associated B cell lymphoproliferative disorders following bone marrow transplantation. *Blood* 71: 1234-1243

Torsteinsdottir S, Brautbar C, Klein G, Klein E, Masucci MG (1988) Differential expression of HLA antigens on B-cell lines of normal and malignant origin: a consequence of immune surveillance or a phenotypic vestige of the progenitor cells? *Int. J. Cancer* 41: 913-919

Young L, Alfieri C, Hennessey K, Evans H, O'Hara C, Anderson KC, Ritz J, Shapiro RS, Rickinson A, Kieff E, Cohen JI (1989) Expression of Epstein-Barr virus transformation-associated genes in tissues of patients with EBV lympho proliferative disease. *N. Eng. J. Med.* 321: 1080-1085

Deregulated c-*myc* Gene Expression and Persistence of EBV are Not Sufficient to Maintain the Malignant Phenotype in Burkitt's Lymphoma x B-Lymphoblastoid Hybrid Cells

J. Wolf, M. Pawlita, J. Bullerdiek[1] and H. zur Hausen

Institut für Virusforschung, Deutsches Krebsforschungszentrum, Im Neuenheimer Feld 280, D-6900 Heidelberg, FRG
[1]Zentrum für Humangenetik der Universität Bremen, Leobenerstraße NW2, D-2800 Bremen, FRG

INTRODUCTION

Evidence for loss of specific genetic information as an essential event in the multistep process of malignant transformation has been provided by somatic cell hybrid studies which demonstrated suppression of the malignant phenotype for a wide spectrum of murine and human tumor cells after fusion with nonmalignant fibroblasts or keratinocytes (Sager 1986, Klein 1987). The model of recessive tumorigenesis seems to be confirmed by the detection of specific allele losses in several human tumors of epithelial and mesenchymic origin (Ponder 1988), however, has not mainly been involved in scenarios for the development of hematopoietic malignancies. Especially malignant lymphomas still are considered to be models for the dominant action of oncogenes activated by specific chromosomal translocations. This concept was supported by transfection experiments which demonstrated tumorigenic conversion of Epstein-Barr virus (EBV)-immortalized human B-lymphocytes after introduction of a constitutively expressed c-myc gene (Lombardi et al. 1987). Accordingly, no tumor suppression has been demonstrated yet in lymphoma/lymphocyte hybrid cells (Harris 1988).
We approached the question of recessive factors in the tumorigenesis of malignant human B-cell lymphomas by establishing lymphoma/lymphocyte hybrids. Therefore, Burkitt's lymphoma cells were fused with EBV-immortalized lymphoblastoid cells (Wolf et al. 1990). Here we summarize and discuss results obtained with these hybrids.

RESULTS AND DISCUSSION

Establishment of Somatic Cell Hybrids Between Burkitt's Lymphoma Cells and EBV-Immortalized Lymphoblastoid Cells

The tumorigenic Burkitt's lymphoma cell line BL 60 (chromosomal translocation (8;22), EBV-positive) and the nontumorigenic EBV-immortalized lymphoblastoid cell line (LCL) IARC 277, which both originate from the same patient were kindly provided by G. Lenoir (IARC, Lyon). BL 60 cells were rendered neomycin resistant (neo^R) by transfection of the plasmid pSV2neo. Subsequently hypoxanthine-guanine-phosphoribosyltransferase deficient (HGPRT$^-$) subclones were selected in 6-thioguanine. A universal fuser subline of BL 60, termed BL60-P7 (neo$^+$HGPRT$^-$) was fused with the IARC 277 cell line, which had not been manipulated experimentally prior to fusion. The seven hybrid clones which were selected for further analysis, demonstrated a stable, near tetraploid karyotype (modal counts: 82-93) including one copy of the BL 60 chromosome 8q$^+$ which is involved in the specific translocation (8;22) versus 3 copies of the normal chromosome 8.
The presence of EBV in the hybrid clones was demonstrated by Southern blot analysis of BamHI digested cellular DNA probed with the BamHI-W fragment of EBV.

Growth Characteristics of the BL/LCL Hybrids in vitro and in vivo

In contrast to the parental BL60-P7 cell line which grows in single cell suspension the parental LCL IARC 277 forms large clumps and shows a lower proliferation rate and saturation density. With regard

Current Topics in Microbiology and Immunology, Vol. 166

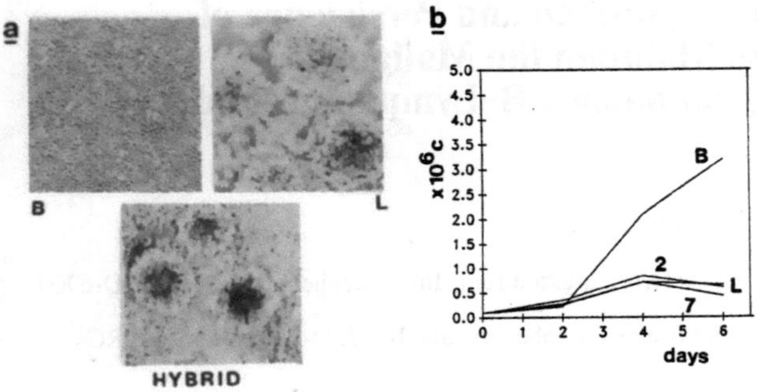

Fig.1: In vitro phenotype of BL/LCL hybrids and parental cell lines.
a. Clumping pattern of BL60-P7 (B), IARC 277 (L) and one representative BL/LCL hybrid clone.
b. Growth kinetics of representative hybrid clones 2 and 7 and the parental BL (B) and LCL (L). $1x10^5$ cells of each cell line were seeded in fresh medium and the number of viable cells was determined at two day intervals.

to these in vitro characteristics the hybrid clones were not distinguishable from the parental LCL (Fig. 1).
To analyze the tumorigenic potential of the cell lines $1x10^7$ cells were inoculated subcutaneously into preirradiated nude mice as described (Gurtsevitch et al. 1987). After a latency period of about 2 weeks the BL60-P7 cells formed large progressively growing tumors reaching a size of about 5 cm diameter within 5 weeks after injection. In these tumors no macroscopic sign of necrosis or regression occured. In contrast the grafts of the parental LCL as well as of the seven hybrid clones showed after a latency period of 2 weeks an initial growth phase up to a size of about 1 cm in diameter, followed by regression (Table 1). Regression took place partly with, partly without a central necroses and was sometimes accompanied by a cachectic state of the animals. However, with the exception of two animals that died in the cachectic state the grafts of all hybrid clones as well as of the parental LCL regressed completely within 9 weeks after injection.

Table 1. Growth of parental and hybrid cell lines in nude mice

	weeks post inoculation				
	2	4	6	8	10
BL60-P7	19/20 (1.5)[a]	19/20 (2.5)	19/20 (5.0)[b]		
IARC277	8/10 (0.6)	7/10 (0.5)	5/10 (0.3)	0/10	
Hybrids 1-7	47/58 (0.5)	34/58 (0.5)	20/56[c](0.5)	4/54[c](0.5)	0/54

[a]number of visible grafts/number injection sites (average diameter in cm)
[b]animals had to be sacrificed due to tumor size
[c]loss of 1 cachectic animal carrying two grafts

When the tumorigenicity assays were performed without preirradiation of the nude mice, the overall take rate of all cell lines was reduced, however the same growth pattern was observed, i.e. unlimited growth potential of the BL cells vs. complete regression after an initial growth phase for the parental LCL and the hybrid clones.

C-myc Transcription Analysis in the BL/LCL Hybrids

C-myc deregulation in BL 60 cells is characterized by highly abundant transcripts originating almost exclusively from the chromosome $8q^{+}$ (Cesarman et al. 1987). Northern blot analysis revealed about the same steady state level of c-myc transcripts in the nontumorigenic hybrids as in the tumorigenic BL60-P7 cell line (Fig. 2a).
To determine the chromosomal origin of the c-myc transcripts in the hybrid cells we made use of a restriction fragment length polymorphism in the c-myc gene (Cesarman et al. 1987). One PvuII restriction site is lost in c-myc exon 1 on chromosome $8q^{+}$ but still present on the three normal chromosomes 8. A cDNA fragment extending from 181bp downstream to 41bp upstream of this PvuII site was generated by first reverse transcription and subsequent amplification in the polymerase chain reaction. PvuII digestion of the 222bp product into two fragments of 181bp and 41bp indicated origin of the c-myc transcripts from the normal chromosomes 8, whereas transcripts from chromosome $8q^{+}$ yielded an undigestible 222bp cDNA fragment. As shown in Fig. 2b, the highly abundant c-myc transcripts of all hybrids originated exclusively from chromosome 8q+, thus indicating the same deregulated c-myc transcription pattern in the hybrids as in the parental BL cell line.

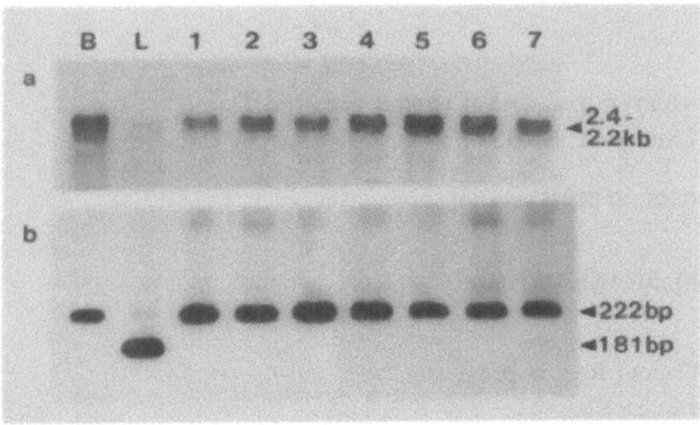

Fig.2. C-myc transcription in BL/LCL hybrids and parental cell lines.
a. Northern blot analysis of equal amounts of total cellular RNA of hybrids 1-7 (lanes 1-7), parental BL60-P7 (lane B) and IARC 277 (lane L) probed with c-myc exon 3.
b. Analysis of the chromosomal origin of the c-myc transcripts by PvuII digestion of c-DNA fragments extending from 181bp upstream to 41bp downstream of the PvuII site which is lost on the BL60 chromosome 8q+. The fragments are identified after electroblotting by hybridization with an internal oligonucleotide probe.

Recessive Tumorigenesis in Endemic Burkitt's Lymphoma?

Our results indicate that the tumorigenicity of a BL cell line can be suppressed by fusion with a nonmalignant EBV-immortalized lymphoblastoid cell line. Persistence of EBV and continued deregulation of the c-myc gene are not sufficient to maintain the malignant phenotype in these hybrids. These observations appear to be controversial to the hypothesis that the BL specific translocation causing the deregulated c-myc expression pattern and thus the outgrowth of a malignant cell clone, takes place in EBV-immortalized lymphoblasts (Klein 1979). This model has been supported by the tumorigenic conversion of EBV-immortalized lymphoblasts after introduction of a constitutively expressed c-myc gene (Lombardi et al. 1987). At present it can not be excluded that specific characteristics of the experimental systems contribute to the apparent discrepancy between the tumorigenic c-myc transfectants and the nontumorigenic hybrids presented here. For instance, a gene dosage effect in the hybrid cells might cause increased production of a factor antagonizing the effects of the c-myc protein. On the other hand malignant transformation in the c-myc transfectants might occur only in a LCL-subset, which had undergone additional, yet unknown genetic changes. In this context it is noteworthy that another group reported altered growth pattern in vitro, but not tumorigenicity in vivo after transfection of a constitutively expressed c-myc gene into EBV-immortalized lymphoblasts (Hotchin et al. 1990).

In analogy to somatic cell hybrids previously established with cell lines derived from tumors of epithelial and mesenchymic origin, our data suggest that the nonmalignant fusion partner contributes specific cellular functions which suppress tumorigenicity and which are absent in the tumorigenic BL parent. In the hybrid model presented here, at present it remains an open question whether the absence of these functions in the BL cells is due to their specific differentiation state or whether it is caused by specific gene loss.
The tumor suppressing functions provided by the parental LCL might exert their activity for instance by directly interfering with cellular growth control, by influencing tumor angiogenesis or resistance to hypoxic environment or by upregulating specific cellular or viral antigens which represent targets for a residual immune response by the nude mouse host.
Further characterization of these putative antioncogenic functions will include analysis of the phenotypic changes which occur in the long-term cultured hybrids. Some of these hybrid clones which have been continuously growing in tissue culture for about 2 years, have lost their initial clumping phenotype and show a higher proliferation rate and higher saturation density. Work is in progress to test the tumorigenicity of these hybrid sublines and to isolate tumorigenic segregants in which reappearance of tumorigenicity might be correlated with specific changes in the karyotype.

REFERENCES

Cesarman E, Dalla-Favera R, Bentley D, Groudine M (1987) Science 238: 1272-1275
Gurtsevitch VE, O'Conor GT, Lenoir GM (1987) Int J Cancer 41: 87-95
Harris H (1988) Cancer Res 48: 3302-3306
Hotchin NA, Allday MJ, Crawford DH (1990) Int J Cancer, in press
Klein G (1979) Proc Natl Acad Science 76: 2442-2446
Klein G (1987) Science 238: 1539-1545
Lombardi L, Newcomb EW, Dalla-Favera R (1987) Cell 49: 161-170
Sager R (1986) Cancer Res 46: 1573-1580
Ponder B (1988) Nature 335: 400-402
Wolf J, Pawlita M, Bullerdiek J, zur Hausen H (1990) Cancer Res, in press

Epstein-Barr Virus Latency and Activation in Vivo

I. Ernberg

Department of Tumor Biology, Karolinska Institute, Box 60 400, 10401 Stockholm, Sweden

INTRODUCTION

Epstein-Barr virus (EBV) infects most human beings in childhood or adolescence. Lifelong latency is established after primary infection.

EBV can infect two cell types in vivo, B-lymphocytes and epithelial cells. The infected B-lymphocytes can be established as EBV positive lymphoblastoid cell lines (LCL) from the peripheral blood of sero-positive donors (Nilsson 1971; Nilsson et al 1971; Lewin et al 1987). EBV genomes and antigens are present in nasopharyngeal carcinoma cells of epithelial origin (zur Hausen et al 1970; Klein et al 1974). EBV replicates in wartlike lesions of the mouth epithelium detected in HIV-carriers, the oral hairy leukoplakias (Greenspan et al 1986). While EBV infected B-lymphocytes of different origins have been studied in vitro for about two decades, no efforts to establish EBV carrying epithelial cell lines in vitro were succesful so far. Thus, studies of the virus-host cell interaction in epithelial cells have been quite difficult.

The normal route of entry of the virus is the mouth (Chang, 1980). The subsequent events are poorly understood as yet. EBV enters B lymphocytes at an early stage after infection (Svedmyr et al 1983), maybe subsequent to a phase of virus replication in oropharyngeal epithelial cells? Infectious mononucleosis (IM) occurs in a large proportion of EBV infected persons post primary infection. It is a disease due to disregulation of the interaction of B- and T-lymphocytes. During this disease, activated, EBV carrying B-lymphocytes can be detected in the peripheral blood (Klein et al 1976; Robinson et al 1980). During convalescence and in healthy carriers no such cells have been observed. Still, B-lymphocytes can be grown out as LCLs from such carriers. No symtoms after primary infection relating to the virus cycle in epithelial cells have been observed.

I will discuss some aspects of the virus-host cell interaction, and the interplay between the epithelial and hematopoetic compartments.

THE CELL PHENOTYPE AND VIRUS GENE EXPRESSION

EBV gene expression varies with the cell phenotype (Table 1). There are five known types of virus gene expression. Burkitt lymphoma (BL) tumors, and some derived cell lines (group I), only express EBNA 1

Current Topics in Microbiology and Immunology, Vol. 166

(Rowe et al 1986). Nasopharyngeal carcinomas (NPC) may show the same type of virus gene expression, but about 65 per cent also express the latent membrane protein (LMP; Fåhraeus et al 1988; Young et al 1988). Lymphoblastoid cell lines (LCL), whether derived spontaneously from peripheral blood B lymphocytes, or by in vitro virus transformation, express all the six EBNAs, EBNA 1-6 and LMP (Rowe et al 1989; Ernberg 1988). This can be considered the growth transformation associated phenotype. Both EBNA 2 and LMP have been shown to induce changes

Table 1. EBV gene expression in different cell types.

Cell type	In vivo/ In vitro	EBV antigen expression	Remarks
Burkitt lymphoma	In vivo	EBNA 1	
Burkitt lymphoma derived cell line, group I	In vitro	EBNA 1	
Burkitt lymphoma derived cell line, group III	In vitro	EBNA 1-6, LMP	Spontaneous activation of lytic cycle in a few cells
Lymphoblastoid cell line	In vitro	EBNA 1-6, LMP	---"---
Nasopharyngeal carcinoma cell	In vivo	EBNA 1, LMP+/-	
Oral hairy leukoplakia	In vivo	Lytic cycle proteins	
Latently infected B cell	In vivo	?	
Lymphproliferative disease lesions in immune deficient patients	In vivo	EBNA 1-6, LMP	Exceptionally other virus protein phenotypes have been found

that are associated with transformation in experimental systems. There is no direct evidence that EBNA 1 is involved in transformation. This protein has been shown to control episome copy number, and to act on a transcriptional regulatory enhancer. Thus, in BLs and LMP-negative NPCs no virus coded transformation associated proteins are expressed. In a few cells in many BL derived cell lines and in most LCLs, the lytic EBV cycle resulting in virus production is induced spontaneously. During virus production some 80 early and late proteins are expressed. The productive cycle has also been detected in biopsies from oral hairy leukoplakias (Greenspan et al 1986).

From these data the EBV host-cell interaction can be classified in four phases (Table 2): 1)tumor phenotype associated, latent, 2) growth transformation associated, 3) early and 3) late productive, lytic infection. After induction of the group I Rael BL derived cell line, it was shown that the lytic cycle could be induced in some 10% of the cells (Masucci et al 1989). It could also be shown that many of the cells that entered the lytic cycle did not express EBNA 2. The virus gene expression seen in LCLs with all the EBNAs is not a prerequisite for the lytic cycle. It has not been established whether EBNAs have to be downregulated before entry of the lytic cycle. At least EBNA 1 is likely to be compatible with the lytic cycle. EBNAs were detected upon double staining in cells entering the lytic cycle (Ernberg et al 1976).

Table 2. EBV gene expression.

Phase of virus infection	Protein/Antigen expression
Latency in B-lymphocytes in vivo	?
Tumor associated, latent	EBNA 1 (LMP in some NPC)
Growth transformation associated (in LCLs)	EBNA1-6, LMP
Immediate, early lytic cycle	BZLF 1, BRLF 1, BMLF 1
Early, lytic cycle	Several early proteins (grouped as early antigens, EA)
Late, lytic cycle	Many late proteins, including structural proteins (VCA)

Like in other herpes-viruses the entry into the lytic cycle requires the expression of 1-3 immediate early proteins (Farrell et al 1989). The productive cycle can be carbon copied on top of that of the strongly lytic herpesviruses, with three phases of the infection (Fig. 1). The tumor- and LCL-types of virus-host cell interactions are unique to EBV.

The phenotype and EBV gene expression in the B lymphocyte which may carry EBV in a latent form in vivo is unknown. If this is a non-proliferating cell there is no need for the known functions of EBNA 1, and it may thus be an EBV-DNA positive, EBNA negative cell.

The two major types of EBV carrying B-cells, which could be studied in vitro, the BL-derived cell and the LCL-type cell show very distinct phenotypes. The LCs express many surface markers characteristic of activated B- blasts from the peripheral blood, "activation markers", including CD 23, CD 39, CD 45R and adhesion molecules (Sakthivel et al 1989). The BL-derived cells of group I do not express these surface molecules, but express CD 10 and CD 77, which refers them to the same group as follicular center cells. Group III BL-cells have drifted towards an LCL-like phenotype upon in vitro culturing (Rowe and Gregory 1989).

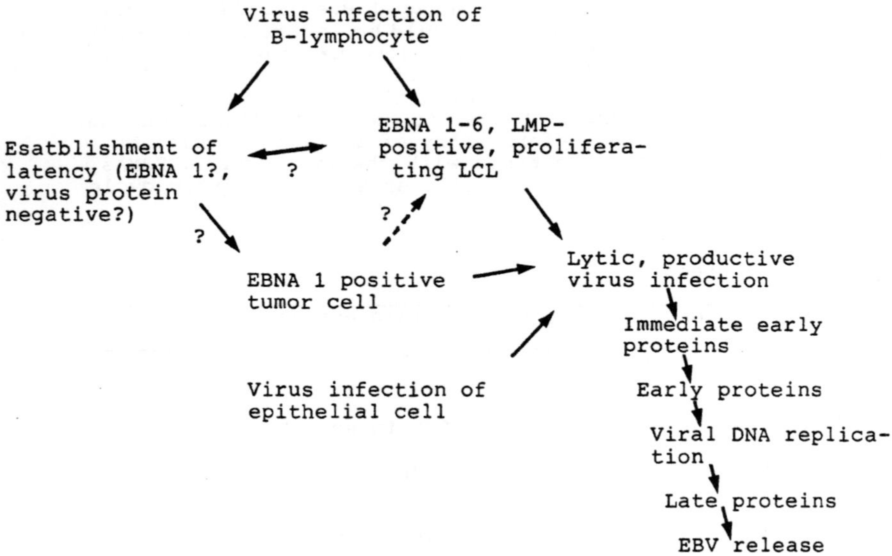

Fig. 1. Schematic representation of different EBV host-cell interactions.

THE SITE OF VIRUS LATENCY?

The interplay between virus in the epithelial cell compartment and in the lymphoid compartment is not accessible to direct studies. It has been assumed on the basis of the findings in hairy leukoplakias, and on the basis of studies of EBV replication in epithelium, that there are foci of epithelial cells in the oropharynx, that can support the productive EBV cycle (Greenspan et al 1986; Sixbey et al 1983). These foci have not been demonstrated directly in healthy carriers. In IM patients, and in the majority of healthy carriers EBV can be isolated from the oropharynx, suggesting at least a constant low level of virus production (Chang 1980; Yao et al 1985; Ernberg and Andersson 1986). The source may be virus production by the oropharyngeal epithelium.

One view has been that virus produced by the epithelium continuously infects B-lymphocytes, which become activated in vivo and behave much like the in vitro proliferating LCLs. As a result there would be a need for continuous elimination of these cells by immunologic and immunoregulatory mechanisms. This view has stimulated a large body of research on cellular immunity against EBV-infected B lymphocytes. There is clear evidence for T cell mediated immunity to the virus infected cells (for review see Rickinson 1986). In healthy carriers these memory cells are not active, but need secondary priming in vitro. Moreover, no one has detected activated B-blasts which express EBNAs in healthy carriers. In contrast, a variable small proportion of such cells were detected in the blood of acute IM patients (Klein et al 1976; Robinson et al 1980). Thus the cellular immunity may operate during the acute primary infection to suppress or eliminate the virus activated B lymphocytes. In healthy carriers the virus may reside in non-proliferating B-lymphocytes in a strictly latent form. Then there is no need for continuous immune elimination of activated B-blasts, nor for continuous infection of B-cells with EBV. In line with this, Lewin et al (1987) have detected spontaneous outgrowth directly from a population of B-lymphocytes with high density from Percoll gradients, i e small, resting B lymphocytes.

Acyclovir in therapeutic doses inhibits the EB-virus replication in the oropharynx completely (Ernberg and Andersson 1986). In several studies of the effect of acyclovir on oropharyngeal EBV production there was no reduction of the frequency of circulating, EBV infected B cells after short or long term treatment (Andersson et al 1985; 1986; Andersson et al 1987; Yao et al 1989). There seems to be no dependence on oropharyngeal virus production to maintain the level of circulating, EBV carrying B cells.

In another study design, where we use the wide variation of molecular weights of EBNAs in immunoblotting (**"EBNotyping"**; Ernberg et al 1989) to monitor the passage of EBV between donors and recipients after bone marrow transplantation, we made the observation that in about 5% of the transplant patients the recipient EBV is lost, eradicated (Gratama et al, 1989). We have taken this as another evidence that EBV establishes long term latency in small, resting B lymphocytes (Gratama et al 1988). In the bone marrow transplant situation the recipient hematopoetic compartment is eliminated with the peri-

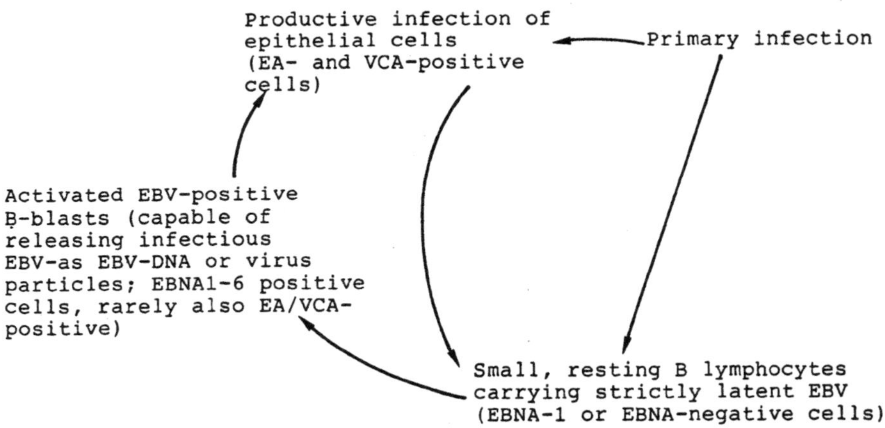

Figure 2. Suggested route of EBV-infection in vivo, based on the recent findings supporting the idea that small, resting B-lymphocytes constitute the reservoir of latent EBV.

transplant cytoreductive therapy. With that goes the latent EBV in the recipient B-lymphocytes. Also, we think this is a strong indication that EBV does not reside in a latent form in the oropharyngeal epithelium. We conclude that the small, resting B lymphocytes are likely to be the site of EBV latency. It is unlikely that EBV also establishes latency in the epithelium. The epithelial cells may be infected by virus from lymphocytes. There is a variable proportion of large B-blasts, from which EBV infected B-cells also can grow out in vitro in a two step process, involving infection and transformation of a new, second B-lymphocyte (Lewin et al 1987). These cells must be able to release virus, which may be the source of infectious virus for the epithelial cells. In contrast to the earlier view, we suggest the B-lymphocytes to be the "reservoir" of EBV (Fig. 2).

THE SIGNIFICANCE OF EBV DNA METHYLATION

We have found that the EBV DNA is extensively methylated in BL-derived cell lines (group I), in BL-biopsies and in NPC-tumors (Ernberg et al 1989; Li Fu et al subm for publication; Minarovits et al subm for publication). In contrast we found that EBV DNA was virtually unmethylated in LCLs. The methylation of CpG-base pairs followed closely the cell phenotype. We have preliminary data that the cellular DNA in general was extensively methylated in NPC-tumors and in the BL-cell lines and -tumors, where the viral DNA was methylated. Methylation of cellular DNA was lower in LCLs. The viral DNA is exposed to the same methylation pressure as the cellular DNA. An analysis of the frequency of CpG-pairs in the EBV DNA, compared to the expected frequency, suggests that EBV normally is exposed to a methylation pressure in vivo in the cell where it is kept latent (Honess et al 1989). EBV DNA contains less CpG than expected, as a result of this "methylation pressure" on the genome. In the tumor cells we have found two "islands" with unmethylated DNA. One is the Bam HI C-region surrounding a promoter for EBNA-transcription, the origin of DNA-replication (ori P) and the genes coding for two small non-translated RNAs (EBERs). The other one is the LMP-regulatory sequence (LRS) 5' of the LMP-gene in those NPC-tumors that expressed LMP (LiFu et al subm for publication).

Klein (1989) has suggested that the BL-tumor cell is a malignant counterpart of the type of B-lymphocyte, which carries latent EBV normally. The cell has lost its ability two return to the resting state due to the activation of c-myc by the chromosome translocation. According to this hypothesis the normal counterpart would be a small, long-lived resting B-lymphocyte, maybe related to follicular center cells (because of the expression of CD 10 and CD 77 on the BL-cell), that expresses EBNA 1 and it has extensively methylated viral and cellular DNA.

VIRUS ACTIVATION IN THE IMMUNODEFICIENT PATIENT

We and others have recently examined the expression of EBV proteins in EBV positive lymphoproliferative lesions and tumors in patients with inherited or acquired immune defects (Young et al 1989; Falk et al subm for publication; Gratama et al subm for publication). With few exceptions, EBV gene expression in most of these lymphoproliferations and tumors was similiar to that of LCLs. EBNA 1, 2 and LMP were expressed, and EBNA 3, 4 and 6 could also be detected by immunoblotting in most of these lesions. Altogether EBV protein expression in tissues from 30 patients have been studied. They include lymphoproliferations from patients with sporadic fatal IM or fatal IM as part of X-linked lymphoproliferative syndrome (XLP), tumors in bone marrow transplant recipients, and a few EBV-positive "AIDS-lymphomas". In a minority of these patients only EBNA 1 could be detected, like in BL. Thus, it is common that LCL-like EBV positive B-blasts proliferate in vivo in immunosuppressed patients to the extent that they give rise to serious clinical symptoms. The analogy to the in vitro proliferating LCLs must be used with caution, however, since in several of these patients the lymphoproliferative disease shows the characteristics of a malignant lymphoma, as e.g. in the bone marrow transplant recipients (Zutter et al 1988), and the tumor cells may have undergone further cellular changes to acquire the more malignant phenotype.

Several studies have shown that the number of EBV carrying cells is increased in the peripheral blood of AIDS-patients (Rinaldo et al 1986; Ragona et al 1986). It is conceivable that "the load" of activated, EBV carrying cells increase in immunosuppressed patients. This may be due to activation of the small, resting B-cells that carry EBV in a latent form, and arise as a result of impaired T-cell dependent regulation of B-cells. This in turn increases the risk of development of lymphoproliferative disease or EBV carrying immunoblastic lymphomas. It may also result in an increase of the number of EBV infected epithelial cells in the oropharynx, and give rise to oral hairy leukoplakias (Fig. 2).

CONCLUSIONS

We have summarized evidence suggesting that EBV is carried under strict latency in small, long-lived, resting B lymphocytes. This cell may be the normal counterpart of the BL cell (Klein, 1989). These cells should be characterized by restricted EBV gene expression, either expressing only EBNA 1 or no EBNAs at all. They may also carry extensively methylated cellular and viral genomes, like the BL-cells. In immune deficient individuals these cells can become activated to B blasts, and acquire a phenotype like that of in vitro transformed lymphoblastoid cell lines. According to this hypothesis EBV does not have a latent phase in oropharyngeal epithelial cells, but the epithelium becomes constantly infected by virus released from activated B blasts.

REFERENCES

Andersson J, Sköldenberg B, Ernberg I, Britton S, Henle W, Andersson U (1985) Acyclovir treatment in primary Epstein-Barr virus infection. A double blind placebo controlled study Scand J Inf Dis, supplement 47:107-115

Anderssson J, Britton S, Ernberg I, Andersson U, Henle W, Sköldenberg B, Tisell A (1986) Effect of acyclovir on infectious mononucleosis. A double-blind, placebo-controlled study J Inf Dis 153:283-290

Andersson J, Sköldenberg B, Henle W, Britton S, Ernberg I (1987) Peroral acyclovir treatment in infectious mononucleosis Infection 15:14-21
Chang RS (1980) Infectious mononucleosis GK Hall Medical Publishers Boston p 130
Ernberg I, Masucci G, Klein G (1976) Persistence of Epstein-Barr virus nuclea antigen (EBNA) in cells entering EB viral cycle Int J Cancer 17:197-203
Ernberg I, Andersson J (1986) Acyclovir efficiently inhibits oropharyngeal excretion of Epstein-Barr virus in patients with acute infectious mononucleosis J Gen Virol 67:2267-2272
Ernberg I, Kallin B, Dillner J, Falk K, Ehlin-Henriksson B, Klein G (1986) Lymphoblastoid cell lines and Burkitt lymphoma derived cell lines differ in the expression of a second Epstein-Barr virus encoded nuclear antigen Int J Cancer 38: 729-737
Ernberg I, Danell E, Gratama JW, Oosterveer MAP, Klein, G (1989). Molecular weight variation of EBNAs as a mechanism to identify different Epstein-Barr virus isolates In Epstein-Barr virus and human disease II (Humana Press, Eds Ablashi) 341-347
Ernberg I (1988) Growth transformation of B lymphocytes by Epstein-Barr virus Cancer Rev 10:92-107
Ernberg I, Falk K, Minarovits J, Busson P, Tursz T, Masuc ci MG, Klein G (1989) The role of methylation in the pheno type dependent modulation of EBNA 2 and LMP genes in latently EBV infected cells J Gen Virol 70:2989-3002
Falk K, Ernberg I, Sakthivel R, Davis J, Christensson B, Luka J, Grierson H, Klein G, Purtilo D. Expression of Epstein-Barr virus encoded proteins and B-cell antigens in fatal infecttious mononucleosis Subm for publication
Farrell PJ, Rowe DT, Rooney CM, Kouzarides T (1989) Epstein-Barr virus BZLF 1 trans-activator specifically binds to a consensus AP-1 site and is related to c-fos EMBO J 8:127-132
Fåhraeus R, Fu Hu-Li, Ernberg I, Finke J, Rowe M, Klein, G, Falk K, Nilsson E, Yadav M, Busson P, Tursz T, Kallin B (1988) Expression of Epstein-Barr virus encoded proteins in nasopharyngeal carcinoma Int J Cancer 42:329-338
Gratama JW,Oosterveer MAP, Zwaan FE, Lepoutre J, Klein G, Ernberg I (1988) Eradication of Epstein-Barr virus by allogeneic bone-marrow transplantation: Implications for sites of viral latency Proc Natl Acad Sci 85:8693-8696
Gratama JW, Oosterveer MAP, Lepoutre J, van Rood JJ, Zwaan FE, Vossen JMJJ, Kapsenberg JG, Klein G, Ernberg I (1989) Persistence and transfer of Epstein-Barr virus after allo geneic bone marrow transplantation Transplantation Proceedings 21:3097-3098
Gratama J, Zutter MM, Minarovits J, Oosterveer MAP, Thomas ED, Klein G, Ernberg I. Expression of Epstein- Barr virus growth transformation-associated proteins in lymphoprolierative lesions of bone marrow transplant recipients Subm for publication
Greenspan JS, Greenspan D, Lennette ET, Abrahams DJ, Conant MA, Petersen V, Freese UK (1985) Replication of Esptein-Barr virus within the epothelial cells of oral "hairy" leukoplakia, an AIDS-associated lesion N Engl J Med 313:1564-1571
Honess RW, Gompels UA, Barrell BG, Craxton M, Cameron KR, Staden R, Chang YN, Hayward GS (1989) Deviation from expected frequencies of CpG dinucleotides in herpesvirus DNAs may be diagnostic of differences in the states of their latent genomes J Gen Virol 70:837-855

Klein G, Giovanella BC, Lindahl T, Fialkow PJ, Singh S, Stehlin J (1974) Direct evidence for the presence of Epstein-Barr virus DNA and nuclear antigen in malignant epithelial cells from patients with anaplastic carcinoma of the nasopharynx Proc Natl Acad Sci 71:4737-4741.

Klein G, Svedmyr E, Jondal M, Persson PO (1976) EBV-determined nuclear antigen (EBNA) positive cells in the peripheral blood of infectious mononucleosis patients Int J Cancer 17:21-26

Klein G (1989) Viral latency and transformation: the strategy of Epstein-Barr virus Cell 58:5-8

Li-Fu H, Minarovits J, Shi-Long C, Contreras-Salazar B, Rymo L, Falk K, Klein G, and Ernberg I.Hypomethylation of BNLF 1 5'-flanking region corelates to EBV-LMP (latent membrane protein) expression in nasopharyngeal carcinoma Subm for publication

Lewin N, Åman P, Masucci MG, Klein E, Klein G, Öberg B, Strander H, Henle W, Henle G (1987) Characterization of EBV-carrying B-cell populations in healthy seropositive individuals with regard to density, release of transforming virus and spontaneous outgrowth Int J Cancer 39:472-476

Masucci MG, Contreras-Salazar B, Ragnar, E, Falk K, Minarovits J, Ernberg I, Klein G.(1989) 5-azacytidine up- regulates the expression of EBNA 2-6 and LMP in EBNA 1 positive, EBNA 2-6 and LMP-negative Burkitt lymphoma line J Virol 63: 3135-3141

Minarovits J, Klein G, Ernberg I (subm for publication) EBV DNA methylation in Burkitt lymphoma tumor biopsies and derived cell lines

Nilsson K (1971) High frequency-establishment of human immunoglobulin producing lymphoblastoid cell lines from normal and malignant tissue and peripheral blood Int J Cancer 8:432-42

Nilsson K Klein G, Henle W, Henle G (1971) The establishment of lymphoblastoid lines from adult and fetal human lymphoid tissue and its dependence on EBV Int J Cancer 8:443-450

Ragona G, Sirianni M, Soddu S, Vercelli B, Piccoli M, Aiuti F (1986) Evidence for disregulation in the control of Epstein-Barr virus latency in patients with AIDS-related complex Clin Exp Immunol 66:17-24

Rinaldo C, Kingsley L, Lyter DW, Rabin B, Atchison RW, Bodner AJ, Weiss SH, Saxinger WC (1986) Association of HTLV III with Epstein-Barr virus infection and abnormalities of T lymphocytes in homosexual men J Infect Dis 154:556-561

Rickinson AB (1986) Cellular immunological responses to the virus infection. In "The Epstein-Barr virus" (Eds Esptein and Achong) William Heineman Med Books, London, p 75

Robinson J, Smith D, Niederman J (1980) Mitotic EBNA positive lymphocytes in peripheral blood during infectious mononucleosis Nature 287:234-235

Rowe DT, Rowe M, Evan G, Wallace LE, Farrel PJ, Rickinson AB (1986) Restricted expression of EBV latent genes and T lymphocyte detected membrane antigen in Burkitt's lymphoma cells EMBO J 5:2599-2607

Rowe M, Gregory C (1989) Epstein-Barr virus and Burkitt's lymphoma In "Adv in Viral Oncology" (Ed G Klein), Raven Press, New York 8:237-259

Sakthivel R, Christensson B, Ehlin-Henriksson B, Klein G (1989) Immunophenotypic characterization of follicle-center-cell-derived non-Hodgkin's lymphomas Int J Cancer 43:624-630

Sixbey JW, Veserinen EH, Nedryd JG, Raab-Traub N, walton LA, Pagano JS (1983) Replication of Epstein-Barr virus in human epithelial cells infected in vitro Nature 306:480-483

Svedmyr ES, Ernberg I, Seeley J, Weiland O, Masucci G, Tsukuda K, Szigeti R, Masucci MG, Blomgren H, Berthold W, Henle W, Klein G (1984) Virologic, immunologic and clinical observations on a patient during the incubation, acute and convalescent phases of infectious mononucleosis Clin Immunol Immunopathol 30:437-450.

Yao QY, Rickinson AB, Epstein MA (1985) A re-examination of the EBV-virus carrier state in healthy seropositive individuals Int J Cancer 35:35-42

Yao QY, Ogan P, Rowe M, Wood M, Rickinson AB (1989) Epstein-Barr virus-infected B cells persist in the circulation of acyclovir treated carriers Int J Cancer 43:67-71

Young L, Alfieri C, Hennessy K, Evans H, O'Hara C, Andersson KC, Ritz J, Shapiro RS, Rickinson AB, Kieff E, Cohen JI (1989) Expression of Epstein-Barr virus transformation associated genes in tisuues of patients with EBV lymphoproliferative disease N Eng J Med 321:1080-1085

zur Hausen H; Schulte-Holthausen H, Klein G, Henle W, Henle G, Clifford P, Santesson L (1970) EBV DNA in biopsies of Burkitts tumours and anaplastic carcinomas of the nasopahrynx Nature 228:1056-1058

Zutter MM, Martin PJ, Sale GE et al.(1988) Epstein-Barr virus lymphoproliferation after bone marrow transplantation Blood 72:520-529

Relative Predispositional Effect of a PADPRP Marker Allele in B-Cell and Some Non B-Cell Malignancies

KISHOR BHATIA, KONRAD HUPPI,
BARRY CHERNEY*, MARK RAFFELD,
MARK SMULSON*, IAN MAGRATH

National Institutes of Health, National Cancer Institute,
9000 Rockville Pike, Bldg 10/13N240,
Bethesda, Maryland, 20892, USA.
*: Dept. of Biochemistry and Molecular Biology,
Georgetown University, Washington, D.C. 20007 U.S.A.

The gene PADPRP codes for a nuclear enzyme Poly (ADP-ribose) polymerase. This enzyme is a DNA binding protein that modulates chromatin structure adjacent to DNA strand breaks (Smulson and Sugimura 1980). The polymerase catalyzes the transfer of the ADP-ribose moiety of NAD to a variety of nuclear proteins and to itself, such that a long chain of ADP-ribose units attached to nuclear proteins is formed. The enzyme also functions as one of the major acceptors for poly ADP-ribose. The principal clue to the cellular function of poly (ADP-ribose) polymerase arises from the observation that the catalytic activity of the enzyme is obligately dependant upon the presence of DNA breaks, both in vivo and in vitro. It has hence been implicated in playing a role in DNA repair (Benjamin and Gill 1980), perhaps stabilizing the broken ends of DNA strands.

Cloning of the PADPRP gene revealed a single long open reading frame extending 3042 bp from the 1st ATG codon to a TAA termination codon at position 3202-3204. The protein deduced from this sequence contains 1014 amino acids, the calculated molecular weight closely approximating the molecular weight estimates of the purified human protein. Two salient features of the sequence information that provide support for a DNA-repair/recombination function for the polymerase include the presence of a putative nucleotide binding fold similar to that present in many DNA repair enzymes. The polymerase also contains regions within the DNA binding domain which are suggestive of a Zn^{2+} binding finger (Cherney et al. 1987). These sequences are of the form cys x_2 cys x_{25} His x_2 His (residues 21-53), His x_2 cys x_5 His x_3 His (residues 53-66) and cys x_2 cys x_{30} His x_2 cys (residues 125-162). Since the NH_2 - terminal portion of the polymerase specifically binds Zn^2, Zn^2 binding fingers may be involved in DNA break recognition events.

Using the 3·7 Kb cDNA as a probe for Southern analysis of EcoRI digests of human rodent somatic cell hybrids, a percent discordancy table was constructed which mapped the gene(s) to chromosome 1, 13, and 14 in humans (Cherney et al. 1987).

Current Topics in Microbiology and Immunology, Vol. 166

The dissection of the molecular mechanisms of B and T-cell carcinogenesis has emphasized the fact that aberrant recombination of antigen receptor genes may play a significant role in promoting tumorigenesis (Cory 1986). The "cut and paste" mechanism involving significantly large segments of chromatin around the immunoglobulin and T cell receptor loci result in localized DNA breaks (albeit transient) at a far greater frequency in lymphoid cells than in non-lymphoid cells. It is, however, safe to assume that the recombinases which catalyse the physiological recombination in specific lymphoid cells may act in concert with a chromatin based DNA repair mechanism activity that maintains the level of aberrant recombination at a minimum. The various properties of the chromatin associated DNA-break-sensitive enzyme poly (ADP-ribose) polymerase make it an attractive candidate for just such a policing activity.

In support of the presence of a chromatin based repair mechanism in B-cell tumors, the following observations from the mouse plasmacytoma models are note worthy.

1) Among a broad panel of Balb/c and $DBA/_2$ congenic strains, those that fortuitously harbor tumor resistant genes of DBA origin have also been found to carry a locus derived from DBA which confers more efficient chromatin repair (Potter et al. 1986).
2) Both the tumor resistance and repair loci cosegregate with the distal end of mouse chromosome 4.
3) Following isolation of the mouse homolog of PADPRP, we obtained preliminary evidence that a 9·oKb Hind III Rf band identified by the murine PADPRP probe also maps to the distal end of mouse chromosome 4 (Huppi et al. 1989).

These observations in mice prompted us to determine whether differences in the human polymerase genes would be present in DNA from Burkitt's lymphoma, as compared to DNA from non-cancer cells.

RESULTS AND DISCUSSION

RFLP ANALYSIS of PADPRP LOCI DEFINES a DELETION on CHROMOSOME 13

We utilized a simple and direct approach of Restriction Fragment Length Polymorphism (RFLP) analysis of Burkitt's tumor DNA and peripheral blood lymphocyte DNA from the general population to reveal the presence of structural rearrangements of the polymerase genes in Burkitt lymphoma cells.

In the course of our work directed toward the chromosomal localization of the PADPRP genes, we also observed several simple two allele polymorphisms in polymerase or polymerase-like sequences (3). These included a Hind III RFLP with alleles of 2·6kb (allele B) and 2·8kb (allele A). RFLP were also detected with Pst I, Sst I, EcoRI, KpnI, Taq I & Msp I. Although the chromosomal localization of these RFLP's was not confirmed.

Three representative DNA samples that provided the three possible genotypes for the Hind III alleles, were digested with various restriction enzymes. When DNA from 2·6kb (BB) and 2·8kb (AA) homozygous individuals, or from a heterozygote (AB) were subjected to EcoRI digestion and polymorphic analysis, the heterozygous sample was resolved into two EcoRI bands (5·1 and 5·3 kb), while a sample homozygous for the 2·6 kb allele gave a single band of 5·1 kb. Collectively these polymorphisms suggested a possible deletion of 200 bp adjacent or within the polymerase like sequences. Furthermore, since the Hind III RFLP segregated with the 5·3 kb EcoRI band, in DNA obtained from somatic cell hybrid data, we concluded that this deletion localizes to chromosome 13.

That this is indeed a deletion, rather than a single base change polymorphism, was confirmed by detecting an extra Pst 1 band of approximately 5·6 kb in DNA obtained from individuals with the AB genotype that was not present in the DNA from AA genotype. From Pst 1 digests of BB genotype DNA, it was evident that this band arose at the expense of the 7kb Pst I band (Fig 1), RFLPS for other enzymes, while including KpnI, MspI also segregate with Hind III, EcoRI and Pst 1 RFLP'S for SstI, BamHI and EcoRV do not cosegregate.

A 200 base pair deletion in the polymerase sequences or linked to polymerase sequences as observed by both Hind III and EcoRI, would be sufficient to account for a 5·6 kb Pst 1 fragment in the AB and the BB genotypes only if a Pst 1 site were created at the break - rejoining point. A more likely possibility is that the 1·4 kb difference in the Pst 1 band is a consequence of a deletion which encompasses regions beyond the 200 base pairs and that since the polymerase sequences are present at the extreme end of the deletion, the apparent size of the deletion as observed using a PADPRP probe is limited to regions spanning the PADPRP. Two hypothetical models for such a deletion are described in fig 1 A and B.

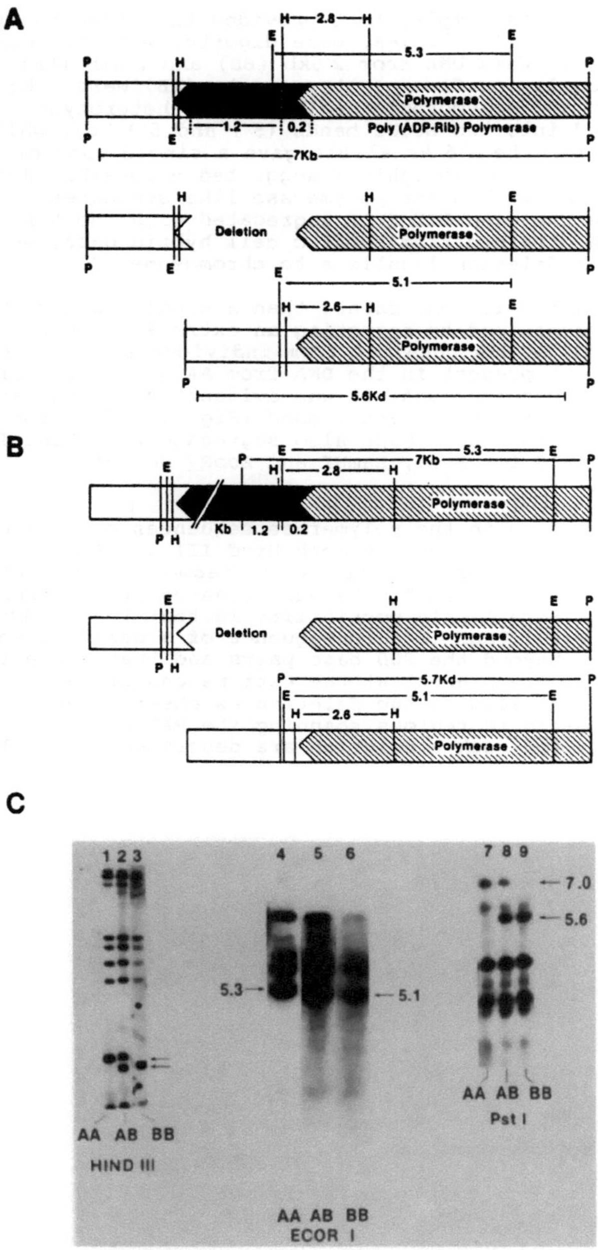
A
H—2.8—H
E
E
5.3
P
H
P
Polymerase
1.2
0.2
Poly (ADP-Rib) Polymerase
P
E
P
7Kb
H
H
E
Deletion
Polymerase
P
E
P
E
5.1
H —2.6— H
E
Polymerase
P
P
5.6Kd
B
E
5.3
E
P
7Kb
P
E
H
2.8
H
Polymerase
Kb
1.2
0.2
P H
E
H
E P
Deletion
Polymerase
P H
P
5.7Kd
P
E
5.1
E
H —2.6— H
Polymerase
C
1 2 3
4 5 6
7 8 9
7.0
5.6
5.3
5.1
AA AB BB
HIND III
AA AB BB
ECOR I
AA AB BB
Pst I

Fig. 1- Schematic representation of the two hypothetical deletion events on chromosome 13 which could lead to the observable *A* and *B* banding patterns derived from genomic DNA Southern analyses using various enzymes [*HindIII (H), Pst I (P)* and *EcoRI (E)]*. Part C shows the various polymorphic bands from the same DNA samples when cut with *HindIII* (lanes 1-3). *EcoRI* (lanes 4-6, or *Pst* (lanes 7-9. In the *HindIII* digests a polymorphic band is at 2.8 kb, and the B polymorphic band is at 2.6 kb. In the *EcoRI* digest, the *A* and *B* bands again have a difference of 200 base pairs (5.3 and 5.1 kb, respectively), while in the *Pst* digest the *A* and *B* bands differ by almost 1.4 kb. Scheme *A* assumes the deletion event does not incorporate a *Pst* I site and does not encompass more than 1.4 kb of the genome. Scheme *B* utilizes the possibility that the deletion event does encompass a *Pst* I site and that the deleted region could be of any length greater than or equal to the largest deletion observed (i.e. 1.4 kb. as observed with *Pst* I digests). Model that would allow for insertion events would be similar and are shown.

FREQUENCY OF THE DELETION IS HIGHER IN BLACK INDIVIDUALS

To establish the frequency of the above noted deletion in a control population, we screened DNA from a large number of randomly chosen individuals. The A allele was found to predominate (0·86) in the control population. The frequency of the B allele was 0·14 in Caucasians. Since our studies were directed towards Burkitt's lymphoma, we also included in the control samples, DNA obtained from Africans. In a series of 87 DNA samples from black individuals African and Americans, we found the frequency of the B allele to be significantly greater than in whites; 0·35 (Table I). The frequency of the B allele amongst black individuals from Africa and America was similar.

Table 1. Increased frequency of B allele in DNA from different population groups

Population	*DNA from normal Cells Genotype Distribution* AA	AB	BB	Total	*B-allele frequency*
Black					
American	*15*	*18*	*4*	*37*	*0.35*
African	*23*	*18*	*9*	*50*	*0.36*
Caucasian					
American	*45*	*12*	*2*	*59*	*0.14*

Distribution of genotypes based upon RFLP banding patterns obtained following restriction of DNA with HindIII (see text for description). DNA was obtained from either Caucasians *(American) or Blacks (American and African). The designation of the genotype and the description of the* A *and* B *bands are provided in Fig. 1 and in the text.*

INCREASED FREQUENCY OF THE B ALLELE IN DNA FROM BURKITT'S LYMPHOMA.

DNA from 45 Burkitt's lymphoma were analyzed for the distribution of the PADPRP B allele (Fig 2). 19 of these lymphoma samples were from endemic Burkitt's lymphoma and were thus derived from African blacks. The other 26 samples were sporadic Burkitt's lymphoma and were obtained from American caucasians. Panel A in Fig 2 is a

Southern blot of a Hind III digest of DNA from 16 cases of the endemic tumors analyzed. The results are particularly striking, in that not a single sample was homozygous for AA.

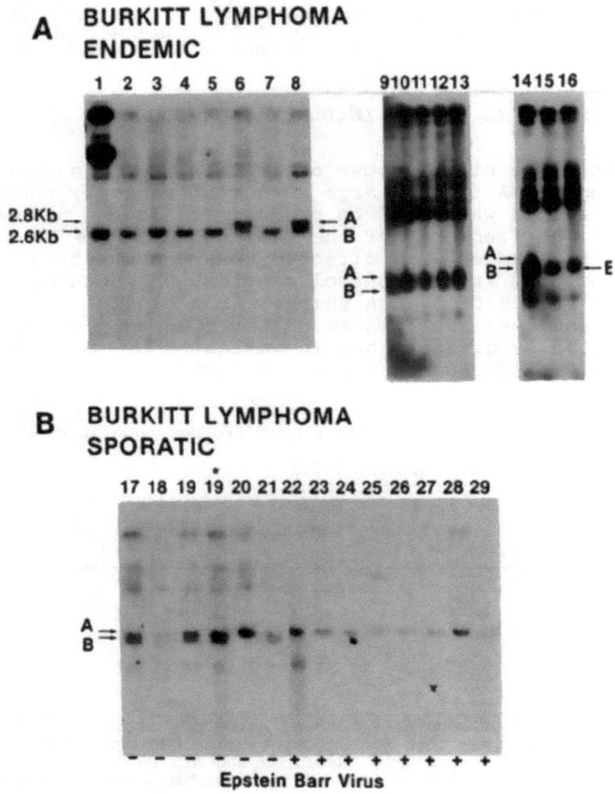

Fig 2. Southern blot of Burkitt lymphoma DNA samples digested with *HindIII* and hybridized with full-length PADPRP probe. A description of the *A* and *B* alleles is provided in the text and in Fig 2. DNA was obtained from endemic Burkitt tumors (lanes 1-8 and lanes 9-18 represent separate gel runs. EBV status is hown for each sporadic tumor (17-29). All endemic cases were EBV +ve (1-16.

In a total of 19 endemic cases of Burkitt's lymphomas, 11 were AB (heterozygous) and 8 were homozygous or hemizygous for the B allele. The pronounced reduction in the homozygosity of the A allele increased the frequency of the B allele to twice that in the control black population (Table 2). It is possible that the marked reduction in the frequency of the A allele in these tumor samples is a result of non-specific chromosomal alteration often observed in tumor progression. However, this is unlikely, since the only differences observed in the banding patterns were in the fragments known to originate from chromosome 13. Other bands which we now know to map to chromosome 14 and to the active gene on chromosome 1 remained unchanged.

Table 2. Increased frequency of B allele in Burkitt DNA from different population groups.

Population	DNA from Tumor Cells Genotype Distribution				B-allele frequency	BB-genotypes over non-cancer control
	AA	*AB*	*BB*	Total		
Black	0	11	8	19	0.70	2.8
Caucasian	12	9	2	23	0.28	2.5

A representative blot of some of these Burkitt lymphoma cases is shown in Fig 1. The designation of the genotype and the description of the *A* and *B* bands are provided in Fig. 2 and in the text. The increase in the frequency of *BB* genotypes in the tumor DNA was calculated by comparison with appropriate racial controls.

Figure 2B is a Southern analysis of DNA derived from sporadic Burkitt's lymphoma. The presence of the AB or BB genotypes approached 100% in all cases that were not associated with EBV. EBV positive tumors however showed only the normally predominant AA allele. Among 23 sporadic Burkitt's lymphomas tested, 8 were EBV +ve and none showed the presence of a B allele, while among 15 that were EB -ve, 4 were homozygous for the A allele, 9 were of AB genotype and 2 were homozygous for the B allele. The frequency of the B allele in this group was thus 3 times that of non-cancer controls, with an 8 fold increase in the homozygous B genotype (Table 2).

Since almost all samples from endemic Burkitt's lymphoma exhibited either a BB or AB genotype, and since endemic Burkitt's lymphoma is nearly always found to be associated with EBV, the presence of the B allele appears to reflect an independent factor involved with the induction or progression of Burkitt's lymphoma. It is possible that the reduced frequency of the A allele along with EBV and malarial infection could contribute to the endemic nature of Burkitt's lymphoma in Africa.

REDUCTIONS IN THE FREQUENCY OF "A" ALLELE IN OTHER TUMORS

We also examined other hematological malignancies for the distribution of the B allele. Thirteen samples of myeloid leukemia showed no significant increase in the frequency of the B allele.

Among a panel of B-cell follicular lymphomas from Caucasian individuals, we again observed an increased frequency of the B allele. Interestingly, in some heterozygous samples we also noted density differences in the A and B band, suggestive of tumor-specific alteration. Among 23 DNA samples from B-cell follicular lymphomas, the B frequency was 0·33 which is close to 3 fold more than in the control population. However, all the B genotypes co-segregated in those tumor samples with demonstrable 14:18 translocations (Table 3).

We wished to determine if the increase frequency of the B allele would be specific only for hematological tumors so we extended our analysis to a number of other malignancies. Among these we found a significantly high frequency of B allele in small lung carcinomas. Our analysis indicated that the frequency of the B allele to be 0·35, again 3 fold higher than controls. Interestingly, on further analysis we found that all of the 6 SCLC samples that contained an L-MYC amplification, possessed at least one B allele (Fig 3, Table 3).

Table 3. Allelic zygosity of the chromosome 13 PADPRP sequences in genomic DNA from tumor samples.

DNA	Genotype AA	AB	BB	Total	B-allele Frequency
Tumor					
B-cell lymphoma	12	7	4	23	0.33
Lung carcinoma	11	9	4	24	0.35

* of the 24 lung tumors, 20 were diagnosed as SCLC and 4 others were classified as non-SCLC

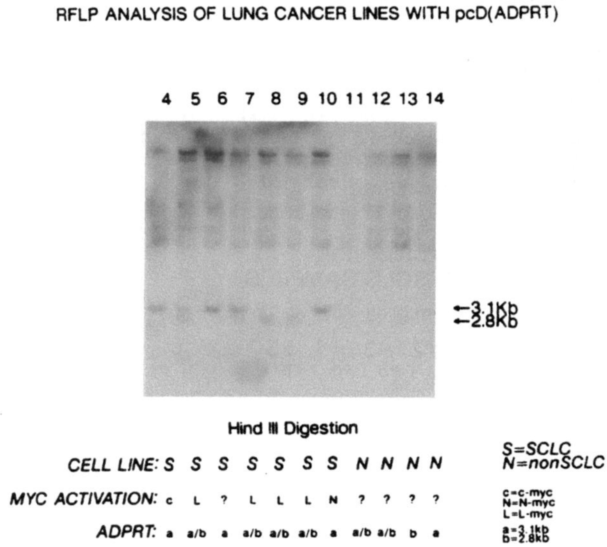

Fig. 3: Southern blot of some representative samples of lung cancer DNA probed with PAPPRP probe.

TUMOR SPECIFIC REDUCTION to HOMOZYGOSITY of the B ALLELE

Growing evidence suggests that various classes of tumors arise by inactivation of both alleles of putative tumor suppressor gene. (Mikkelsen 1990). Since the PADPRP B allele arises as a result of a deletion in the A allele, we have analyzed matched normal and tumor DNA samples to determine if instances of tumor specific reduction to homozygosity occur. As is seen in Fig. 4, in a limited panel of matched DNA from SCLC tumors, we observed 2 cases of tumor specific loss of the A allele.

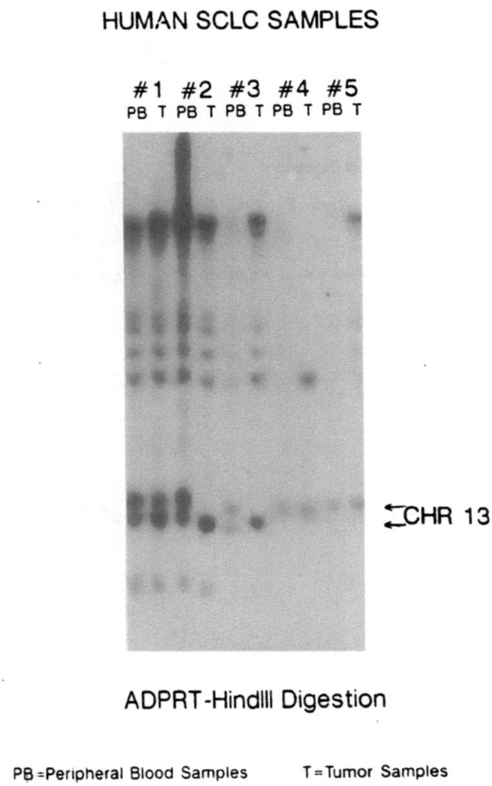

Fig 4. Southern blot of matched normal and tumor DNA from SCLC samples, probed with PADPRP probe.

It thus appears that the increased frequency of the B allele in Burkitt's lymphoma, and B follicular lymphomas, detected by the polymerase probe, could be a result of tumor specific loss of a gene linked to these sequences. Such a hypothesis would also predict that the A allele in the tumor DNA with AB genotype is inactive as a consequence of mutation. Cloning and characterization of this putative suppressor region is now in progress.

REFERENCES

Benjamin RC, Gill DM, (1980) J Biol Chem 255:10502-10508
Cherney B, McBride OW, Chen D, Alkhatib M, Bhatia K, Hensley P, Smulson, ME (1987) Proc Natl Acad Sci (USA) 84:8370-8374
Cory S, Activiation of cellular oncogenes in hemopietic cells by chromosome translocations. Adv Can Res 47: 198-234
Huppi K, Bhatia K, Siwarski D, Klinman D, Cherney B, Smulson M (1989) Sequence and organizations of the mouse poly (ADP-ribose) polymerase. Nucleic Acids Res pp 3387-3401
Mikkelsen T, Cavenee W (1990) Suppressors of the malignant phenotype. Cell Growth and Differentiation 1:201-207
Potter M, Sanford KK, Parshad R, Tarone, RE, Price SM, Mock B, Huppi K (1988) Genes on chromosomes 1 and 4 in the mouse are associated with repair of radiation induced chromatin damage. Genomics Vol 2:257-62
Smulson ME, Sugimura T, eds (1980) Novel ADP-ribosylations of regulatory enzymes and proteins. Elsevier/North Holland, New York

ACKNOWLEDGEMENTS

This work was carried out at the Dept. Biochemistry and Molecular Biology, Georgetown University and was supported by grants from the National Cancer Institute (CA 25344; CA 13195)

Epstein-Barr Virus Terminal Protein Gene Transcription is Dependent on EBNA2 Expression and Provides Evidence for Viral Integration into the Host Genome

U. Zimber-Strobl, KO. Suentzenich, M. Falk, G. Laux, M. Cordier [1], A. Calender [1], M. Billaud [1], G.M. Lenoir [1], G.W. Bornkamm

Institut für Klinische Molekularbiologie und Tumorgenetik, Hämatologikum der GSF, Marchioninistraße 25, 8000 München 70, FRG

INTRODUCTION

Burkitt's lymphoma is a particularly aggressive, extremely rapidly growing human B cell malignancy. It is frequent in countries with holoendemic and hyperendemic malaria (i.e. in tropical Africa and New Guinea) and occurs with 20 to 50 fold lower incidence in all parts of the world. Burkitt's lymphoma is characterized by chromosomal translocations involving the c-myc locus on the long arm of chromosome 8 and one of the immunoglobulin gene bearing heavy or light chain loci on chromosome 14, 2 or 22. Activation of the c-myc oncogene is a crucial and probably rate limiting step in the development of Burkitt's lymphoma, it is, however, not sufficient to fully induce the malignant phenotype. Other factors have therefore to be regarded as cooperative steps in the development of the tumor. Almost all african and about 15% of cases in caucasians are associated with EBV. The tumor cells harbor the EBV DNA in high copy number, usually as episomes and express at least one viral antigen, EBNA1. Therefore EBV is a prime candidate for providing cooperative functions in the multistep development of cancer.

EBNA2 IS ESSENTIAL FOR IMMORTALIZATION OF PRIMARY B LYMPHOCYTES IN VITRO, BUT IS NOT EXPRESSED IN BURKITT'S LYMPHOMA CELLS IN VIVO

Infection of primary B lymphocytes by EBV in vitro induces unlimited proliferation and allows establishment of cell lines in culture, a process called immortalization or transformation. EBV immortalized cells express at least nine viral proteins, six nuclear (EBNA1, 2, 3A, 3B, 3C and LP) and three membrane antigens, LMP, TP1 and TP2. Biological functions of some of these proteins start to be elucidated. EBNA1 is required for episomal replication of EBV by binding to an element called oriP (origin of plasmid replication) (Rawlins et al. 1985; Yates et al. 1985). EBNA2 is absolutely necessary for immortalization of primary B lymphocytes, since deletion of the EBNA2 gene in the viral mutant P3HR1 abolished the immortalizing capacity of the virus (Bornkamm et al. 1982, Hammerschmidt and Sugden 1989). Expression of EBNA2 is responsible for induction of cellular activation markers such as CD21 (the receptor for the complement component C3d) and CD23 (the low affinity Fc_E receptor), as shown by the analysis of EBV negative Burkitt's lymphoma cells which have been converted in vitro by either the B95-8 or P3HR1 strain of EBV (Calender et al. 1987). Apo 1, a cell surface antigen defined by a monoclonal antibody inducing apoptosis (Trauth et al., 1989), shows a similar

[1]International Agency for Research on Cancer, 150 Cours Albert Thomas, 69372 Lyon, Cedex 08, France

Current Topics in Microbiology and Immunology, Vol. 166

pattern of expression and is present on B95-8-, but not P3HR1-EBV converted cell lines (unpublished observation). LMP transforms rodent fibroblasts (Wang et al. 1985, Hammerschmidt et al. 1989) and is involved in modulating the expression of adhesion molecules (Wang et al.1988).

The great potential of EBNA2 and LMP to alter the phenotype of cells and to change the pattern of gene expression has raised a dilemma regarding the role of EBV in Burkitt's lymphoma. On one hand, looking at tumor biopsies, EBV positive and EBV negative Burkitt's lymphoma are virtually indistinguishable with regard to histological presentation, cell morphology and phenotype. On the other hand, wild type EBV induces substantial changes in the growth pattern, morphology and cell surface phenotype of EBV negative Burkitt's lymphoma cells infected in vitro. Furthermore , Burkitt's lymphoma patients have a normal T cell response towards autologous B lymphocytes which were immortalized by EBV in vitro; yet, the EBV carrying tumor cells proliferate in vivo, apparently without any evidence of cytotoxic activity directed against them. This dilemma was recently resolved by the finding of Rickinson and coworkers (Rowe et al. 1986, 1987) who showed that cells obtained from tumor biopsies differ significantly in morphology, cell surface phenotype, growth pattern and viral gene expression from those of Burkitt's lymphoma cells established as cell lines.

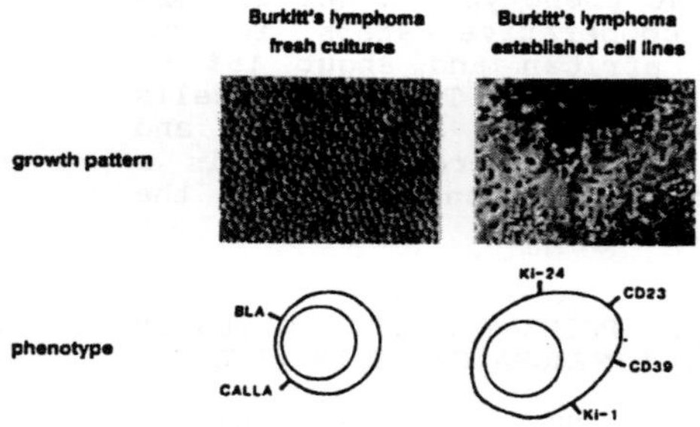

	Burkitt's lymphoma fresh cultures	Burkitt's lymphoma established cell lines
EBNA1	+	+
EBNA2	-	+
LMP	-	+
EBNA3 A,B,C	-	+
LP	?	+
TP	?	+
HLA restricted T cell killing	-	+

Fig. 1.
Phenotype and antigen expression pattern of Burkitt's lymphoma cells according to Rowe et al. 1987 and Murray et al. 1988.

Cells obtained from biopsies apparently do not express EBNA2 and LMP and are positive for CALLA (CD10) and the Burkitt's lymphoma associated glycolipid antigen CD77. They show low levels or no expression of CD21, CD23, CD30 (Ki 1) and grow in single cell suspension. Most EBV positive Burkitt's lymphoma cell lines, however, express EBNA2 and LMP, grow in large clumps and show high expression of B cell activation markers similarly to cells immortalized by EBV in vitro. Only few Burkitt's lymphoma cell lines maintain their original surface pattern and do not shift to the lymphoblastoid penotype. These cell lines may be taken as a model reflecting the pattern of viral gene expression in Burkitt's lymphoma cells in vivo. The lack of EBNA2 and LMP expression in the tumor cells in vivo can also explain the missing T cell response in Burkitt's lymphoma patients, since EBNA2 and LMP are involved in T cell defense mechanisms either by serving as targets for cytotoxic T cells directly, by inducing antigens recognized by T cells, or by mediating the contact between target and effector cells (Moss et al. 1988, Murray et al. 1988).

TERMINAL PROTEIN GENE EXPRESSION IS DEPENDENT ON EBNA2

Regarding the role of EBV in the development of Burkitt's lympoma, the lack of EBNA2 and LMP expression in biopsy cells has turned the interest onto the other viral genes expressed in B lymphocytes immortalized by EBV. We have focussed on the analysis of terminal protein (TP) gene expression in Burkitt's lymphoma cells. This gene is composed of exons located at the right and left hand end of the linear viral genome generating a functional transcription unit only if the viral genome is circularized. TP1 and TP2 transcripts share exons 2 to 9 but are initiated at different promoters and have distinct first exons (Fig. 2).

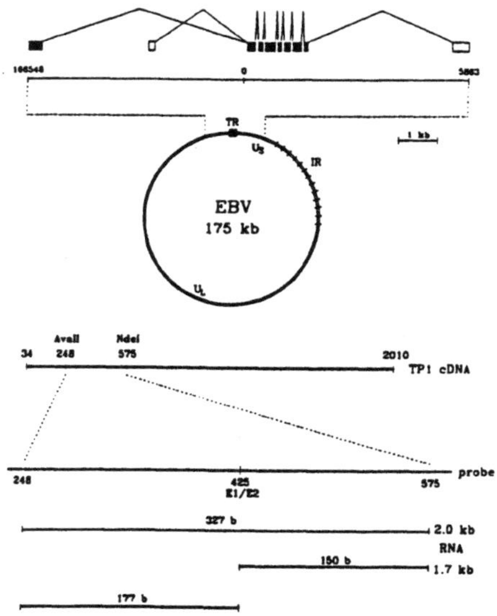

Fig. 2. Schematic representation of the EBV genome circularized through the terminal repeats (TR) and the terminal protein gene. The internal repeats (IR) divide the genome into a short unique (U_S) and a long unique region (U_L). The TP gene gives rise to a 2.0 kb TP1 mRNA initiated at position 166526 and a 1.7 kb TP2 mRNA initiated at position 169795 of the B95-8 EBV sequence (Baer et al. 1984). Translated and non-translated exons are indicated by solid and open boxes, respectively.

The first exon of TP2 RNA does not seem to have coding capacity. TP2 RNA has a translation initiation codon at the beginning of exon2 and codes for an extremely hydrophobic protein lacking the hydrophilic N-terminus encoded by the first exon of TP1 RNA (Laux et al. 1988, 1989, Sample et al. 1989).

We have studied TP expression by S1 analysis using a single stranded TP1 cDNA probe cloned in M13 which spans the junction between exon1 and exon2, with 177b derived from exon1 and 150b from exon2.

TP expression varied greatly in different Burkitt's lymphoma lines. The 327b fragment corresponding to TP1 RNA was present in most of the lines tested, whereas the 150b fragment corresponding to TP2 RNA was detected in only a minority of cell lines (Fig. 3).

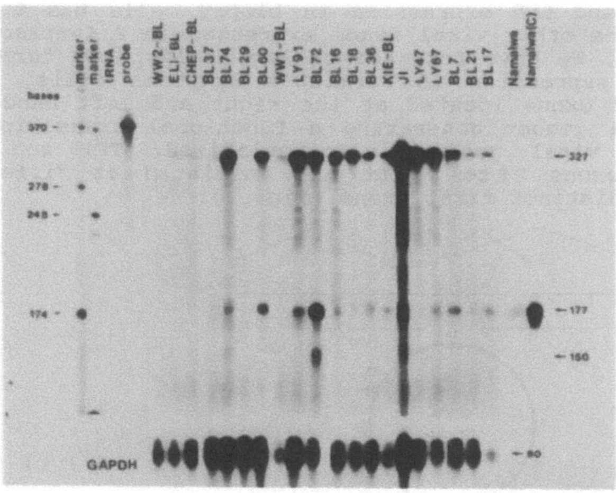

Fig. 3. S1 analysis of TP expression in Burkitt's lymphoma cell lines.
50 μg RNA of each cell line was hybridized to a single stranded probe which spans the junction between exon1 and exon2 of a TP1 cDNA. After S1 digestion TP1 RNA gives rise to a fragment of 327b and TP2 RNA of 150b . Protected fragments were separated on a 5% denaturating polyacrylamide gel. The sizes are given in bases. As internal control a GAPDH probe was added to the hybridization mixture.

Attempts to correlate TP expression with the phenotype revealed high TP expression in cell lines shifted to a more lymphoblastoid phenotype as well as in EBV immortalized cell lines. TP expression was low or absent in cells which have maintained the phenotype of Burkitt's lymphoma cells in vivo.

The dependence of TP transcription on EBNA2 expression became apparent when the incidence of TP RNA was analysed in EBV negative Burkitt's lymphoma cell lines converted with immortalization-defective or -competent virus strains. TP RNA was expressed in all B95-8 or AG876 virus convertants, whereas TP RNA was absent from P3HR1 virus converted cell lines (Fig. 4).

Stable transfection of EBNA2 into P3HR1 virus converted BL41 cells induced significant levels of TP specific transcripts. Transient transfection experiments with plasmid constructs containing the TP promoters in front of the chloramphenicol acetyltransferase gene indicated that EBNA2 is increasing the level of TP RNA by driving the TP promoters (data not shown).

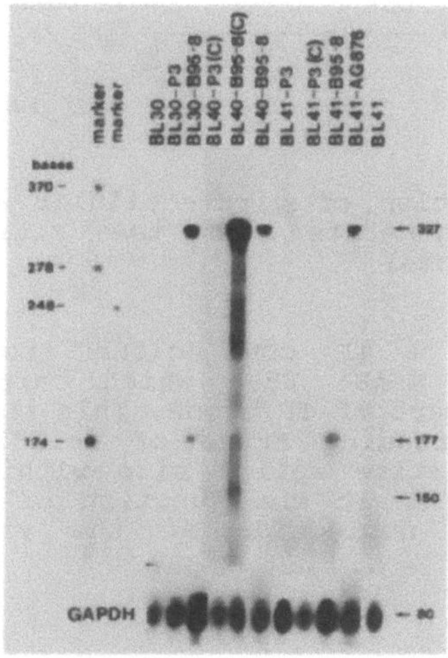

Fig. 4. TP expression in EBV converted cell lines.
TP expression in BL30-, BL40- and BL41-converted cells. The indicated EBV negative cell lines were infected <u>in vitro</u> by the EBV-strains P3HR1, B95-8 or AG876. S1 analysis was performed as described in Figure 3. Cells indicated with (C) were treated with 50 μg/ml cycloheximide for 16 hours.

ABERRANT TP TRANSCRIPTS PROVIDE EVIDENCE FOR VIRAL INTEGRATION INTO THE HOST GENOME

A novel TP RNA species giving rise to a protected fragment of 177b was detected in most of the Burkitt's lymphoma cell lines examined and in about 50% of the EBV immortalized cell lines. This fragment corresponds in size exactly to the part of the probe derived from exon1, but does neither represent TP1 nor TP2 RNA. Northern blot and S1 analysis using a number of additional probes revealed unambiguously that the novel RNA species is initiated at the TP1 promoter and contains the first exon spliced to unknown sequences. The novel TP RNA was the only TP specific transcript in Namalwa cells (Fig. 3, rightmost lane). Since Namalwa cells contain only two viral copies integrated into the host genome via the terminal repeats and no episomal viral DNA, it is concluded that the novel TP RNA represents a viral-cellular fusion transcript generated by integration of the viral DNA into the host genome.

ACTIVATION OF TERMINAL PROTEIN GENE EXPRESSION BY INSERTIONAL MUTAGENESIS?

RNA giving rise to the 177b fragment is initiated at the viral TP1 promoter and transcribed into the cellular genome across the integration site. The reciprocal site of viral integration might generate a cellular-viral transcription unit with a cellular promoter inserted in front of the coding part of TP2 . Integration might thus render TP expression independent of EBNA2.

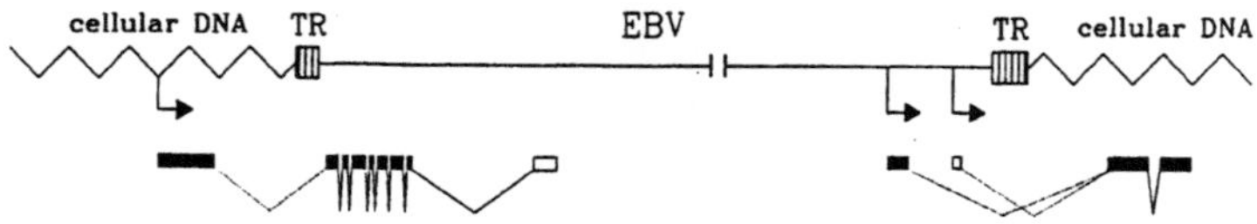

Fig. 5. Schematic representation of viral-cellular and cellular-viral fusion transcripts generated at the sites of viral integration into the host genome.

We have in fact isolated a TP cDNA clone from primary B lymphocytes immortalized by M-ABA EBV, which carries cellular sequences in front of the 3'part of TP exon2. This cDNA represents a transcript with an open reading frame of cellular sequences fused in frame to an alternative splice site within the second exon of the TP gene (Fig. 6). At the junction of cellular and viral sequences, the first nucleotide at the viral site is mutated.

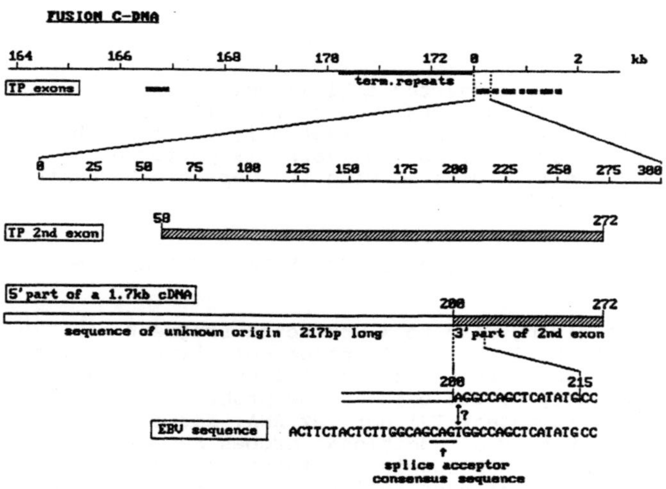

Fig. 6. Fusion cDNA of cellular sequences at the 5' and viral sequences at the 3'end. The cellular sequence is spliced to a normally not used additional splice acceptor site within the second exon of TP RNA at position 200 of the B95-8 EBV sequence.

With regard to the role of EBV in the development of Burkitt's lymphoma it will be particularly interesting to see whether such cellular-viral fusion transcripts do consistently exist in Burkitt's lymphoma cells, and if so, what their biological function might be in the development of the tumor. It is, of course, also mandatory, to learn more about the biological function of the TP gene products and their role in the process of immortalization of normal cells.

ACKNOWLEDGEMENT

This work was supported by the Deutsche Forschungsgemeinschaft (SFB31) and Fonds der chemischen Industrie.

REFERENCES

Bornkamm GW, Hudewentz J, Freese UK, Zimber U (1982) Deletion of the nontransforming Epstein-Barr virus strain P3HR-1 causes fusion of the large internal repeat to the DSL region. J. Virol. 43:952-968

Calender A, Billaud M, Aubry JP, Banchereau M, Vuillaume M, Lenoir GM (1987) Epstein-Barr virus (EBV) induces expression of B-cell activation markers on in vitro infection of EBV-negative B-lymphoma cells. Proc. Natl. Acad. Sci. U. S. A. 84:8060-8064

Cordier M, Calender A, Billaud M, Zimber U, Rousselet G, Pavlish O, Banchereau J, Tursz T, Bornkamm GW, Lenoir GM (1990) Stable transfection of Epstein-Barr virus (EBV) nuclear antigen 2 in lymphoma cells containing the EBV P3HR1 genome induces expression of B-cell activation molecules CD21 and CD23. J. Virol. 64(3): 1002-1013.

Hammerschmidt W, Sugden B (1989) Genetic analysis of immortalizing functions of Epstein-Barr virus in human B lymphocytes. Nature 340:393-397

Hammerschmidt W, Sugden B, Baichwal VR (1989) The transforming domain alone of the latent membrane protein of Epstein-Barr virus is toxic to cells when expressed at high levels. J. Virol. 63:2469-2475

Laux G, Perricaudet M, Farrell PJ (1988) A spliced Epstein-Barr virus gene expressed in immortalized lymphocytes is created by circularization of the linear viral genome. EMBO J. 7:769-774

Laux G, Economou A, Farrell PJ (1989) The terminal protein gene 2 of Epstein-Barr virus is transcribed from a bidirectional latent promoter region. J. Gen. Virol. 70(11): 3079-3084.

Moss DJ, Misko IS, Burrows SR, Burman K, McCarthy R, Sculley TB (1988) Cytotoxic T-cell clones discriminate between A- and B-type Epstein-Barr virus transformants. Nature 331:719-721

Murray RJ, Wang D, Young LS, Wang F, Rowe M, Kieff E, Rickinson AB (1988) Epstein-Barr virus-specific cytotoxic T-cell recognition of transfectants expressing the virus-coded latent membrane protein LMP. J. Virol. 62:3747-3755

Rawlins DR, Milman G, Hayward SD, Hayward SD (1985) Sequence-specific DNA binding of the Epstein-Barr virus nuclear antigen (EBNA-1) to clustered sites in the plasmid maintenance region. Cell 42:859-868

Rowe DT, Rowe M, Evan GI, Wallace LE, Farrell PJ, Rickinson AB (1986) Restricted expression of EBV latent genes and T-lymphocyte-detected membrane antigen in Burkitt's lymphoma cells. EMBO J. 5:2599-2607

Rowe M, Rowe DT, Gregory CD, Young LS, Farrell PJ, Rupani H, Rickinson AB (1987) Differences in B cell growth phenotype reflect novel patterns of Epstein-Barr virus latent gene expression in Burkitt's lymphoma cells. EMBO J. 6:2743-2751

Sample J, Liebowitz D, Kieff E (1989) Two related Epstein-Barr virus membrane proteins are encoded by separate genes. J. Virol. 63:933-937

Trauth BC, Klas C, Peters AMJ, Matzku S, Möller P, Falk W, Debatin KM, Krammer PH (1989) Monoclonal antibody-mediated tumor regression by induction of apoptosis. Science 245:301-304

Wang D, Liebowitz D, Kieff E (1985) An EBV membrane protein expressed in immortalized lymphocytes transforms established rodent cells. Cell 43:831-840

Wang D, Liebowitz D, Wang F, Gregory C, Rickinson A, Larson R, Springer T, Kieff E (1988). Epstein-Barr virus latent infection membrane protein alters the human B-lymphocyte phenotype: deletion of the amino terminus abolishes activity. J. Virol. 62:4173-4184

Yates JL, Warren N, Sugden B (1985) Stable replication of plasmids derived from Epstein-Barr virus in various mammalian cells. Nature 313:812-815

Genomic Integration as a Novel Mechanism of EBV Persistence

E. A. Hurley*, , J. A. McNeil[1], J. B. Lawrence[1] and D. A. Thorley-Lawson*.

Dept. of Pathology, Tufts University School of Medicine, 136, Harrison Ave., Boston, MA 02111*.

INTRODUCTION

EBV is associated with several malignancies including nasopharyngeal carcinoma, Burkitt's and other B cell lymphomas arising in immunosuppressed and AIDS patients. It infects resting B cells <u>in vitro</u> causing them to become activated (Thorley-Lawson and Mann, 1985) while a single genome becomes joined at the ends to generate a covalently closed circle (CCC) that eventually amplifies (Sugden et al., 1979). We suggested that the CCC forms only during the G_1 stage of activation and is necessary to obtain the correct viral transcription required to push the cell on to proliferate (Hurley and Thorley-Lawson, 1988). This hypothesis predicts that CCC would not form upon infection of cells that are already proliferating.
Early work, suggesting that integration by EBV may occur, was based on density gradient fractionation of cellular and viral DNA (Andersson-Anvret et al., 1978). The only detailed molecular analysis of EBV integration was in the BL line Namalwa (Matsuo et al., 1984). Here we show that when EBV infects activated or proliferating cells it forms CCC infrequently or not at all. Rather, it persists in these cells as a single integrated copy. This is the first system which allows the reproducible integration of EBV to be studied.

MATERIALS AND METHODS

<u>Cells, Cell Lines and Viruses.</u>

The EBV converted BL lines used in this study have been described in detail elsewhere (Calendar et al., 1987).Peripheral blood B cells

1-Dept.of Cell Biology, University of Massachusetts Medical Center.

were purified and infected as described previously (Hurley and Thorley-Lawson, 1988).

Gardella Gels

The technique used was based on that originally described by Gardella et al with modifications (Gardella et al., 1984, Hurley and Thorley-Lawson, 1988).

Probes

The Mlu 1 (right hand) probe in the vector pHSI-LM and the EcoR1 I (left hand) probe in the SP65 vector were kind gifts from Dr. Elliott Kieff.

Fluorescence in situ Hybridization

The methodology used has been described previously (Lawrence et al 1988)

RESULTS

Infection of EBV^- BL Cells

We performed a time course study of the appearance of CCC upon infection of several EBV^- BL lines compared to normal resting B cells. As we have described previously (Hurley and Thorley-Lawson, 1988) CCC were detected in the resting population by 24 hrs. post-infection and the signal increased in intensity thereafter. In comparison, no CCC were detected at any time in the infected BL cells even though, at early time points, 10-50% of the cells were EBNA positive and, therefore, contained and expressed EBV DNA.

Form of the genome in EBV converted lines analyzed by Gardella gels.

Although it was not possible to detect CCC in the newly infected EBV^-BL cells, EBV DNA must be retained at some frequency since stably converted sublines have been derived from such infected cultures

(Calendar et al.,1987).We have analyzed the status of the viral genome in a panel of 16 such converted cell lines and the results are summarised in Table 1. In the first experiment the sublines were analyzed by the Gardella gel technique. The results from a representative sample of 7 such lines is shown in Fig 1. CCC could

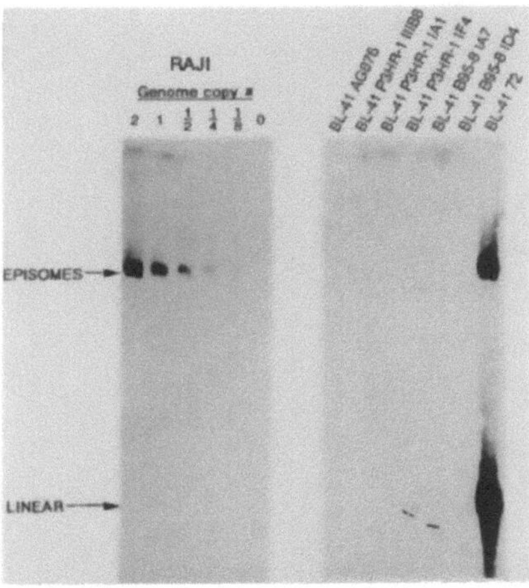

Fig.1 Analysis of EBV Converted Sublines of BL-41 for the presence of CCC.
The cell lines were analyzed by the Gardella gel technique which allows the resolution of the linear and circular forms of the virus. The relative migration of the different forms of the genome is marked. Note that only BL-41 72 contains detectable CCC. To the left is the RAJI cell standard. RAJI cells were serially diluted with EBV negative cells (RAMOS) to yield mixtures corresponding to 1/4-8 circular EBV genomes average per cell. Note that as few as 0.1 genomes per cell can be detected.

only be detected in 6/16 of the sub-lines. The remaining, independently derived, lines all contained no detectable CCC. Essentially 100% of the cells in the converted lines are EBNA positive and contain, on the average, 0.5-2 genomes per cell as assessed by quantitative dot blots (not shown). Thus, they all contain viral genetic information but not in the form of intact CCC. This suggests the possibility that the virus could be integrated.

<u>Form of the genome in converted cell lines analyzed by Southern blot.</u>

An alternate approach which can be used to determine the status of the genome in an infected cell is to perform Southern blot analysis with probes derived from unique sequences at the ends of the linear viral genome (Matsuo et al., 1984; Raab-Traub and Flynn, 1986). The results for seven representative lines are presented in Fig 2 and for all the lines are summarised in Table I. As expected the cell lines

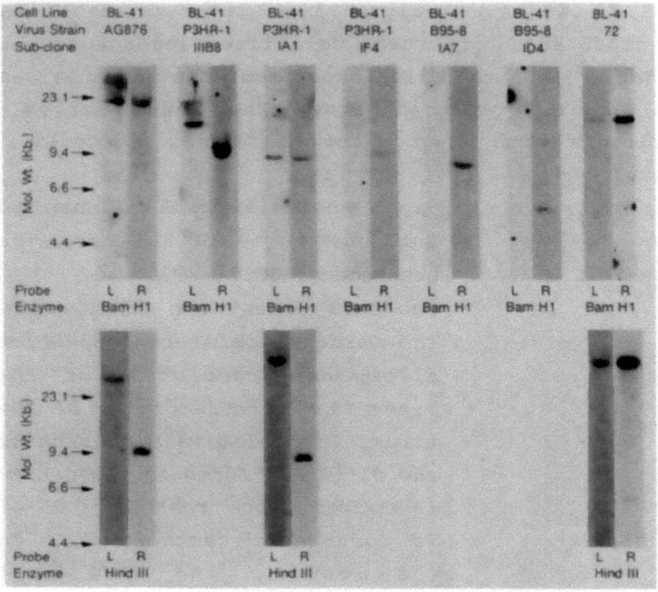

Fig. 2. Southern Blot Analysis of the Converted Sublines of BL-41. *Whole genomic DNA from each of the cell lines was digested and fractionated on agarose gels. The digested DNA was then transferred in two directions to generate duplicate blots. The blots were hybridized with either right (R) or left end (L) probes. The resulting films were sliced and remounted so that R and L probes could be readily compared for each cell line. DNA's that gave similar size R and L fragments after digestion with Bam H1 were reanalyzed after digestion with Hind III. Note that only BL-41 72 gives identical R and L fragments after digestion with either enzyme consistant with the presence of CCC.*

that contain CCC revealed the same size fragment when hybridized with either left or right end probes. Of the lines that had no detectable CCC all showed clear evidence of integration with the two probes failing to hybridize to the same size fragments in either Bam H1 or Hind III digests. Interestingly, three had deleted the left end of their genomes. More precise analysis revealed that all three lines had deleted all of the Bam H1 C fragment (detectable down to about 150 bp, not shown) but had retained W repeats. Thus, in all three lines the origin of replication, which lies in Bam C, has been deleted.

Structure of the viral genome by in situ hybridization

To provide an independent verification of integration, six of the cell lines were analyzed by fluorescence in situ hybridization with biotinated probes (Lawrence et al., 1988). This technique can readily demonstrate single copy integrated genomes in metaphase cells due to the identical labeling of the sister chromatids in the same chromosome. In contrast, cells containing episomes, which are always present in multiple copies, should demonstrate multiple single hybridization singles, randomly associated with the chromosomes (Harris et al., 1985). The six lines analyzed included two with episomes, three with intact integrated DNA and one with integrated DNA deleted at the left end. The results are summarised in Table I.

Table I Summary of the Cell lines Studied and the Form of the Persistent Genome.

	Status of the Genome		
Cell Line	Gardella	Southern	in situ
BL-30 B95-8	E	E	E
BL-30 P3HR-1	E	E	
BL-31 B95-8 B	E	E	
BL-31 P3HR-1	E	E	
BL-40 B95-8	E	E	E
BL-41 72	E	E	
BL-2 B95-8	I	I	I
BL-2 P3HR-1	I	I	I
BL-31 B95-8 A	I	I	
BL-41 P3HR-1 IA1	I	I	
BL-41 P3HR-1 IIIB8	I	I	
BL-41 AG876	I	I	
BL-70 B95-8	I	I	I
BL-41 B95-8 ID4	I	I	I
BL-41 B95-8 IA7	I	I	
BL-41 P3HR-1 IF4	I	I	

I = Integrated
E = Episome

Form

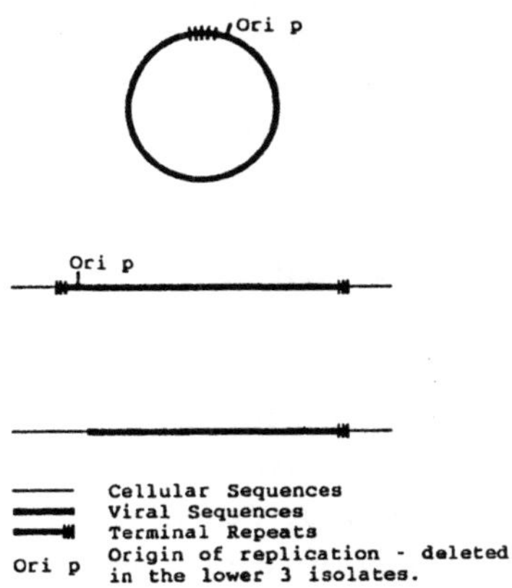

The BL-30 B95-8 and BL40 B95-8 cell lines showed multiple hybridization signals consistent with the 15-25 CCC they are known to contain. By comparison the other four cell lines all demonstrated a single site of hybridization visible as an intense signal at the same position on each of the sister chromatids. The same site of fluorescence was seen in approximately 90% of the cells in any given cell line and no other signals were seen. Furthermore the signal was located on a different chromosome for each of the lines studied, implying that integration is not occuring at preferential sites.

DISCUSSION

When EBV successfully infects a cell that is already proliferating it does not form stable CCC, in most cases, rather it has to integrate in order to persist. This may reflect on an unfavorable environment for CCC formation in an activated cell. Thus proteases and nucleases could be more active in these cells causing degradation of the incoming capsid and its DNA. Furthermore, integration may be more likely in an activated cell since the DNA may be more accessible.

The form of the persisting genome may reflect on heterogeneity in the target cell population. BL cells in vivo are thought to be related to the germinal center centroblast (Gregory et al., 1987), an immature proliferating B cell. However, the phenotype of BL cells is known to drift towards an activated cell phenotype in culture (Rooney et al., 1986). Thus, cultured EBV^{-}BL cells are heterogeneous.

In every case integration ocurred, via repeat sequences, as an intact linear genome excepting three lines that deleted sufficient of the left end of the genome to remove the origin of replication. It is apparent that, if there is anything in common between the integration sites, it was not revealed at the level of analysis presented here.

Our studies do not directly reflect on the possible biological significance of integration in vivo. However, an important issue that arises from these studies is to what extent integration could be playing a role in neoplasia? The first question is, does integration occur in vivo, either as a single linear molecule or in the presence of multiple CCC.Our demonstration of reproducible integration by EBV suggests a reanalysis of these possiblities is now warranted.

REFERENCES

Andersson-Anvret, M., and Lindahl, T. (1979) Integrated viral DNA sequences in Epstein-Barr virus-converted human lymphoma lines. J. Virol 25: 710-718

Calender, A., Billaud, M., Aubrey, J.-P., Banchereau, J., Vuillaume, M., and Lenoir, G.M. (1987) Epstein-Barr virus (EBV) induces expression of B-cell activation on in vitro infection of EBV-negative B-lymphoma cells. Proc. Natl. Acad. Sci. 84: 8060-8064

Gardella, T., Medveczky, P., Sairenji, T., and Mulder, C. (1984) Detection of circular and linear herpesvirus DNA molecules in mammalian cells by gel electrophoresis. J. Virol. 50: 248-254.

Gregory, C.D., Tursz, T., Edwards, C.F., Tetaud, C., Talbot, M., Caillou, B., Rickinson, A.B., and Lipinski, M. (1987) Identification of a subset of normal B cells with a Burkitt's lymphoma (BL)-like phenotype. J. Immunol. 139: 313-318.

Harris, A., Young, B.D., and Griffin, B.E. (1985) Random association of Epstein-Barr virus genome with host cell metaphase chromosomes in Burkitt's lymphoma-derived cell lines. J. Virol. 56: 328-332.

Hurley, E.A., and Thorley-Lawson, D.A. B cell activation and the establishment of Epstein-Barr virus latency. J. Exp. Med. 168: 2059-2075.

Lawrence, J., Villnave, C., and Singer, R. (1988) Sensitive, high-resolution chromatin and chromosome mapping in situ: Presence and orientation of two closely integrated copies of EBV in a lymphoma line. Cell 52: 51-61.

Matsuo, T., Heller, M., Petti, L., O'Shiro, E., and Kieff, E. (1984) Persistence of the entire Epstein-Barr virus genome integrated into human lymphocyte DNA. Science 226: 1322-1324.

Raab-Traub, N., and Flynn, K. (1986) The structure of the termini of the Epstein-Barr virus as a marker of clonal cellular proliferations. Cell 47: 883-889.

Rooney, C.M., Gregory, C.D., Rowe, M., Finerty, S., Edwards, C., Rupani, H., and Rickinson, A.B. (1986) Endemic Burkitt's lymphoma: Phenotypic analysis of tumor biopsy cells and of derived tumor cell lines. J. Natl. Cancer Institute 77: 681-687.

Sugden, B., Phelps, M., and Domoradzki, J. (1979) Epstein-Barr virus DNA is amplified in transformed lymphocytes. J. Virol. 31: 590-595.

Thorley-Lawson, D.A., and Mann, K.P. (1985) Early events in Epstein-Barr virus infection provide a model for B cell activation. J. Exp. Med. 162: 45-59.

Effect of TGF-beta on the Proliferation of B Cell Lines and on the Immortalisation of B Cells by EBV

A. Altiok, M.T. Bejarano and E. Klein

Department of Tumor Biology, Karolinska Institutet, S-104 01 Stockholm, Sweden

INTRODUCTION

The ubiquitous TGF-beta peptides exhibit a multitude of biological functions and act on many type of cells. (For review see Roberts and Sporn 1988, Wakefield and Sporn 1990, Massagué et al. 1990). The designation was coined on the basis of the results obtained in the system in which they were first encountered, transformation of rat fibroblasts. It was however soon discovered that TGF-beta has a variety of other type of actions, including inhibition of cell proliferation. Among the 5 known peptides in the group, TGF-beta1 and TGF-beta2 are most frequently found and are best characterised. They have 70-80% amino acid sequence identity and their functions are similar. They are conserved across species.

Almost all cells have TGF-beta receptors with variable affinities for the active form of the peptides and almost all cells produce TGF-beta, but in a biologically inactive, latent form. Only few cell lines, notably those derived from retinoblastoma were shown to be devoid of TGF-beta receptors. Two types of receptors, I and II differ in molecular weight, 65 and 85 kd respectively. These bind TGF-beta 1 with higher affinity. The type III receptor, 280 kd, binds both peptides with similar affinities. Studies with mutants of mink lung epithelial cells indicated that the type I receptor is the functionally important one. Usually, the type III receptor is expressed in highest quantity. The secreted TGF-beta molecules are in a complex, which has to be cleaved to exhibit receptor binding capacity.Processing of the molecule can occur by the secreting cells or by neighbouring cells. It is assumed that the latter event is more common. Through the tuning of secretion, activation and expression of receptors of variable affinities the TGF-beta system can regulate cell behaviour, including proliferation and differentiation, with high flexibility. TGF-beta also provides messages for cell recruitment in response to tissue injury, and regulates extracellular matrix formation.

TGF-beta is the best characterised cytokine which can inhibit cell proliferation.Whether it is inhibitory or stimulatory in a given assay system depends on the cell type and the conditions. It is likely that the stimulatory effects are indirect, mediated by induction of growth promoting factors. TGF-beta changes the expression of several gene products involved in cell cycle regulation. When T cells are stimulated in the presence of TGF-beta the upregulation of IL-2 and transferrin receptors is inhibited.

TGF-beta can control the expression of c-myc; c-jun/AP-1; PDGF-A and PDGF-B, and the group of cell adhesion molecules. According to Ignotz and Massagué (1987) the regulation of integrins is of great significance and some of the other effects may be secondary to it. The mode of signal transmission by TGF-beta is still not established, it does not correspond to the known pathways.

In the bone marrow TGF-beta inhibits the proliferation of stem cells. When erythropoetic and myelopoietic cells differentiate they become refractory to TGF-beta. This change does not seem to occur through downregulation of receptor expression. Both B and T lymphocytes retain sensitivity when they mature, their

Current Topics in Microbiology and Immunology, Vol. 166

proliferative responses and functions were shown to be inhibited by TGF-beta in several systems (Wahl et al 1989).

The growth inhibitory effect of TGF-beta was shown to be lost when fibroblast lines are established and it occurs concurrent with immortalisation, the proliferation of these cells was even stimulated by TGF-beta (Sorrentino and Bandyopadhyai 1989).

We have investigated the effect of TGF-beta on the proliferation of B cell lines. Ex vivo B cells and their established lines provide almost unique possibilities for studies of activation, transformation and malignancy related characteristics. The panel was selected to reveal whether the sensitivity of EBV carrying ones derived from normal and malignant cells differ and whether in the lines established from malignancies the presence of the EBV genome influences responsivness. The characteristics of B cell lines have been extensively studied (for review see Klein 1989, Nilsson and Klein 1982, Rowe and Gregory 1989).

The cells of the polyclonal LCLs - lymphoblastoid cell lines derived from normal B cells - have an irregular shape, grow in tight aggregates, express activation markers and 8 EBV coded proteins, high level of MHC Class I antigens and adhesion molecules, they do not grow in nude mice and do not clone in agarose. Burkitt lymphomas (BL) are monoclonal, their cells are round, they grow dispersed in culture, they have markers characterising resting germinal center B cells, they have no activation markers, express often only one of the EBV encoded proteins - EBNA-1 -, have often lower level of the MHC antigens with differences between the expression of the alleles, lower level of adhesion molecules, they grow in nude mice and clone well in agarose. Usually the BL cells change in vitro and acquire some degree of similarity with the LCLs. On the basis of cell morphology and social organisation in the cultures and the expression of cell surface markers, the BL lines are categorised as Type I-III. EBV negative BL lines and freshly explanted EBV positive BL cells exhibit the Type I character. Type III is represented by cells which have marked similarities with LCL. Acquisition of LCL-like traits, the change in the BL lines in vitro from Type I to III is imposed by the EBV genome, as shown by the comparison between the EBV positive and negative BL lines and between the latter and their in vitro converted sublines. It is noteworthy that while these EBV carrying tumor derived lines can acquire similarities with the LCLs, their phenotypes still remain different and even when the BLs express the EBV encoded proteins, the levels are always lower compared to the LCLs. Studies with cell hybrids suggest that the regulation of EBV gene expression in the latent infection is an inherent capacity of the maturation stage of the B cell (Contreras-Salazar et al. 1989).

We tested the effects of TGF-beta 1 and 2 on the growth of B cell lines. Furthermore we investigated the influence of TGF-beta 1 on the in vitro EBV induced B cell immortalisation system. The results will be published in detail elsewhere.

PROLIFERATION OF B CELL LINES

The sensitivity of cells for the growth modifying effect of TGF-beta was determined by culturing them for 72 hours with variable concentrations of TGF-beta, in the range of 0.04-10 ng/ml. The ^{3}H Thy incorporation was measured during the last 7-10 hours. In accordance with the experience mentioned in the various reports on TGF-beta effects, the results depended strongly on the conditions of the test. Though the results with TGF-beta 1 on the same cells were qualitatively similar in repeat tests, particularly with the higher TGF-beta concentrations, the levels of growth inhibition and the threshold dose to which they responded differed. The cell density seeded was an important factor. E.g. BL41, one EBV negative BL line, was seeded in concentrations 5,10 and 20 x 10^6 cells/well in 0.2 ml volume.The cell growth in the cultures of the two lower concentrations was inhibited by 0.04 ng/ml TGF-beta 1. The cultures containing the higher cell number was growth inhibited first with 1 ng/ml. The survey of a group of lines was performed with cultures containing 5 x 10^3 cells/well.

We tested two types of LCLs. Conventional ones originating from blood-derived B lymphocytes and sublines of one proB line established from cells of fetal liver (Altiok E and al. 1989). The latter did not have rearranged Ig genes. None of these lines were inhibited by TGF-beta 1, tested up to 10 ng/ml.

Among the EBV carrying BL lines Raji,Daudi,Rael and Jijoye, the growth of Raji was not inhibited by TGF-beta 1, (0.2-12 ng/ml) and it was regularly stimulated by 10ng/ml TGF-beta 2. This panel was selected to comprise cells differing in a number of characteristics. E.g. the virus carried by Daudi does not have transforming capacity, Daudi does not express MHC Class I antigens and Rael is a Type I BL (Masucci et al. 1989). Raji expresses at high level the assorted features defining similarities with LCL, such as relatively high level of EBV encoded proteins, adhesion molecules, MHC Class I antigens, and CD21 (C3d and EBV receptor). We could not define however one particular phenotypic trait which could be correlated with its different behaviour with regard to the response to TGF-beta.

We posed the question whether the EBV carrying sublines of the EBV negative BLs show alterations in response to TGF-beta. Groups of cells derived from Ramos, Bjab and BL41 were studied. The sublines were infected in vitro either with the non transforming P3HR1 or with the transforming B958 virus strain. The growth of the two P3HR1 virus carrying sublines of Ramos was not influenced by TGF-beta 1. Similar results were reported by Blomhoff et al. (1987). The original line and two B958 virus carrying sublines were inhibited. This pattern did not hold however for the Bjab and BL41 groups, within these, all lines, even the P3HR1 virus carrying ones, were sensitive.

Altogether the effects of TGF-beta 2 were different, though the pattern showed a relationship with the effect of TGF-beta 1. Some of the cell lines were stimulated by TGF-beta 2, others were inhibited. The inhibitions were weaker than that imposed by TGF-beta 1. The effects of the two TGF types correlated, in that some lines which were not inhibited by TGF-beta 1 were stimulated by TGF-beta 2. Raji was one typical example. The conditions of our test with TGF-beta 2 detected the capacity of TGF-beta both to inhibit or enhance proliferation. It is possible that the difference in affinities between TGF-beta 1 and 2 to the Type I receptor led to these differences in the growth modifiyng effects. Moreover TGF-beta may signal more than one cellular growth regulating systems and the end result is the balance of these effects.

The results with TGF-beta 2 on one converted cell line suggested that the responsiveness is coupled to the differentiation or maturation state of the B cell. One of the EBV carrying subline, BL41/95 acquired most markedly the characteristics of LCL among all the converted lines studied. It has the chromosomal translocation typical for BL. This line is one good example on the impact of EBV for the phenotypic change of lymphoma derived cells concurrent with the adaptation to the in vitro conditions. In this line the resemblance to the LCL reached the highest degree, because even the malignancy related traits, growth in agarose and in nude mice were reduced (Torsteindottir et al. 1989). We have regarded therefore BL41/95 as a partial "revertant" (Klein G 1989). It is noteworthy that down regulation of the malignancy related parameters was imposed by the EBV genome. This line was in all experiments less sensitive to TGF-beta 2 compared to the sublines in this group and thus this parameter classified it also to be closer to the LCL type.

The responsivness to TFG-beta is an additional parameter in which the two main categories of EBV carrying B cells, LCL and BL, differ. While in the BL lines the presence of EBV genome did not seem to influence directly the growth modifying effect of TGF-beta, it could do so indirectly by inducing the changes which rendered the cells similar to LCL. Altogether our results suggest that the phenotype representing the in vitro immortalised B cell is not growth inhibited by TGF-beta and the BL lines show some differences probably correlated with the acquision of LCL likeness in culture.

EBV INDUCED IMMORTALIZATION OF B CELLS.

When the total lymphocyte population is infected with EBV the presence of T cells inhibits the growth of B cells (for review see Tosato 1987, Wallace and Murray 1988). The efficiency of inhibition depends on the cell density seeded. TGF-beta inhibits T and B cell proliferation, therefore it is expected to influence B cell immortalisation depending on the relative sensitivity of the two cell compartments. We found that 10ng/ml TGF-beta 1 enhanced the outgrowth of B cells due to its efficient inhibition of T cells.

When purified B cell populations were infected, TGF-beta 1 inhibited proliferation if applied at the initiation of the culture. In that, the effect corresponded to the results obtained on B cells stimulated with mitogens. Added a few days later, the inhibitory effect gradually decreased, and it gave place in some experiments to stimulation of B cell growth.

The infection of B cells by EBV can be monitored by the expression of the nuclear antigen EBNA. Usually EBNA positive cells can be detected in the cultures within 8-10 hours. However not all EBNA carrying cells proceed to immortalisation. In our cultures the proportion of EBNA carrying cells increased between 3-10 days and was not influenced by the presence of TGF-beta. Thus, the absolute numbers of EBNA positive cells were lower in the treated cultures, they were reduced parallelly with the total cell counts. Therefore at this stage there seemed to be no difference between the EBNA positive and negative cells with regard to TGF-beta sensitivity. Between 14 and 21 days the proportion of EBNA positive cells increased in the treated cultures and the immortalised lines could be easier established. The morphology of the cultures differed. Compared to the untreated cultures, the cell aggregates were larger in the presence of TGF-beta. In some cultures, mainly in those in which viral infection was less efficient TGF-beta seemed to enhance the proliferation of EBNA positive B cells.

In order to ascertain whether the growth of recently immortalised cells are stimulated by TGF-beta we seeded these cells in limiting dilutions, 2×10^3 - 1×10^2 / wells. We observed no influence on the plating efficiency by 10ng/ml TGF-beta 1.

We interpreted the results as follows: The inhibitory effect of T cells on the EBV induced immortalisation is abrogated by TGF-beta 1. Activated B cells are also inhibited by it. Immortalisation of B cells is accompanied by loss of sensitivity to TGF-beta, therefore a proportion of B cells escape the growth inhibitory effect. These cells are saved by TGF-beta from the T lymphocytes and therefore the final result is enhancement of immortalisation in the treated cultures. According the proposed scheme by Thorley-Lawson (1985), EBV infection of B cells acts initially like a B cell mitogen. At this stage the EBV infected B cell is, similarly to mitogen stimulated cells, growth inhibited by TGF-beta. In several experiments, when immortalisation was established, the cells were stimulated by TGF-beta. This effect may have been indirect. At this stage the accompanying, non transformed B cells in the culture may have been inhibited and when the immortalised cells acquired autonomy, help of stimulatory factors produced by the activated cells become superfluous. On the contrary, such cells may compete with the transformed cells in the culture. Thus through the achievment of the pure immortalised population early in the culture, the events may appear as direct growth stimulation of the immortalised cells by TGF-beta.

The results with the B lines and the freshly immortalised B cells are similar to those obtained with fibroblast cultures. Thus the loss of the growth inhibitory effects of TGF-beta concurrent with immortalisation occurs also in B cells. On the other hand, in accordance with the behaviour of several malignancies, the lymphoma derived B cell lines were sensitive to TGF-beta.

ACKNOWLEDMENTS

This work was supported by PHS grant 5R01 CA25250-06 awarded by the National Cancer Institute, DHHA and by the Swedish Cancer Society.

We thank Michael Sporn, National Cancer Institute, Bethesda, Md; Rolf Ohlsson Center for Biotechnology, Stockholm and L. Ellingsworth, Collagen Company, Palo Alto, Calif. for suggestions and for the TGF-beta preparations.

REFERENCES

Altiok E, Klein G, Zech L, Uno M, Ehlin-Henriksson B, Battat B, Ono Y and Ernberg I. Epstein-Barr virus transformed pro-B cells are prone to illegitimate recombination between the switch region of the μ chain gene and other chromosomes. Proc Natl Acad Sci. USA, 86:6333-6337 (1989).

Blomhoff HK, Smeland E, Mustafa AS, Godal T and Ohlsson R. Epstein-Barr virus mediates a switch in responsiveness to transforming growth factor, type beta, in cells of the B cell lineage. Eur J Immunol. 17:299-301 (1987).

Contreras-Salazar B, Klein G and Masucci MG. Host cell-dependent regulation of growth transformation associated Epstein-Barr-virus antigens in somatic cell hybrids. J Virol. 63:2768-2772 (1989).

Ignotz RA and Massagué J. Cell adhesion protein receptors as targets for transforming growth factor-beta action. Cell 51:189-197 (1987).

Klein G. Reversion of tumorigenicity in an EBV-converted Burkitt's lymphoma line. In: Genetic analysis of the tumor suppression. Ciba Foundation symp. 142:36-53 (1989).

Klein G. Viral latency and transformation: the strategy of Epstein-Barr virus. Cell 58:5-8 (1989).

Masucci MG, Contreras-Salazar B, Ragnar E, Falk K, Minarovits J, Ernberg I and Klein G. 5-Azacytidine upregulates the expression of EBNA 2-6 and LMP in an EBNA-1 positive, EBNA 2-6 and LMP-negative Burkitt lymphoma line. J Virol. 63:3135-3141 (1989).

Massagué J, Cheifetz S, Boyd TB and Anders JL. Annals of the New York Academy of Sciences in press.

Nilsson K and Klein G. Phenotypic and cytogenetic characteristics of human B lymphoid cell lines and their relevance for the etiology of Burkitt's lymphoma. Adv Cancer Res. Acad Press 37:319-380 (1982).

Roberts AB and Sporn MB. Transforming growth factor beta. Adv Cancer Res. Acad. Press 51:107-145 (1988).

Rowe M and Gregory C. Epstein-Barr virus and Burkitt's lymphoma. Adv Viral Oncology. Acad. Press 8:237-239 (1989).

Sorrentino V and Bandyopadhay S. TGF beta inhibits Go/s phase transition in primary fibroblasts. Loss of response to the antigrowth effect of TGF beta is observed after immortalization. Oncogene 4:569-574 (1989)

Thorley-Lawson DA and Mann KP. Early events in Epstein-Barr virus infection provide a model for B-cell activation. J. Exp. Med. 162:45-59 (1985).

Torsteindottir S, Colombani C, Brautbar C, Klein E and Klein G. Down-regulation of Class I HLA antigens and of the Epstein Barr virus encoded latent membrane protein in Burkitt lymphoma lines. Proc Natl Acad Sci.84:3283-3286 (1974)

Torsteindottir S, Andersson M, Avila-Carino J, Ehlin-Henriksson B, Masucci MG, Klein G and Klein E. Reversion of tumorigenicity and decreased agarose clonability after EBV conversion of an IgH/myc translocation carrying Bl line. Int.J Cancer 43:273-278 (1989)

Tosato G. The Epstein-Barr virus and the immune system. Adv Cancer Res. 49:75-125 (1987).

Wahl SM, McCartney-Francis N, Mergenhagen SE. Inflammatory and immunomodulatory roles of TGF beta. Immunol today 10:258-261 (1989).

Wakefield LM and Sporn MB. Suppression of carcinogenesis: A role for TGF-beta and related molecules in prevention of cancer. In: Tumor Suppressor Genes. M Dekker, New York, in press.

Current Topics in Microbiology and Immunology

Volumes published since 1984 (and still available)

Vol. 113: **Potter, Michael; Melchers, Fritz; Weigert, Martin (Ed.):** Oncogenes in B-Cell Neoplasia. Workshop at the National Cancer Institute, National Institutes of Health, Bethesda, MD March 5–7, 1984. 1984. 65 figs. XIII, 268 pp. ISBN 3-540-13597-9

Vol. 115: **Vogt, Peter K. (Ed.):** Human T-Cell Leukemia Virus. 1985. 74 figs. IX, 266 pp. ISBN 3-540-13963-X

Vol. 116: **Willis, Dawn B. (Ed.):** Iridoviridae. 1985. 65 figs. X, 173 pp. ISBN 3-540-15172-9

Vol. 122: **Potter, Michael (Ed.):** The BALB/c Mouse. Genetics and Immunology. 1985. 85 figs. XVI, 254 pp. ISBN 3-540-15834-0

Vol. 124: **Briles, David E. (Ed.):** Genetic Control of the Susceptibility to Bacterial Infection. 1986. 19 figs. XII, 175 pp. ISBN 3-540-16238-0

Vol. 125: **Wu, Henry C.; Tai, Phang C. (Ed.):** Protein Secretion and Export in Bacteria. 1986. 34 figs. X, 211 pp. ISBN 3-540-16593-2

Vol. 126: **Fleischer, Bernhard; Reimann, Jörg; Wagner, Hermann (Ed.):** Specificity and Function of Clonally Developing T-Cells. 1986. 60 figs. XV, 316 pp. ISBN 3-540-16501-0

Vol. 127: **Potter, Michael; Nadeau, Joseph H.; Cancro, Michael P. (Ed.):** The Wild Mouse in Immunology. 1986. 119 figs. XVI, 395 pp. ISBN 3-540-16657-2

Vol. 128: 1986. 12 figs. VII, 122 pp. ISBN 3-540-16621-1

Vol. 129: 1986. 43 figs., VII, 215 pp. ISBN 3-540-16834-6

Vol. 130: **Koprowski, Hilary; Melchers, Fritz (Ed.):** Peptides as Immunogens. 1986. 21 figs. X, 86 pp. ISBN 3-540-16892-3

Vol. 131: **Doerfler, Walter; Böhm, Petra (Ed.):** The Molecular Biology of Baculoviruses. 1986. 44 figs. VIII, 169 pp. ISBN 3-540-17073-1

Vol. 132: **Melchers, Fritz; Potter, Michael (Ed.):** Mechanisms in B-Cell Neoplasia. Workshop at the National Cancer Institute, National Institutes of Health, Bethesda, MD, USA, March 24–26, 1986. 1986. 156 figs. XII, 374 pp. ISBN 3-540-17048-0

Vol. 133: **Oldstone, Michael B. (Ed.):** Arenaviruses. Genes, Proteins, and Expression. 1987. 39 figs. VII, 116 pp. ISBN 3-540-17246-7

Vol. 134: **Oldstone, Michael B. (Ed.):** Arenaviruses. Biology and Immunotherapy. 1987. 33 figs. VII, 242 pp. ISBN 3-540-17322-6

Vol. 135: **Paige, Christopher J.; Gisler, Roland H. (Ed.):** Differentiation of B Lymphocytes. 1987. 25 figs. IX, 150 pp. ISBN 3-540-17470-2

Vol. 136: **Hobom, Gerd; Rott, Rudolf (Ed.):** The Molecular Biology of Bacterial Virus Systems. 1988. 20 figs. VII, 90 pp. ISBN 3-540-18513-5

Vol. 137: **Mock, Beverly; Potter, Michael (Ed.):** Genetics of Immunological Diseases. 1988. 88 figs. XI, 335 pp. ISBN 3-540-19253-0

Vol. 138: **Goebel, Werner (Ed.):** Intracellular Bacteria. 1988. 18 figs. IX, 179 pp. ISBN 3-540-50001-4

Vol. 139: **Clarke, Adrienne E.; Wilson, Ian A. (Ed.):** Carbohydrate-Protein Interaction. 1988. 35 figs. IX, 152 pp. ISBN 3-540-19378-2

Vol. 140: **Podack, Eckhard R. (Ed.):** Cytotoxic Effector Mechanisms. 1989. 24 figs. VIII, 126 pp. ISBN 3-540-50057-X

Vol. 141: **Potter, Michael; Melchers, Fritz (Ed.):** Mechanisms in B-Cell Neoplasia 1988. Workshop at the National Cancer Institute, National Institutes of Health, Bethesda, MD, USA, March 23–25, 1988. 1988. 122 figs. XIV, 340 pp. ISBN 3-540-50212-2

Vol. 142: **Schüpbach, Jörg:** Human Retrovirology. Facts and Concepts. 1989. 24 figs. 115 pp. ISBN 3-540-50455-9

Vol. 143: **Haase, Ashley T.; Oldstone, Michael B. A. (Ed.):** In Situ Hybridization. 1989. 33 figs. XII, 90 pp. ISBN 3-540-50761-2

Vol. 144: **Knippers, Rolf; Levine, A. J. (Ed.):** Transforming Proteins of DNA Tumor Viruses. 1989. 85 figs. XIV, 300 pp. ISBN 3-540-50909-7

Vol. 145: **Oldstone, Michael B. A. (Ed.):** Molecular Mimicry. Cross-Reactivity between Microbes and Host Proteins as a Cause of Autoimmunity. 1989. 28 figs. VII, 141 pp. ISBN 3-540-50929-1

Vol. 146: **Mestecky, Jiri; McGhee, Jerry (Ed.):** New Strategies for Oral Immunization. International Symposium at the University of Alabama at Birmingham and Molecular Engineering Associates, Inc. Birmingham, AL, USA, March 21–22, 1988. 1989. 22 figs. IX, 237 pp. ISBN 3-540-50841-4

Vol. 147: **Vogt, Peter K. (Ed.):** Oncogenes. Selected Reviews. 1989. 8 figs. VII, 172 pp. ISBN 3-540-51050-8

Vol. 148: **Vogt, Peter K. (Ed.):** Oncogenes and Retroviruses. Selected Reviews. 1989. XII, 134 pp. ISBN 3-540-51051-6

Vol. 149: **Shen-Ong, Grace L. C.; Potter, Michael; Copeland, Neal G. (Ed.):** Mechanisms in Myeloid Tumorigenesis. Workshop at the National Cancer Institute, National Institutes of Health, Bethesda, MD, USA, March 22, 1988. 1989. 42 figs. X, 172 pp. ISBN 3-540-50968-2

Vol. 150: **Jann, Klaus; Jann, Barbara (Ed.):** Bacterial Capsules. 1989. 33 figs. XII, 176 pp. ISBN 3-540-51049-4

Vol. 151: **Jann, Klaus; Jann, Barbara (Ed.):** Bacterial Adhesins. 1990. 23 figs. XII, 192 pp. ISBN 3-540-51052-4

Vol. 152: **Bosma, Melvin J.; Phillips, Robert A.; Schuler, Walter (Ed.):** The Scid Mouse. Characterization and Potential Uses. EMBO Workshop held at the Basel Institute for Immunology, Basel, Switzerland, February 20–22, 1989. 1989. 72 figs. XII, 263 pp. ISBN 3-540-51512-7

Vol. 153: **Lambris, John D. (Ed.):** The Third Component of Complement. Chemistry and Biology. 1989. 38 figs. X, 251 pp. ISBN 3-540-51513-5

Vol. 154: **McDougall, James K. (Ed.):** Cytomegaloviruses. 1990. 58 figs. IX, 286 pp. ISBN 3-540-51514-3

Vol. 155: **Kaufmann, Stefan H. E. (Ed.):** T-Cell Paradigms in Parasitic and Bacterial Infections. 1990. 24 figs. IX, 162 pp. ISBN 3-540-51515-1

Vol. 156: **Dyrberg, Thomas (Ed.):** The Role of Viruses and the Immune System in Diabetes Mellitus. 1990. 15 figs. XI, 142 pp. ISBN 3-540-51918-1

Vol. 157: **Swanstrom, Ronald; Vogt, Peter K. (Ed.):** Retroviruses. Strategies of Replication. 1990. 40 figs. XII, 260 pp. ISBN 3-540-51895-9

Vol. 158: **Muzyczka, Nicholas (Ed.):** Viral Expression Vectors. 1990. approx. 20 figs. approx. XII, 190 pp. ISBN 3-540-52431-2

Vol. 159: **Gray, David; Sprent, Jonathan (Ed.):** Immunological Memory. 1990. 38 figs. XII, 156 pp. ISBN 3-540-51921-1

Vol. 160: **Oldstone, Michael B.A.; Koprowski, Hilary (Ed.):** Retrovirus Infections of the Nervous System. 1990. 16 figs. XII, 176 pp. ISBN 3-540-51939-4

Vol. 161: **Racaniello, Vincent R. (Ed.):** Picornaviruses. 1990. 12 figs. X, 194 pp. ISBN 3-540-52429-0

Vol. 162: **Roy, Polly; Gorman, Barry M. (Ed.):** Bluetongue Viruses. 1990. 37 figs. X, 200 pp. ISBN 3-540-51922-X

Vol. 163: **Moyer, Richard W.; Turner, Peter C. (Ed.):** Poxviruses. 1990. 23 figs. X, 210 pp. ISBN 3-540-52430-4

Vol. 164: **Bækkeskov, Steinunn; Hansen, Bruno (Ed.):** Human Diabetes. 1990. 9 figs. X, 198 pp. ISBN 3-540-52652-8

Vol. 165: **Bothwell, Mark (Ed.):** Neuronal Growth Factors. 1990. 14 figs. approx. X, 200 pp. ISBN 3-540-52654-4

MIX
Papier aus verantwortungsvollen Quellen
Paper from responsible sources
FSC® C105338
FSC
www.fsc.org

If you have any concerns about our products,
you can contact us on
ProductSafety@springernature.com

In case Publisher is established outside the EU,
the EU authorized representative is:
Springer Nature Customer Service Center GmbH
Europaplatz 3, 69115 Heidelberg, Germany

Printed by Libri Plureos GmbH
in Hamburg, Germany